Chromatographic Analysis of Environmental and Food Toxicants

CHROMATOGRAPHIC SCIENCE SERIES

A Series of Monographs

Editor: JACK CAZES
Cherry Hill, New Jersey

Chromatographic Analysis of Environmental and Food Toxicants

edited by

Takayuki Shibamoto
University of California
Davis, California

CRC Press
Taylor & Francis Group
Boca Raton London New York

CRC Press is an imprint of the
Taylor & Francis Group, an **informa** business

CRC Press
Taylor & Francis Group
6000 Broken Sound Parkway NW, Suite 300
Boca Raton, FL 33487-2742

First issued in paperback 2019

© 1998 by Taylor & Francis Group, LLC
CRC Press is an imprint of Taylor & Francis Group, an Informa business

No claim to original U.S. Government works

ISBN-13: 978-0-367-40057-6

Library of Congress Cataloging-in-Publication Data

Chromatographic analysis of environmental and food toxicants / edited
    by Takayuki Shibamoto.
            p.   cm. -- (Chromatographic science series ; 77)
        Includes bibliographical references and index.

        1.Chromatographic analysis. 2. Environmental toxicology.
    3. Food--Toxicology.   I. Shibamoto, Takayuki.   II. Series:
    Chromatographic science ; v. 77.
    QD79.C4C477   1998
    615.9' 07--dc21   97-46799
                                                                                    CIP

**Visit the Taylor & Francis Web site at
http://www.taylorandfrancis.com**

**and the CRC Press Web site at
http://www.crcpress.com**

Preface

Throughout history people have learned how to avoid poisons by trial and error. There was probably a tremendous sacrifice of human lives before people leraned to identify and subsequently avoid toxic substances. The nature of "poison" has been understood since ancient times. Areolus Phillipus Theophrastus Bombastus von Hohenheim Paracelsus (1493–1541) defined the fundamental concept of poison as a function of dosage: "All substances are poisons; there is none which is not a poison. The right dose differentiates a poison from a remedy." Therefore, two of the primary objectives of toxicology are to qualitate and quantitate poisonous substances. For centuries, only acutely toxic substances were known as poisons. Recently, however, chronically toxic substances have been discovered. Beginning with the industrial revolution and accelerated by the progress of science and technology, this planet has been contaminated with tremendous numbers of man-made chemicals, some of which possess significant toxicities. Consequently, substances with chronic toxicities such as carcinogenicity, mutagenicity, and teratogenicity have begun to receive much attention.

Analytical techniques are critical to assess to toxicity of substances in the environment and in food. The presence of certain toxicants has recently been identified and recognized using state-of-the-art techniques. Consistent and continuous intake of a certain substance can induce chronic toxicity, even if it is consumed at a trace level. As mentioned above, toxicity depends upon the dose of a substance. Therefore, it is extremely important to know the exact amount of a substance present in the environment and in food to assess its adverse effect on humans.

Today, if a pure compound is obtained, identification is not difficult because advanced instrumentation such as mass spectrometry, NMR, and electron microscopy (solid) provide significant information for elucidation of its structure. Therefore, purification techniques play a more important role in identification of chemicals at present. Chromatographic methods have been the most effective techniques to isolate and purify chemicals. The development of gas chromatography and high-performance liquid chromatography has contributed significantly to the discovery of toxic contaminants in the environment and in foods. However, conventional column and thin-layer chromatographies (TLC) are also powerful techniques to separate or to purify substances. Certain materials still can be isolated only by TLC exclusively. Chromatography thus has the advantage of accomodating both the clean-up and the separation of substances of interest. Chromatographic techniques can be applied to isolate diverse kinds of toxic substances ranging from extremely volatile chemicals such as volatile aldehydes to less volatile chemicals such as pesticides. It is important to chose the right technique to analyze toxicants of interest.

Every month, numerous articles and reports on analytical methods and their applications for toxic chemicals appear in scientific journals. Additionally, many reference books for analytical methodologies have been published. However, these publications focus on specific areas such as food, ground water, air, or soil. There is virtually no other reference book that covers chromatographic methods for a wide range of toxic chemicals found in the environment and in food.

Takayuki Shibamoto

Contents

Contributors

Bruce N. Ames Department of Molecular and Cell Biology, University of California, Berkeley, California

Michael D. David University Center for Environmental Sciences, University of Nevada, Reno, Nevada

Joe W. Dorner National Peanut Research Laboratory, Agricultural Research Service, United States Department of Agriculture, Dawson, Georgia

John D. Ebeler Department of Viticulture and Enology, University of California, Davis, California

Susan E. Ebeler Department of Viticulture and Enology, University of California, Davis, California

Jo A. Engebretsen Department of Environmental Toxicology, University of California, Davis, California

James S. Felton Biology and Biotechnology Research Program, Lawrence Livermore National Laboratory, Livermore, California

Guillermina Font Laboratory of Food Chemistry and Toxicology, Faculty of Pharmacy, University of Valencia, Valencia, Spain

Kenneth G. Furton Department of Chemistry, Florida International University, Miami, Florida

Harold J. Helbock   Department of Molecular and Cell Biology, University of California, Berkeley, California

Mark G. Knize   Biology and Biotechnology Research Program, Lawrence Livermore National Laboratory, Livermore, California

Jordi Mañes   Laboratory of Food Chemistry and Toxicology, Faculty of Pharmacy, University of Valencia, Valencia, Spain

Juan-Carlos Moltó   Laboratory of Food Chemistry and Toxicology, Faculty of Pharmacy, University of Valencia, Valencia, Spain

Charles R. Mourer   Department of Environmental Toxicology, University of California, Davis, California

Gretchen Pentzke   Department of Chemistry, Florida International University, Miami, Florida

Yolanda Picó   Laboratory of Food Chemistry and Toxicology, Faculty of Pharmacy, University of Valencia, Valencia, Spain

Norimitsu Saito   Iwate Prefectural Institute of Public Health, Morioka, Japan

James N. Sieber   University Center for Environmental Sciences, University of Nevada, Reno, Nevada

Takayuki Shibamoto   Department of Environmental Toxicology, University of California, Davis, California

James E. Woodrow   University Center for Environmental Sciences, University of Nevada, Reno, Nevada

Kenji Yamaguchi   Yokogawa Analytical Systems, Inc., Tokyo, Japan

Geoffrey Yeh   Department of Viticulture and Enology, University of California, Davis, California

Helen C. Yeo   Department of Molecular and Cell Biology, University of California, Berkeley, California

1

Polycyclic Aromatic Hydrocarbons

Kenneth G. Furton
and Gretchen Pentzke
Florida International University, Miami, Florida

I. INTRODUCTION

In this chapter, we survey some of the numerous methods available for the chromatographic analysis of PAHs and highlight some of the recent advances. We discuss only chromatographic and related techniques and do not cover methodology focusing on the distribution, biotransformation, flux, and spectroscopic properties of PAHs. The analysis techniques have been grouped into sample preparation (including new extraction techniques and hyphenated techniques), supercritical fluid chromatography and extraction, gas chromatography, liquid chromatography, and miscellaneous techniques including thin layer chromatography and micellar electrokinetic capillary chromatography. Readers are referred to more comprehensive publications, if more thorough background material or historical information is required [1,2], including a recent review of air sampling and analysis of PAHs [3]. PAHs comprise the largest class of known chemical carcinogens and are produced during the combustion, pyrolysis, and pyrosynthesis of organic matter. PAHs are ubiquitous in air, water, soil, and food, and their accurate identification and determination continues to be an important analytical problem.

In fact, PAHs are even present in outer space and are probably more abundant than all other known interstellar polyatomic molecules combined [4]. It has been suggested that interstellar PAHs on meteorites may represent the starting material for the synthesis of complex molecules including amino acids [5] and primitive pigments in the prebiotic environment [6]. Readers interested in the subject of interstellar PAHs are referred to Ref. 7.

The earliest reports of the carcinogenic properties of PAHs were made by Percival Pott in 1775 based on studies of combustion products such as soot [8]. The first identification of a specific chemical carcinogen was made by Kennaway and Hieger in 1930 who identified dibenz(a,h)anthracene as the first chemical recognized to have carcinogenic activity [9]. Samples typically contain an extremely complex mixture of many different PAHs including isomers, alkylated, and nonalkylated forms of PAHs. The structures of the 16 PAHs identified as priority pollutants by the EPA are shown in Figure 1 and include the first carcinogenic PAH isolated, dibenz(a,h)anthracene. By far, a person's greatest exposure to carcinogenic PAHs in the environment comes from food. The average estimated intake of carcinogenic PAHs by nonsmokers in the US is 1–5 µg/day with 96.2% coming from food, 1.9% from soil, 1.6% from air, and 0.2% from water [10]. Smokers increase their exposure to carcinogenic PAHs 2–5 µg/day per pack of cigarettes smoked. Additional exposure can also come from occupational environments, cosmetics applied to the skin, and asphalic materials applied to roofs or driveways. The estimated potential doses of carcinogenic PAHs from water, soil, air, and food are shown in Figure 2, and the median concentrations and ranges of PAHs in environmental media are given in Table 1 [10]. Exposure risks are greatest from surface waters, indoor air with smokers, and urban soil/road dust. Foods with the greatest average PAH levels tend to be charcoal broiled or smoked meats, green leafy vegetables, and fats and oils, as seen in Figure 3. The average dietary intake of PAHs in Finland is reported to be approximately 18 µg/day, with the greatest levels found in samples collected from industrialized areas where large-scale manufacturing industries employ fossil fuels [11].

The cooking mode as well as cooking time has been shown to affect greatly the levels of PAHs in foods, as seen in Table 2 [12]. The levels of PAHs increased from pressure cooker to microwave to pan frying and again with cooking time, as seen in Table 2 for the production of benzo[a]pyrene in mutton and chicken as a function of pan frying time. Additionally, many deep-fried South Indian food dishes, particularly those deeply charred, showed high quantities of PAHs and were suggested to play a role in the high incidence of gastric cancers in South India [13]. Although charbroiled and smoked meats generally have the highest reported levels of PAHs, there may be methods to reduce these levels. The use of a vertical barbecue where fat drips away sufficiently distant from the heat source during grilling to prevent its pyrolysis resulted in 10–30 times lower PAHs levels, as shown in Table 3 for sardines [14]. The use of liquid smoke flavors has been

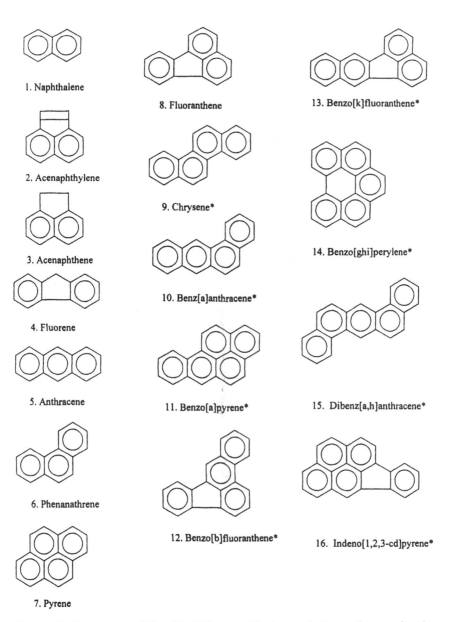

1. Naphthalene

2. Acenaphthylene

3. Acenaphthene

4. Fluorene

5. Anthracene

6. Phenanathrene

7. Pyrene

8. Fluoranthene

9. Chrysene*

10. Benz[a]anthracene*

11. Benzo[a]pyrene*

12. Benzo[b]fluoranthene*

13. Benzo[k]fluoranthene*

14. Benzo[ghi]perylene*

15. Dibenz[a,h]anthracene*

16. Indeno[1,2,3-cd]pyrene*

Figure 1 Structures of the 16 PAHs identified as priority pollutants by the US EPA.

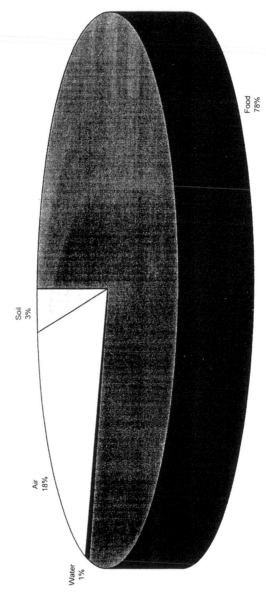

FIGURE 2 Maximum potential doses of carcinogenic PAHs from water, soil, air, and food.

Table 1 Carcinogenic PAH Concentrations
in Environmental Media

Media	Median concentration	Concentration range
Ground water	1.2 ng/L	0.2–6.9
Drinking water	2.8 ng/L	0.1–62
Surface water	8.0 ng/L	0.1–830
Outdoor air	5.7 ng/m^3	0.2–65
Indoor air in homes (Ohio)	8 ng/m^3	0.6–29
Homes with tobacco smoke	13 ng/m^3	7–29
Rural soil	0.07 mg/kg	0.01–1.3
Urban soil	1.10 mg/kg	0.6–5.8
Road dust	137.0 mg/kg	8–336

Source: Data from Ref. 10.

demonstrated to lower PAH concentrations by two orders compared to traditional smoking, possibly due to the sorption process of PAHs into polyethylene packaging materials [15]. Recently, PAH concentrations of 91 μg/kg in liquid smoke flavor filled into low density polyethylene bottles were observed to decrease to zero in 164 hours, offering a potential solution to the problems caused by the presence of PAHs in foods or food additives [16]. The mutagenicity of wood smoke condensates typically experienced by people in developing nations has been shown to be more mutagenic than cigarette smoke condensate, depending on the type of wood and combustion conditions and allowing for the possibility of minimizing exposure to the more carcinogenic PAHs by careful selection of wood combinations and combustion conditions [17]. Gas-fire drying of wheat grain does not appear to increase PAH levels [18].

The mechanisms of action of food-associated PAH carcinogens have become clearer recently, and it appears that they require metabolism to dihydrodiol epoxide metabolites in order to express their biological activities. Each PAH-reactive metabolite is unique in its interactions with cells but as sufficient to shift the reaction with DNA from the amino group of guanine residues to the amino group of adenine residues [19]. It has also been hypothesized that food restriction enhances detoxification of PAHs in the rat and may be an important mechanism involved in the observed protective action of reduced food intake [20].

II. SAMPLE PREPARATION

The PAHs from solid samples such as air particulates, soils, sediments, and food are traditionally extracted by Soxhlet extraction using a variety of organic solvents

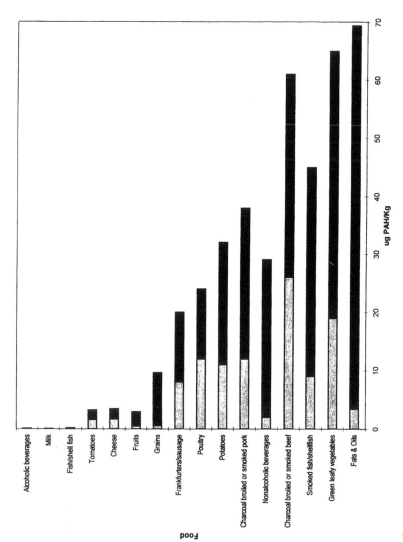

FIGURE 3 Average minimum and maximum reported concentrations (μg/kg) of carcinogenic PAHs in food items.

Table 2   Effect of Cooking Mode and Cooking Time on the Level of
Benzo[a]pyrene (μg/kg) Found in Foods

Cooking time (minutes)	Cooking mode	Level found in mutton (μg/kg)	Level found in chicken (μg/kg)
15	Pressure cooker	17	7
4	Microwave	63	47
10	Pan frying	102	not detected
15	Pan frying	195	118
20	Pan frying	332	192
25	Pan frying	471	272

Source: Data from Ref. 12.

including acetone, benzene, toluene, and methylene chloride. Although Soxhlet extraction is efficient for many samples, it requires large volumes of solvents (ca. 300 ml), is time consuming (ca. 6–24 hours), and may yield incomplete recovery of higher molecular weight PAHs from materials on which they are strongly adsorbed (i.e., carbon black or coal fly ash). For these reasons, alternatives to Soxhlet extraction have been developed recently including ultrasonic methods, supercritical fluid extraction (SFE), accelerated solvent extraction (ASE), and microwave assisted extraction (MAE). The main advantages of these new methods are reduced extraction time, reduced solvent usage, and greater sample throughput/reduced sample handling, due, in part, to the ease of automation of these new techniques. The recovery of PAHs from aqueous solutions has traditionally involved liquid–liquid extraction (LLE) methods with organic solvents (i.e., hexane, dichloromethane, chloroform, etc.). Since environmental samples

Table 3   Effect of Barbecue Geometry and Cooking Time on PAH Levels (μg/kg)
in Sardines

PAH	Raw	Vertical, medium (2 × 4 min)	Horizontal, rare (2 × 2 min)	Horizontal, medium (2 × 4 min)	Horizontal, well done (2 × 8 min)
Benzo[a]pyrene	<0.1	0.2 ± 0.1	1.0 ± 0.2	30 ± 2	32 ± 2
Fluoranthene	<0.1	1.1 ± 0.3	1	70 ± 30	60 ± 10
Benzo[b&k]fluoranthene	<0.1	0.4 ± 0.2	1.3 ± 0.4	30 ± 2	30 ± 6
Indeno [123cd]pyrene and benzo[ghi]perylene	<0.1	<0.1	2.3 ± 0.5	55 ± 2	60 ± 5
Total	<0.4	1.7	5.6	185	182

Source: Data from Ref. 14.

generally contain interferents and trace amounts of PAHs of interest, concentration and cleanup procedures are usually required prior to the final chromatographic analysis. In many cases, the sample pretreatment procedure is the critical step in achieving reliable quantitative results. The most promising alternatives to LLE include solid phase extraction cartridges and disks and solid phase microextraction (SPME).

PAH concentration and cleanup is increasingly being performed by solid phase extraction (SPE). For preconcentration of PAHs from drinking water samples, best results were obtained for combined octadecylsilane (C18)/ammonia (NH$_2$) solid phase cartridges, whereas the enrichment of PAHs from soil samples was best achieved with silica (Si)/cyano (CN) or C18/CN combinations [21]. Saponification and silica gel cleanup of Soxhlet extracted PAHs from sewage sludge–amended soils have been reported to be superior to other methods including XAD-2 cleanup [22]. A newly developed SPE method for the analysis of the 16 EPA priority PAHs recommends C$_8$ SPE columns [23]. The choice of SPE sorbent type is often dictated by the chromatographic method to be subsequently used for PAH separation and identification. For example, a recent study showed that for the determination of PAHs in lake sediments, C18 and silica columns could be satisfactorily used to clean up extracts for subsequent HPLC analysis with fluorescence detection; however, they could not be used for GC-MS for PAHs greater than chrysene due to interferences from aliphatic waxes. Fully activated silicic acid and neutral alumina columns were recommended [24]. Silica cleanup was also recommended recently for the analysis of PAHs from freeze-dried sediment samples after Soxhlet extraction [25]. A standard leaching test employing SPE with C18 packings has proven to be a fast, reliable method for determining the PAH leachability from waste materials [26]. Florisil (SiO$_2$ and MgO) cartridges have yielded rapid and efficient recovery of PAHs for petroleum and sediment extracts [27]. Extraction and concentration of PAHs in oils was achieved by charge-transfer liquid chromatography on an improved tetrachlorophthalimidopropyl-bonded silica [28]. A quantitative procedure for the determination of PAHs in biomass tar has been described using SPE with aminopropylsilane packings [29]. Chromosorb T and XAD-2 have been compared for the in situ extraction of PAHs from fresh- and seawater. Neither sorbent was useful for PAHs with molecular weights less than that of phenanthrene due to low recoveries or PAH contaminants and were comparable for the study of three-ring and higher PAHs [30]. A convenient method for the separation and preconcentration of traces of PAHs in aqueous solutions has been achieved by adsorption on cobalt phthalocyanine and barium salts of sulphophthalocyanines followed by thermal desorption gas chromatography [31]. A procedure for the size-selective extraction of PAHs from oil-in-water microemulsions has been reported using cyclodextrins [32]. A recent study illustrates the difficulties in interpreting results from samples spiked with PAHs in the laboratory. The recovery of PAHs from waste incinerator fly ash has been shown to be influenced by the spiking method, PAH concentration, molecular size, storage time, and carbon content of the matrix [33].

Numerous studies have dealt with the sampling procedures for airborne PAHs, including the combination of a Teflon filter for particulates with polyurethane foam (PUF) for gaseous compounds [34]. A comparison of filter material used in high-volume sampling of PAHs revealed that glass-fiber filters yielded substantially higher concentrations of the lower molecular weight PAHs compared to Teflon filters [35]. A new technique for controllable vapor phase deposition of PAHs onto particulate matter was developed to provide particle-bound radiolabeled substrate for use in metabolism and toxicological studies [36]. The sorption and desorption properties of PAH-coated ultrafine particles were studied with a photoelectric sensor revealing the following ranking for desorption temperatures: Aerosil 200 > aluminum oxide > carbon > sodium chloride [37]. The concentrations of PAHs adsorbed onto air particulates was found to correlate negatively with temperature during the sampling, due to volatilization, photodegradation, and seasonal modifications of emissions from urban traffic [38]. Storage stability tests revealed that PAHs (fluoranthene/pyrene to coronene) from ambient air collected on Teflon/PUF can be stored in the dark in closed vessels at room temperature for up to 118 days without observable losses [39].

Details of the preparation and analysis of a new National Institute of Science and Technology (NIST) Standard Reference Material (SRM) 1941 have been described. SRM 1941, Organics in Marine Sediment, has been certified for concentrations of 11 PAHs and provides noncertified values for 24 additional PAHs using results from gas chromatography (GC) with flame ionization detection, GC/mass spectrometry (MS), and liquid chromatography with fluorescence detection [40]. Improvement in the precision and accuracy of the analytical procedures used by fourteen European laboratories should now permit the certification of coconut oil reference materials for low concentrations of PAHs [41]. A simplified version of an HPLC method has been described for the determination of PAH in suspended particles from small air volumes collected indoors, outdoors, and in personal exposure measurements. A comparison of procedures for the determination of PAHs in low-volume samples has recently appeared [42]. A simple low-pressure liquid chromatography procedure has been developed for the isolation of PAHs from shale oil followed by gas chromatographic analysis [43].

Alternatives to classical Soxhlet extraction include ultrasonication, which has been demonstrated to reduce analysis time and solvent use with comparable efficiencies for a variety of samples including PAHs in soil [44]. Microwave-assisted extraction (MAE) of PAHs from six certified reference materials has shown the technique to reduce analysis time to less than one hour and reduce solvent use to 30 ml with recoveries for eleven PAHs in the 65–85% range, and three PAHs (acenaphthalene, benzo[a]pyrene, and fluorene) had recoveries of ca. 50%, possibly due to degradation [45]. Other studies have confirmed the potential for incorporating MAE into rapid, low solvent use, PAH analysis methods with comparable recoveries to Soxhlet for marine sediments [46] and highly contaminated soils [47]. A new alternative technique for the extraction of PAHs from solid matrices has recently been introduced called accelerated solvent extraction (ASE) combining elevated temperatures (50–200°C) and pressures (500–3000 psi) with liquid solvents and short extraction times (ca. 10 minutes). ASE has been demonstrated to yield comparable recoveries to Soxhlet extraction for PAHs from Urban Dust Standard Reference Material 1649 and will likely successfully compete with MAE and SFE for replacement of Soxhlet extraction in the near future [48]. A promising new extraction technique called solid phase microextraction (SPME) has been introduced and refined by Pawliszyn and coworkers for the recovery of compounds, including PAHs, from liquids and headspaces above solids, that is fast, inexpensive, solventless, and readily automated [49]. PAHs have been analyzed by SPME from a variety of samples and in different modes including from the headspace of solid environmental samples [50] and forensic samples [51,52] and directly from water [53]. Recently, subcritical water (250–300°C, 50–350 bar) has been demonstrated to yield efficient extraction of PAHs from environmental solids in 15 minutes due to the decreased dielectric constant of water at elevated temperatures [54]. Subcritical water extraction has also recently been combined with SPME yielding an inexpensive method for the efficient extraction of PAHs from environmental solids using no organic solvents [55].

III. SUPERCRITICAL FLUID EXTRACTION AND CHROMATOGRAPHY

Supercritical fluid extraction (SFE) has proven to be a powerful alternative to conventional liquid extraction methods used in environmental analysis [56,57]. PAHs have been extracted directly from endogenous solid and liquid matrices, as well as trapped onto solid adsorbents with subsequent recovery by SFE [58,59]. Solid sorbents can also be used for the efficient trapping of PAHs extracted via SFE from soils and sediments [60,61]. One major advantage of SFE is the relative ease with which it can be coupled to any chromatographic technique including gas chromatography (GC), high-performance liquid chromatography (HPLC), and supercritical fluid chromatography (SFC). Hyphenated SFE-CG [62–64] and SFE-SFC [65–67] techniques have been applied for the determination of PAHs from environmental samples. SFE-HPLC has been applied more recently for the determination of PAHs from a variety of matrices ranging from urban dust particulates [68] to smoked and broiled fish [69]. Comparisons of SFE to Soxhlet, sonication, and MAE indicate that further optimization and recipe-type procedures for SFE are needed for its acceptance [70,71].

    Carbon dioxide is the primary fluid used in most SFE applications because it has low critical points ($T_c = 31.3°C$, $P_c = 1070$ psi) and is nontoxic, nonflammable, ordorless, readily available in high purity, and inexpensive; and it eliminates solvent waste disposal problems. Unfortunately, the nonpolar nature of carbon dioxide has hindered its application for the recovery of higher molecular PAHs or those strongly adsorbed to (or trapped in) the environmental matrix. Workers have striven to overcome this limitation in recent years. Alternative fluids such as N_2O and $CHClF_2$ (Freon-22) yield higher recovery of PAHs from petroleum waste sludge and railroad bed soil, compared to CO_2 [72]. Alternatively, the use of organic solvent modifiers (i.e., toluene, methanol) or in situ chemical derivitization has been shown to improve the recovery of PAHs while still employing the preferred supercritical fluid, carbon dioxide [73,74]. Most moderate temperature (ca. 90–120°C) SFE methods developed for the recovery of PAHs from environmental solids have relied on the addition of modifiers including water, methanol, dichloroethane, trifluoroacetic acid, and triethylamine [75–79]. Other studies have focused on optimizing the major controllable SFE variables and minimizing problems including restrictor plugging, particularly when extracting high molecular weight PAHs or employing samples with a high sulfur content. One approach to minimizing restrictor blocking is to employ a copper scavenger column placed after the sample cell. By this technique, SFE of PAHs was accomplished for high sulfur content samples without restrictor blocking [80]. Restrictor plugging while extracting PAHs has also been minimized by nebulizing an organic solvent with the restrictor effluent or by simply heating the restrictor from 50 to 200°C, depending on the analyte and the sample matrix [81].

Models for dynamic SFE have been proposed and applied to real systems including the SFE of the PAH, phenanthrene, from railroad bed soil with generally good agreement [82]. Models such as this are useful as they provide an extrapolation method for obtaining quantitative analytical extractions in the shortest analysis time. A dynamic tracer response technique has been applied for simultaneous measurement of equilibrium and rate parameters for the dynamic extraction of analytes from solid matrices. The technique has allowed adsorption equilibrium constants, effective diffusivities, and axial dispersion coefficients to be determined for the system naphthalene-alumina-supercritical CO_2 [83]. We have reported the measurable effect of the extraction vessel dimensions (id : length) on the elution of PAHs from octadecylsilane SPE sorbents [84], and the relative effects compare to the two major controllable variables, namely, temperature and density [85]. These results from SPE sorbents differ from those seen for the SFE of PAHs directly from environmental solids where no effect is observed [86]. The effect of SFE variables for the elution of PAHs from different SPE sorbents indicates that this process is more like SFC, whereas the extraction of PAHs from real samples is generally matrix dependent [87–89]. The extremely different extraction behaviors of spiked versus native PAHs from heterogeneous environmental samples has been demonstrated for SFE and sonication [90]. The strong influence of the matrix on the kinetics of dynamic SFE of analytes has been discussed and modeled [91]. Recent studies have indicated that temperature is more important than pressure (density) for the SFE of PAHs from octadecyl sorbents [85] and from real matrices [92]. High temperature (HT) SFE has recently been shown to yield much faster extraction of PAHs from geological samples than conventional Soxhlet extraction. Figure 4 compares the recovery of PAHs from New Albany shale samples by 30 minute HT-SFE (350°C, 5000 psi CO_2) compared to 48 hours Soxhlet extraction with methylene chloride [93]. With the exception of the recoveries of high molecular weight PAHs ($M \geq 252$) from a soot sample, SFE at 200°C with pure CO_2 yielded comparable recoveries to 18 hour Soxhlet extraction for real-world environmental samples [94]. Temperature has been shown to increase the solubility of PAHs in supercritical CO_2 greater than that seen with increasing pressure (at constant temperature) using on-line flame ionization detection [95,96]. Increasing the SFE temperature has been shown to increase the desorption kinetics of all of the analyte–matrix combinations studied, including PAHs from marine sediments and railroad bed soil [97]. Increasing SFE temperature from 80°C to 200°C increased recoveries of PAHs from marine sediment, diesel soot, and air particulates, and recoveries were further enhanced with the addition of 10% (v/v) diethylamine [98]. Stepwise high-temperature SFE has recently been applied to study the speciation of PAHs in geochemical samples. Figure 5 shows the extraction of naphthalenes and alkylnaphthalenes from Posidonia shale with 1 hour supercritical carbon dioxide at successively increasing temperatures from 50°C to 350°C with overall recoveries much higher than that observed employing Soxhlet extraction [99]. Stepwise HT-SFE is a useful tool in the study of weakly versus strongly bound (or trapped) PAHs in environmental and geological samples.

Supercritical fluid chromatography (SFC) has been applied to extend the molecular weight range of PAHs normally separated by gas chromatography (GC) while reducing analysis times typical with HPLC. The modification of a GC/MS to SFC/MS mode has been described and used to separate PAHs with molecular weights up to 532 [100]. The retention behavior of PAHs in SFC for various stationary phases has been shown to be controlled by molecular size, as in liquid chromatography, but also influenced by additional parameters including solute dipole moments, solubilities, and volatility [101]. A molecular theory of chromatography based on mean-field statistical thermodynamics has been developed to describe the partitioning of blocklike molecules such as PAHs between an isotropic mobile phase and an anisotropic stationary phase. The theory was qualitatively applied to the interpretation and analysis of experimental data in gas, liquid, and supercritical fluid chromatography [102]. The supercritical fluid retention of PAHs on a polymeric smectic phase has been compared to theoretical predictions using this molecular theory of chromatography with encouraging results [103]. The potential for predicting the utility of SFE from existing supercritical fluid chromatographic retention data has recently been addressed [104]. In

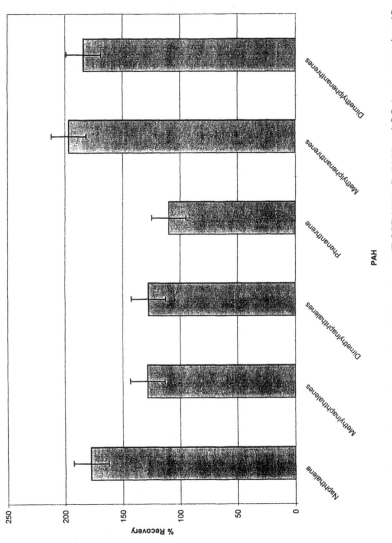

FIGURE 4 Recoveries and precision of 30 minute HT-SFE (350°C, 5000 psi CO_2) compared to 48 hour methylene chloride Soxhlet extractions.

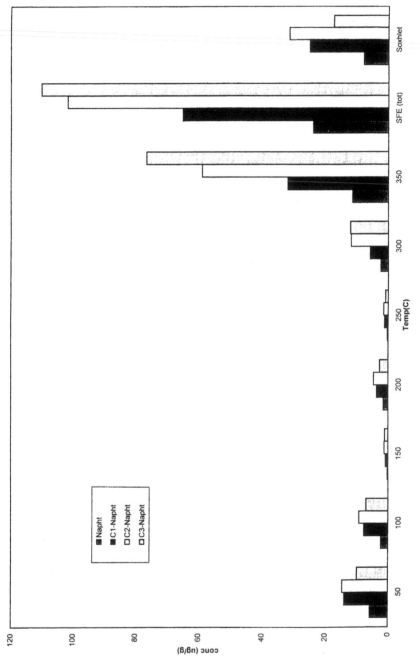

FIGURE 5 Naphthalene and alkylnaphthalene extraction recoveries obtained during stepwise supercritical carbon dioxide extractions (1 hour at each temperature followed by 30 minutes at 50°C) compared to 24 hour methylene chloride Soxhlet extractions.

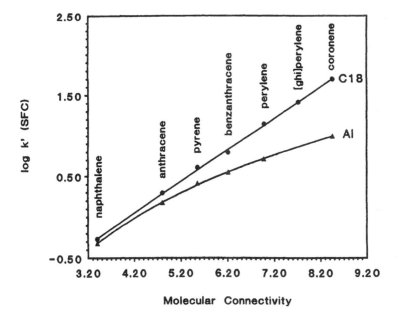

FIGURE 6 Plot of SFC log k' versus the molecular connectivity for PAHs. C18 = octadecyl column, 100°C, 300 atm, supercritical carbon dioxide mobile phase; Al = alumina column, 245°C, 47 atm, supercritical isopropanol mobile phase.

addition, we have found that of numerous physical and molecular descriptors studied, the molecular connectivity correlates best with SFC retention data for normal and reversed phase systems. The excellent correlation between molecular connectivity and the logarithm of the capacity factor for different SFC systems is illustrated in Figure 6 with linear correlation for the reversed phase octadecysilane column and logarithmic for the normal phase alumina column. A recent SFC method has been published demonstrating the separation of the 16 EPA PAHs in 6 minutes compared to 24 minutes by HPLC, although acenaphthylene and ace-naphthene could not be resolved, and benzo[ghi]perylene and indeno[1,2,3-cd]pyrene eluted in reverse order compared to HPLC [105]. A SFC method with CO_2/acetonitrile using UV has been shown to be faster, and it reduced solvent consumption by 56% compared to HPLC for the separation of the 16 EPA PAHs [106].

IV. GAS CHROMATOGRAPHY

A review of gas and high-performance liquid chromatographic techniques for the analysis of PAHs in airborne particulates has recently been published [107]. Gas chromatography is the method of choice for high-resolution separation of complex PAH mixtures with moderate to low molecular weights. Gas chromatographic analysis of high molecular weight PAHs (exceeding 500 amu) has been performed; however, these analyses have traditionally involved very short columns with correspondingly low resolution. Improved phases for high-temperature use continue their development and will allow even higher molecular weight PAHs to be separated while maintaining high resolution. A recent comparison of four high-temperature GC columns illustrated their utility for the high-resolution separation of PAHs with a molecular weight of 328 with seven-ring PAHs including dinaphtho(2,1-a:2,1-h)anthracene eluting in less than 35 minutes [108]. A column coated with a biphenyl-substituted silarylene-siloxane copolymer has been demonstrated to be stable up to 400°C, allowing GC analysis of PAHs up to mass 450 with much higher separation efficiency compared to commercially available 50% methyltrifluoropropyl-substituted polysiloxane or 5% diphenyl-substituted methylpolysiloxane phases [109]. The utility of selective liquid crystalline phases has been demonstrated for the determination of bioconcentration factors of PAHs in polychaete worms [110]. GC using 1,1'-binaphthyl as an internal standard has been used to quantitate the PAH exposure of fish in a laboratory flow-through system containing sediments contaminated with coal-tar creosote [111]. At concentrations of 16 to 320 μg/liter fish refused food, severe fin erosion occurred above 76 μg/liter, and hepatic microsomal ethoxyresorufin *O*-deethylase (EROD) activity increased in the first 2 days of exposure and then declined at concentrations above 35 μg/liter.

Another recent study praised the long column lifetime (more than six years) and short analysis time (ca. 15 minutes) of a liquid-crystalline stationary phase for the analysis of PAHs in carbochemical products [112]. A double internal standard procedure has been described to increase the precision and accuracy of PAH determinations by GC [113].

Several recent studies have focused on relationships between GC retention of PAHs and molecular properties. Regularities of GC retention behavior have been described by structural models containing Van der Waals volume and molecular connectivity indices of different levels depending on the class of PAHs [114]. The heats of adsorption for PAHs on macroporous silicas has been studied by GC in the range 80–200°C [115]. Investigation of the relationship between GC retention indexes and computer-calculated physical properties of PAHs revealed that molecular polarizability was the most important property [116]. A study of the relationship between GC retention and thermal reactivity for PAHs in coal tar pitch indicates that, generally, those that are retained on OV-1701 more strongly than on SE-54 stationary phase are more thermally stable [117]. Although flame ionization detection and mass spectrometry (MS) remain the methods of choice for PAH detection, other methods continue to show promise. GC/Matrix isolation infrared spectrometry (MI-IR) can identify PAHs difficult to distinguish by EI-MS [118]. The GC/Fourier transform IR spectra of 33 PAHs have been measured and interpreted [119]. Cryotrapping GC-FTIR of soil extracts allowed for the conformational analysis of PAHs at a level of 1–4 ng per component injected, and previously undetermined PAHs were identified [120]. A highly sensitive method for determining PAHs using GC-ECD after derivatization with bromine has been presented [121]. Pentafluorobenzyl bromide has been used for the derivatization of hydroxy-PAHs followed by GC with electron capture detection or negative ion chemical ionization mass spectrometry with 0.01 to 3.3 pg detection limits [122]. Other selective detectors, including the photoionization detector (PID), continue to find application particularly when attempting to eliminate interferences such as aliphatic hydrocarbons [123].

Analysis techniques for PAHs by GC and GC/MS continue their development. A complete method for the determination of PAHs in soil by GC has been described [124]. Thermal desorption GC/MS has been applied to the analysis of PAHs in contaminated soils [125,126]. The importance of solvent choice and initial column temperature has been investigated for PAH determination by GC/MS with splitless injection [127]. The use of toluene and xylenes gave enhanced signals up to 100 times greater compared to other solvents for the higher molecular weight PAHs through benzo[ghi]perylene. Peak splitting observed in the capillary GC separation of oxo- and nitro-PAHs can be eliminated by careful selection of injection solvent (acetonitrile is found to be best), initial column temperature (60°C optimal), and the use of a retention gap [128]. Complete analytical methodologies including GC/MS analysis have recently been described for the determination of PAHs in sediments [129], in glass manufacturing oils [130], in soils [131], and in soot produced by combustion of plastics and wood [132]. A recent method for the analysis of PAHs in vegetable oils and fish includes a gel permeation chromatography cleanup step prior to final analysis by GC/MS [133]. A GC/MS method based on microextraction and large volume injection of a toluene extract allows for the determination of ppt levels of PAHs in drinking and drainage water [134]. A method for the determination of low levels of 1-nitropyrene and 2-nitrofluorene in low-volume air samples has been developed based on acetone extraction by sonication, silica gel column fractionation, nitroreduction, and derivatization with heptafluorobutyric anhydride followed by GC/MS analysis [135]. A routine method for the analysis of PAHs in street dust has been used based on SFE-GC/MS with supercritical CO_2 [136]. GC with ion-trap mass detection yielded higher sensitivities for a wide range of PAHs with full-scan spectrum than those previously obtained by MS with quadrupole detection using selected ion monitoring or by HPLC with fluorescence detection [137].

V. LIQUID CHROMATOGRAPHY

Since its inception more than twenty five years ago, high-performance liquid chromatography (HPLC) has been applied to the separation of PAHs. A recent review of the determination of PAHs by liquid chromatography has recently appeared [138]. HPLC is the method of choice for moderate to high molecular weight PAHs and PAH metabolites. Although HPLC still cannot compete with GC in terms of high efficiency and short analysis times, it does offer numerous advantages, including very sensitive and selective detectors and the ability to be used as a fractionation method for other chromatographic or spectroscopic techniques. The application of HPLC for PAH fractionation has become very popular due to its high efficiency, ease of automation, and potential for column switching techniques and on-line coupling with other techniques including gas chromatography. The recoveries of PAHs from a soot sample were compared for ultrasonic ether, Soxhlet toluene, and Soxhlet extraction with liquid CO_2 followed by HPLC fractionation [139]. The results showed liquid CO_2 Soxhlet extraction to be superior for lower molecular weight PAHs (to chrysene). A routine method for the analysis of mononitro-PAHs in environmental samples has been developed based on micro-scale liquid–liquid partition (dimethylformamide/cyclohexane) and silica column HPLC fractionation prior to GC/ECD and GC/MS [140]. HPLC fractionation followed by GC analysis has been applied to the determination of PAHs in urban street dusts with primary components found to range from phenanthrene (three aromatic rings) to benzo[ghi]perylene (six aromatic rings) [141]. A column switching technique utilizing a silica gel and an aminosilane-bonded silica gel column has been used to separate PAHs in lubricating oil base stocks into compound class fractions followed by GC-MS analysis [142]. A similar HPLC column switching technique with silica gel and aminosilane-bonded silica gel columns has been used to fractionate monomethylated PAHs from heavy oil followed by GC analysis [143]. Column-switching HPLC techniques have also been developed for the analysis of PAHs in petroleum products [144] and the group separation of PAHs and nitrogen containing PAHs [145]. A fully automated column switching HPLC method has been developed for the determination of 1-hydroxypyrene in urine of subjects exposed to PAHs [146]. A new on-line concentrator has been developed and applied to the analysis of PAHs in soot by on-line HPLC-GC [147]. On-line HPLC fractionation-GC analysis has also been applied to the separation and identification of PAHs in heavy oil [148] and urban air samples and automobile lubricating oil [149]. On-line LC-GC-MS has been demonstrated for the analysis of PAHs in vegetable oils with detection limits down to 1 pg with selective ion monitoring [150] and for chlorinated PAHs in urban air down to 250 fg [151].

The most popular method of PAH separation is reversed phase HPLC with octadecylsilica phases dominating. Notable exceptions include anthyrl-modified silica phases used to separate PAHs and nitro-PAHs [152], and phenyl-modified silica gel column used for improved separation of ^{32}P-labeled nucleoside 3′,5′-bisphosphate adducts of PAHs [153]. Multidentate phenyl-bonded phases have been shown to provide higher nonplanarity recognition of PAHs than that typically seen for octadecylsilica phases [154]. Tetraphenylporphyrin-based stationary phases have been shown to provide shape-selective separation of PAHs [155]. Reversed phase HPLC provides unique selectivity for the separation of PAH isomers and particularly alkyl-substituted PAHs. Anomalous retention behavior of methyl-substituted PAHs on polymeric C18 phases was found to be related to the nonplanarity of the PAHs due to the presence of the methyl group in the so-called "bay-region" of the PAH structure [156]. A recent comparison of C18 packing indicated that oligomeric not endcapped C18 packing materials are optimal for the separation of PAHs [157]. Microcolumn C18 HPLC with 200,000 theoretical plates was used to separate a standard mixture containing 16 PAHs employing selective fluorescence quenching [158].

Numerous standard methods for the analysis of PAHs employing C18 columns have recently appeared, including the determination of PAHs in air particulate samples [159], diesel soot [160], biomass emissions [161], mineral waters [162], sea mussels [163], and oyster tissues [164]. A variety of applications have involved the separation and detection of PAH metabolites. The determination of 1-hydroxypyrene in human urine has been developed as an indicator of exposure to PAHs [165,166]. Similarly, hydroxy-phenanthrene has been detected after intake of PAHs [167]. Details of the metabolism [168], as well as the HPLC separation of nitro-PAHs, has been described [169]. A combination of both reversed and normal phase HPLC was used to provide efficient separation of the ring-oxidized derivatives of nitro-PAHs. A method using three C18 columns in tandem has been described for the separation of fish biliary PAH metabolites [170]. Additional methods employing C18 HPLC separation have been developed for use in a variety of applications, including the study of PAHs originating from cooking food [12–14] and burning wood [171], and from sewage sludge–amended agricultural soil [172]. A method for derivatizing PAHs to quinones for C18 HPLC with selective electrochemical detection has been applied to the detection of selective PAHs in tap water and motor oil [173]. Recent C18 HPLC methods for the determination of PAHs in water have used SPME collection and fractionation including XAD-2 sorbents [174] and a procedure demonstrating that PAHs sampled via SPME on-site are stable for at least one week prior to elution and determination at ng/L-levels in drinking and surface waters [175]. On-line reduction followed by C18 HPLC has been applied to the determination of mono- and di-nitro PAHs from diesel exhaust particulates using chemiluminescence detection [176].

More HPLC/MS methods have appeared in recent years, including the analysis of hydroxy-PAHs by C18 HPLC followed by pneumatically assisted electrospray mass specrometry [177]. C18 HPLC/particle beam mass spectrometry (PBMS) has been used to detect and identify high molecular weigh PAHs in soils [178], PAHs and oxygenated metabolites in sediment and water samples from the Exxon Valdez oil spill in Alaska [179], and metabolites in biologically treated hazardous waste [180]. Although HPLC/PBMS has powerful possibilities for determining the identities of PAHs and metabolites it also has limitations including time consuming data interpretation, often with ambiguous results, and sensitivity problems for metabolites requiring sample enrichment prior to analysis [180]. Micellar liquid chromatography (MLC), in which critical micelle concentrations of surfactants are present in the mobile phases of reversed phase HPLC systems, has been applied to the analysis of PAHs with limited success due to its reduced efficiency compared to conventional systems. Correlations between retention data for PAHs in MLC and various molecular descriptors have been demonstrated [181], and a comparison of retention models has recently appeared [182]. PAH retention in MLC has been correlated to octanol–water partition coefficients [183], to the number of carbons in the PAH [184], and to molecular descriptors when organic modifiers are present [185]. Lower fluorescence detection limits have been reported for the MLC of PAHs using 3% 2-propanol and 0.035 M SDS compared to conventional reversed phase HPLC, attributed to the higher fluorescence intensity of PAHs in the presence of the micellar mobile phase [186].

The retention mechanism in reversed phase liquid chromatography for large PAHs has been investigated by Fourier transform infrared spectroscopy, nuclear magnetic resonance spectroscopy, and differential scanning calorimetry. The results indicate that a change in mobile phase from methanol to dichloromethane induces further nonplanarity in nonplanar solutes, whereas increases in column temperature drastically change the structure of the stationary phase from solidlike to liquidlike, with subsequent losses in planarity recognition [187]. The use of subambient temperatures (i.e., 0 to $-20°C$) can significantly enhance shape selectivity of PAHs for polymeric C18 phases and result in a phase with liquid-crystal-like retention properties [188]. Results for a novel wide-pore C18 bonded phase packed into a 3.0 mm id column showed that lower temperatures (down to 10°C) enhanced the selectivity of critical PAH isomer pairs [189]. A recent study of PAH retention time reproducibility on C18 columns showed the importance of thermostating the HPLC column with indeno[1,2,3-cd]pyrene, diben[a,h]anthracene, and benzo[ghi]perylene, especially in case of small variations in column temperature [190]. The use of microcolumns packed with monomeric and polymeric C18 phases allowed for temperature programming without the radial thermal gradient problems seen across conventional columns, and all sixteen EPA PAHs could be separated without applying gradient analysis using temperature optimization [191]. Retention characteristics of nitrated PAHs on C18 columns demonstrated a linear dependence of the logarithm of the capacity factor versus both the organic modifier concentration and the reciprocal of the absolute column temperature, allowing thermodynamic variables to be evaluated [192]. The retention in nonaqueous reversed phase HPLC has been studied with the use of large PAHs and correlated to the amount of red shift in the fluorescence spectra for 11 common HPLC solvents [193]. The elution order of 10 PAHs up to coronene on

C18 columns could not be satisfactorily explained with retention models based on molecular weight and length-to-breadth ratio alone, but improvement was made by taking into account the effects of intramolecular steric strain and the resulting degree of nonplanarity [194]. The separation of large PAHs (benzo[ghi]perylene to ovalene) from a diesel particulate extract by reversed phase HPLC with photo-diode array detection revealed that all of the PAHs of six or more rings were highly fused; no linear or nonalternate types were seen [195]. Relationships of structures of forty-six related nitro-PAHs and their corresponding parent PAHs with C18 HPLC retention order suggests that the polarity of the PAH or nitro-PAH is the principal factor determining its HPLC retention time [196].

Although reversed phase HPLC has dominated PAH separations, numerous normal phase and specialty columns have been investigated to improve the selectivity of PAH separations. The separation of amino- and acetylamino-PAHs using six different reversed and normal phase columns has been compared with a Pirkle-type chiral phase exhibiting the best separation [197]. Chiral stationary phases derived from nitrated fluorenylideneaminooxy carboxylic acids covalently bonded to silica gel via aminopropyl spacers have been shown to yield enantiomeric resolution of dihydrodiols of PAHs by π-donor-acceptor interactions [198]. Derivatization of dihydrols of PAHs to methyl ethers has been shown to improve enantiomeric separation on Pirkle-type chiral stationary phases [199]. A Liquid-crystal-bonded phase has been shown to possess a planarity recognition power higher than that seen for typical polymeric C18 phases [200]. A comparison of several normal phase packings demonstrated the advantage of a cyanopropyl-dimethyl-bonded silica gel packing for the group separation of chloro-added and chloro-substituted PAHs [201]. The effect of polar mobile phase modifiers on the retention of various classes of PAHs and the selectivity of their separation by normal phase (silica gel) HPLC has been studied. The linear dependence of the logarithm of the capacity factor versus the number of carbon atoms of the sorbate has been studied for unsubstituted PAH, monoalkylbenzene, and monoalkylnaphthalenes with different mobile phases [201]. The effect of the column material and mobile phase solvents on retention of PAHs in size exclusion chromatography has been investigated. The use of sulfonated poly(divinylbenzene) packings has been shown to improve the performance of this technique, whose application to PAH analysis has been problematic [203]. The incorporation of cyclodextrins has received attention recently as a way of improving the selectivity of PAH separations. The unique shape selectivity of a cyclodextrin phase towards eleven five-ring PAHs has been compared to C18 columns. Although the polymeric C18 column demonstrated the highest overall selectivity for the PAH isomers, it was suggested that the lack of retention dependence on molecular weight could be advantageous for the separation of PAHs of different molecular weights [204]. Bonded β-and γ-cyclodextrin phases showed relatively low efficiency but high selectivity, allowing the separation of different classes of isomeric compounds including PAHs that were difficult to separate on conventional LC stationary phases [205]. β-Cyclodextrin has also been used as a selective inclusion reagent in reverse phase HPLC separation of PAHs. Molecular interactions of PAHs were determined to be mostly electrostatic, and retention order was strongly influenced by PAH molecular shape [206,207].

VI. MISCELLANEOUS TECHNIQUES/APPLICATIONS

Thin-layer chromatography (TLC) is often used as a sample preparation method in PAH analysis, particularly when oil samples are involve. Standardized gas-chromatographic methods employing TLC separation have been developed for the determination of PAHs in petrochemical plants and oil refineries [208] and in olive oil [209]. The utility of urea-solubilized β-cyclodextrin TLC mobile phases was demonstrated by the resolution of a variety of compounds, including four PAHs, on a polyamide stationary phase [210]. The analysis of PAHs in spa waters has been accomplished by two-dimensional TLC using plates containing a mixture of aluminum oxide, silica gel, and acetylated cellulose with detection by spectrofluorometry [211]. Laser mass spectrometry has been used to detect separated PAHs directly from polyamide TLC plates [212]. On-line coupled HPLC-TLC allowed for the successful separation of PAHs in marine sediment using a simple isocratic microbore HPLC system and a conventional fluorescence spectrometer for detection [213].

The effectiveness of the anionic surfactant dodecylsulfate in solubilizing various PAHs has been studied for several different sediment and soil solid phases and aqueous phases [214]. The solubilization and partitioning of PAHs between micelle-phase/aqueous phase has also been determined for nonionic polyoxyethylene surfactants [215]. This data can be combined with additional information on surfactant and PAH sorption on soil/sediment to understand mechanisms effecting the behavior of PAHs in soil/sediment–water systems in which surfactants play a role in contaminant remediation or facilitated transport. The effect of varying sodium dodecylsulfate concentration and the introduction of γ-cyclodextrin in the micellar electrokinetic capillary chromatographic separation of PAHs investigated with an average theoretical plate number of more than 160,000 [216, 217]. Micellar extraction of PAHs from aqueous media using different nonionic surfactants allowed for the enrichment and the simultaneous detection of extracted PAHs by synchronous fluorescence in the micellar phase while eliminating the need for large volumes of organic solvents typical with liquid–liquid extractions [218]. On-line micelle-mediated preconcentration on selective sorbents using Brij-35 surfactant has been used to analyze PAHs from surface waters at the low-to sub-ng/L level [219].

Direct experimental determinations of Henry's law constants for 9 PAHs using a wetted-wall column technique and gas chromatographic analysis compared favorably to other calculated and measured values [220]. Molecular topology has been used to model *n*-octanol/water partition coefficients of PAHs and their alkyl derivatives [221]. PAH partitioning mechanisms with activated sludge have been studied and an equation developed from thermodynamic principles to estimate lipid–waste water distribution coefficients [222]. PAH-chlorobutane and PAH-dichlorobutane equilibrium constants have been calculated from solubility data [223]. The potential health hazard from water chlorination due to formation of chlorinated PAH has been investigated. Chlorination of PAH-contaminated humus-poor lake water was found to result in formation of chlorinated derivatives for some PAHs. However, in the presence of high concentrations of humic substance, no chlorinated PAHs were detected [224]. The extent of reaction of PAHs with hypochlorite has been shown to depend on the chlorine dose, the solution pH, the concentration of both compounds (high values studied), and the structures [225].

REFERENCES

1. K. D. Bartle, M. L. Lee, and S. A. Wise, *Chem. Soc. Rev. 10*: 113 (1981).
2. L. B. Ebert, ed., *Polynuclear Aromatic Hydrocarbons*, ACS Symposium Series 217, American Chemical Society, Washington, D.C., 1988.
3. K. Peltonen and T. Kuljukka, *J. Chromatogr. A 710*: 93 (1995).
4. L. J. Allamandola, in *Advances in the Theory of Benzenoid Hydrocarbons* (I. Gutman and S. J.Cyvin, eds., Springer-Verlag, Berlin, 1990.
5. E. L. Shock and M. D. Schulte, *Nature 343*: 728 (1990).
6. D. W. Deamer, *Adv. Space Res. 12*(4):183 (1992).
7. L. J. Allamandola, A. G. G. M. Tielens, and J. R. Barker, *Astophys. J. Suppl. Ser. 71*: 733 (1989).
8. P. Pott, reprinted in 1963 in *Natl. Cancer Inst. Monogr. 10*: 7 (1775).
9. E. L. Kennaway and I. Hieger, *Br. Med. J. 1*: 1044 (1930).
10. C. A. Menzie, B. B. Potocki, and J. Santodonato, *Environ. Sci. Technol. 26*: 1278 (1992).
11. V. Hietaniemi and E.-L. Kupila, *Analytical Sciences 7*: 979 (1991).
12. S. N. Sivaswamy and B. Nagarajan, *Med. Sci. Res. 19*: 289 (1991).
13. S. N. Sivaswamy, B. Balachandran, and V. M. Sivaramakrishnan, *Indian J. Exp. Biol. 29*: 611 (1991).
14. B. Saint-Aubert, J. F. Cooper, C. Astre, J. Astre, J. Spiliotis, and H. Joyeux, *J. Food Composition Analysis 5*: 257 (1992).
15. P. Simko and B. Brunckova, *Food Addit. Contam. 10*: 257 (1993).
16. P. Simko, P. Simon, V. Khunova, B. Brunckova, and M. Drak, *Food Chem. 50*: 65 (1994).
17. A. Asita, M. Matsui, T. Nohmi, A. Matsuoka, M. Hayashi, M. Ishidate Jr., T. Sofundi, M. Koyano and H. Matsushita, *Mutation Research 264*: 7 (1991).
18. B. K. Larsson, S. Regner, and P. Baeling, *J. Sci. Food Agric. 56*: 373 (1991).
19. A. Dipple and C. A. H. Bigger, *Mutation Research 259*: 263 (1991).

20. C. L. Wall, W. Gao, W. Qu, G. Kwie, F. C. Kauffman, and R. G. Thurman, *Carcinogenesis 13*: 519 (1992).
21. H. G. Kicinski, S. Adamek, and A. Kettrup, *Chromatographia 28*: 203 (1989).
22. G. Codina, M. T. Vaquero, L. Comellas, and F. Broto-Puig, J. *Chromatogr. 673*: 21 (1994).
23. P. R. Kootstra, M. H. C. Straub, G. H. Stil, E. G. van der Velde, W. Hesselink, and C. C. J. Land, *J. Chromatogr. 697*: 123 (1995).
24. R. Leeming and W. Maher, *Org. Geochem. 15*: 469 (1990).
25. I. Holoubek, J. Paasivirta, P. Maatela, M. Lahtipera, I. Holoubkova, P. Korinek, Z. Bohacek, and J. Caslavsky, *Toxicol. Environ. Chem. 25*: 137 (1990).
26. D. H. Bauw, P. G. M. de Wilde, G. A. Rood, and Th. G. Aalbers, *Chemosphere 22*: 713 (1991).
27. Ph. Garrigues and J. Bellocq, *J. High Resolut. Chromatogr. 12*: 400 (1989).
28. P. Jadaud, M. Caude, R. Rosset, X. Duteurtre, and J. Henoux, *J. Chromatogr. 464*: 333 (1989).
29. C. Brage and K. Sjostrom, *J. Chromatogr. 538*: 303 (1991).
30. M. B. Yunker, F. A. McLauglin, R. W. Macdonald, W. J. Cretney, B. R. Fowler, and T. A. Smyth, *Anal. Chem. 61*: 1333 (1989).
31. K. Kusuda, K. Shiraki, and T. Miwa, *Anal. Chim. Acta 224*: 1 (1989).
32. N. B. Elliott, A. J. Prenni, T. T. Ndou, and I. M. Warner, *J. Colloid Interface Sci. 156*: 359 (1993).
33. R. Fischer, R. Kreuzig, and M. Bahadir, *Chemosphere 29*: 311 (1994).
34. R. Niehaus, B. Scheulen, and H. Durbeck, *Sci. Total Environ. 99*: 163 (1990).
35. W. K. Raat, G. L. Bakker, and F. A. Meijere, *Atomspheric Environ. 24A*: 2875 (1990).
36. S. V. Lucas, K. W. Lee, C. W. Melton, B. A. Peterson, J. Lewtas, L. C. King and L. M. Ball, *Aerosol Sci. Technol. 14*: 210 (1991).
37. R. Niessner and P. Wilbring, *Fresenius' Z. Anal. Chem. 333*: 439 (1989).
38. F. Valerio and M. Pala, *Fresenius' J. Anal. Chem. 339*: 777 (1991).
39. G. Kloster, R. Niehaus, and H. Stania, *Fresenius' J. Anal. Chem. 342*: 405 (1992).
40. M. M. Schantz, B. A. Benner, S. N. Chesler, B. J. Koster, K. E. Hehn, S. F. Stone, W. R. Kelly, R. Zeisler, and S. A. Wise *Fresenius' J. Anal. Chem. 338*: 501 (1990).
41. H. A. M. G. Vaessen, P. J. Wagstaffe, and A. S. Lindsey, *Fresenius' J. Anal. Chem. 336*: 503 (1990).
42. A. Sisovic and M. Fugas, *Environ. Monitor. Assess. 18*: 235 (1991).
43. C. E. Rovere, J. Ellis, and P. T. Crisp, *Fuel 68*: 249 (1989).
44. U. Hechler, J. Fischer, and S. Plagemann, *Fresenius' J. Anal. Chem. 351*: 591 (1995).
45. V. Lopez-Avilla, R. Young, and W. F. Beckert, *Anal. Chem. 66*: 1097 (1994).
46. K. K. Chee, M. K. Wong, and H. K. Lee, *J. Chromatogr. 723*: 259 (1996).
47. I. J. Barnabuas, J. R. Dean, and S. P. Owen, *Analyst 120*: 1897 (1995).
48. B. R. Richter, B. A. Jones, J. L. Ezzell, N. L. Porter, N. Advalovic, and C. Pohl, *Anal. Chem. 68*: 1033 (1996).
49. C. L. Arthur, L. M. Killam, K. D. Buchholz, and J. Pawliszyn, *Anal. Chem. 64*: 1960 (1992).
50. Z. Zhang and J. Pawliszyn, *J. High Resolut. Chromatogr. 16*: 689 (1993).
51. K. G. Furton, J. Bruna, and J. R. Amirall, *J. High Resolut. Chromatogr. 18*: 625 (1995).

52. K. G. Furton, J. R. Amirall, and J. C. Bruna, *J. Forensic Sci. 41*: 12 (1996).
53. D. W. Potter and J. Pawliszyn, *Environ. Sci. Technol. 28*: 298 (1994).
54. S. B. Hawthorne, Y. Yang, and D. J. Miller, *Anal. Chem. 66*: 2912 (1994).
55. K. J. Hageman, L. Mazeas, C. B. Grabanski, D. J. Miller, and S. B. Hawthorne, *Anal. Chem. 68*: 3892 (1996).
56. S. B. Hawthorne, *Anal. Chem. 62*: 633A (1990).
57. V. Lopez-Avila, N. S. Dodhiwala, and W. F. Beckert, *J. Chromatogr. Sci. 28*: 468 (1990).
58. S. B. Hawthorne, M. S. Krieger, and D. J. Miller, *Anal. Chem. 61*: 736 (1989).
59. J. M. Wong, N. Y. Kado, P. A. Kuzmicky, H.-S. Ning, J. E. Woodrow, D. P. H. Hsieh, and J. N. Seiber, *Anal. Chem. 63*: 1644 (1991).
60. A. Meyer and W. Kleibohmer, *J. Chromatogr. 657*: 327 (1993).
61. A. Meyer, W. Kleibohmer and K. Cammann, *J. High Resolut. Chromatogr. 16*: 491 (1993).
62. S. B. Hawthorne, D. J. Miller, and J. J. Langenfeld, *J. Chromatogr. Sci. 28*: 2 (1990).
63. M. D. Burford, S. B. Hawthorne, and D. J. Miller, *J. Chromatogr. 685*: 79 (1994).
64. M. D. Burford, S. B. Hawthorne, and D. J. Miller, *J. Chromatogr. 685*: 95 (1994).
65. J. M. Levy, R. A. Cavalier, T. N. Bosch, A. F. Rynaski, and W. E. Huhak, *J. Chromatogr. Sci. 27*: 341 (1989).
66. C. P. Ong, H. K. Lee, and S. F. Y. Li, *Environ. Monitoring Assessment. 19*: 63 (1991).
67. A. Honer, M. Arnold, N. Husers, and W. Kleibohmer, *J. Chromatogr. 700*: 129 (1995).
68. A. C. Lewis, D. Kupiszewska, K. D. Bartle, and M. J. Pilling, *Atomspheric Environ. 29*: 1531 (1995).
69. E. Jarvenpaa, R. Huopalahti, and P. Tapanainen, *J. Liq. Chromatogr. & Rel. Technol. 19*(9):1473 (1996).
70. J. R. Dean, I. J. Barnabas, and I. A. Fowlis, *Anal. Proc. Anal. Communic. 32*: 305 (1995).
71. V. Lopez-Avila, R. Young, and N. Teplitsky, *J. AOAC Int. 79*: 142 (1996).
72. S. B. Hawthorne, J. J. Langenfeld, D. J. Miller, and M. D. Burford, *Anal. Chem. 64*: 1614 (1992).
73. S. B. Hawthorne, D. J. Miller, D. E. Nivens, and D. C. White, *Anal. Chem. 64*: 405 (1992).
74. T. Paschke, S. B. Hawthorne, and D. J. Miller, *J. Chromatogr. 609*: 333 (1992).
75. H.-B. Lee, T. E. Peart, R. L. Hong-You, and D. R. Gere, *J. Chromatogr. 653*: 83 (1993).
76. J. Dankers, M. Groenenboom, L. H. A. Scholtis, and C. Van der Heiden, *J. Chromatogr. 641*: 357 (1993).
77. I. J. Barnabas, J. R. Dean, W. R. Tomlinson, and S. P. Owen, *Anal. Chem. 67:*2064 (1995).
78. G. Reimer and A. Suarez, *J. Chromatogr. 699*: 253 (1995).
79. C. Friedrich, K. Cammann, and W. Kleibohmer, *Fresenius' J. Anal. Chem. 352*: 730 (1995).
80. S. M. Pyle and M. M. Setty, *Talanta 38*: 1125 (1991).
81. S. B. Hawthorne, D. J. Miller, and M. D. Burford *J. Chromatogr. 642*: 301 (1993).
82. K. D. Bartle, A. A. Clifford, S. B. Hawthorne, J. J. Langenfeld, D. J. Miller, and R. Robinson, *J. Supercritical Fluids 3*: 143 (1990).

83. C. Erkey and A. Akgerman, *Amer. Inst. Chem. Eng. J. 36*: 1715 (1990).
84. K. G. Furton and J. Rein, *Chromatographia 31*: 297 (1991).
85. K. G. Furton and J. Rein, in *Supercritical Fluid Technology* (ACS Symposium Series 488), F. V. Bright and M. E. P. McNally, eds., Washington, D.C., 1992.
86. J. J. Langenfeld, M. D. Burford, S. B. Hawthorne, and D. J. Miller, *J. Chromatogr. 594*: 297 (1992).
87. K. G. Furton and J. Rein, *Anal. Chim. Acta 248*: 263 (1991).
88. K. G. Furton and Q. Lin, *Chromatographia 34*: 185 (1992).
89. K. G. Furton, E. Jolly, and J. Rein, *J. Chromatogr. 629*: 3 (1993).
90. M. D. Burford, S. B. Hawthorne, and D. J. Miller, *Anal. Chem. 65*: 1497 (1993).
91. A. Clifford, M. D. Burford, S. B. Hawthorne, J. J. Langenfeld, and D. J. Miller, *J. Chem. Soc. Faraday Trans. 91*(9):1333 (1995).
92. J. J. Langenfeld, S. B. Hawthorne, D. J. Miller, and J. Pawliszyn, *Anal. Chem. 65*: 338 (1993).
93. K. G. Furton, C.-W. Huang, R. Jaffe, and M. A. Sicre, *J. High Resolut. Chromatogr. 17*: 679 (1994).
94. S. B. Hawthorne and D. J. Miller, *Anal. Chem. 66*: 4005 (1994).
95. D. J. Miller, S. B. Hawthorne, A. A. Clifford, and S. Zhu, *J. Chem. Eng. Data. 41*: 779 (1996).
96. D. J. Miller and S. B. Hawthorne, *Anal. Chem. 67*: 273 (1995).
97. J. J. Langenfeld, S. B. Hawthorne, D. J. Miller, and J. Pawliszne, *Anal. Chem. 67*: 1727 (1995).
98. Y. Yang, A. Gharaibeh, S. B. Hawthorne, and D. J. Miller, *Anal. Chem. 67*: 641 (1995).
99. D. Diaz, in *Hydrocarbon Speciation Studies in Ancient Sediments by High Temperature Supercritical Carbon Dioxide Extraction*, M.S. thesis, Florida International University, 1996.
100. S. B. Hawthorne and D. J. Miller, *J. Chromatogr. 468*: 115 (1989).
101. K. Jinno and S. Niimi, *J. Chromatogr. 455*: 29 (1988).
102. C. Yan and D. E. Martire, *J. Phys. Chem. 96*: 3489 (1992).
103. C. Yan and D. E. Martire, *J. Phys. Chem. 96*: 3505 (1992).
104. J. Rein, C. M. Cork, and K. G. Furton, *J. Chromatogr. 545*: 149 (1991).
105. D. M. Heaton, K. D. Bartle, A. A. Clifford, P. Myers, and B. W. King, *Chromatographia 39*: 607 (1994).
106. B. W. Wenclawiak and T. Tees, *J. Chromatogr. 660*: 61 (1994).
107. H. K. Lee, *J. Chromatogr. A 710*: 79 (1995).
108. A. Bemgard, A. Colmsjo, and B.-O. Lundmark, *J. Chromatogr. 595*: 247 (1992).
109. A. Bemgard, A. Colmsjo, and B.-O. Lundmark. *J. Chromatogr. 630*: 287 (1995).
110. P. Fernandez and J. M. Bayona, *J. High Resolut. Chromatogr. 12*: 802 (1989).
111. D. W. Sved, P. A. Van Veld, and M. H. Roberts, *Marine Environ. Res. 34*: 189 (1992).
112. K. Gorczynska, T. Kreczmer, D. Ciecierska-Stoklosa, and A. Utnick, *J. Chromatogr. 509*: 53 (1990).
113. V. M. Pozhidaev and K. A. Pozhidaeva, *J. Anal. Chem. USSR. 44*: 1716 (1989).
114. V. A. Gerasimenko and V. M. Nabivach, *J. Chromatogr. 498*: 357 (1990).
115. V. I. Dernovaya and Yu. A. Eltekov, *J. Chromatogr. 520*: 47 (1990).
116. V. S. Ong and R. A. Hites, *Anal. Chem. 63*: 2829 (1991).

117. M. D. Guillen, M. J. Iglesias, A. Dominquez, and C. G. Blanco, *J. Chromatogr. 591*: 287 (1992).
118. J. W. Childers, N. K. Wilson, and R. K. Barbour, *Appl. Spectrosc. 43*: 1344 (1989).
119. J. Semmler, P. W. Yang, and G. E. Crawford, *Vibrational Spectrosc.* 2: 189 (1991).
120. T. Visser, M. J. Vredenbregt, and A. P. J. M. de Jong, *J. Chromatogr. 687*: 303 (1994).
121. M. Seym and H. Parlar, *J. Chromatogr. 464*: 227 (1991).
122. M. T. Galceran, E. Moyano, and J. M. Poza, *J. Chromatogr. 710*: 139 (1995).
123. J. A. Lebo, J. L. Zajicek, T. R. Schwartz, L. M. Smith, and M. P. Beasley *J. Assoc. Off. Anal. Chem. 74*: 538 (1991).
124. W. Auer and H. Malissa, *Anal. Chim. Acta 237*: 451 (1990).
125. W. H. McClennen, N. S. Arnold, K. A. Roberts, H. L. C. Meuzelaar, J. S. Lighty, and E. R. Lindgren, *Combust. Sci. Tech. 74*: 297 (1990).
126. A. Robbat, T.-Y. Liu, and B. M. Abraham, *Anal. Chem. 64*: 1477 (1992).
127. I. D. Brindle and X.-F. Li, *J. Chromatogr. 498*: 11 (1990).
128. M. T. Galceran and E. Moyano, *J. Chromatogr. 607*: 287 (1992).
129. F. I. Onuska and K. A. Terry, *J. High Resolut. Chromatogr. 12*: 362 (1989).
130. E. Menichini, L. Bonanni, and F. Merli, *Toxicol. Environ. Chem. 28*: 37 (1990).
131. F. J. Gonzalez-Vila, J. L. Lopez, F. Martin, and J. C. del Rio, *Fresenius' J. Anal. Chem. 339*: 750 (1991).
132. M. Elomaa and E. Saharinen, *J. Appl. Polymer Sci. 42*: 2819 (1991).
133. K. Speer, E. Steeg, P. Horstmann, Th. Kuhn, and A. Montag, *J. High Resolut. Chromatogr. 13*: 104 (1990).
134. R. Kubinec, P. Kuran, I. Ostrovsky, and L. Sojak, *J. Chromatogr. A. 653*: 363 (1993).
135. P. T. J. Scheepers, D. D. Velders, M. H. J. Martens, J. Noordhoek, and R. P. Bos. *J. Chromatogr. A. 677*: 107 (1994).
136. Y. Yang and W. Baumann. *Analyst 120*: 243 (1995).
137. G. Castello and T. C. Gerbino. *J. Chromatogr. 642*: 351 (1993).
138. S. A. Wise, L. C. Sander, and W. E. May. *J. Chromatogr. 642*: 329 (1993).
139. H. A. Claessens, M. M. Rhemrev, J. P. Wevers, A. A. Janssen, and L. J. Brasser, *Chromatographia 31* :569 (1991).
140. E. Schneider, P. Krenmayr, and K. Varmuza, *Monatshefte für Chemie 121*: 393 (1990).
141. H. Takada, T. Onda, and N. Ogura, *Environ. Sci. Technol. 24*: 1179 (1990).
142. J.-P. F. Palmentier, A. J. Britten, G. M. Charbonneau, and F. W. Karasek, *J. Chromatogr. 469*: 241 (1989).
143. S. Matsuzawa, P. Garrigues, O. Setokuchi, M. Sato, T. Yamamoto, Y. Shimizu, and M. Tamura, *J. Chromatogr. 498*: 25 (1990).
144. A. J. Packham and P. R. Fielden, *J. Chromatogr. 552*: 575 (1991).
145. C. Grosse-Rhode, H. C. Kicinski, and A. Kettrup, *J. Liq. Chromatogr. 13*(17):3415 (1990).
146. K.-S. Boos, J. Lintelmann, and A. Kettrup, *J. Chromatogr. 600*: 189 (1992).
147. G. S. Heo and J. K. Suh, *J. High Resolut. Chromatogr. 13*: 748 (1990).
148. S. Matsuzawa, P. Garrigues, O. Setokuchi, M. Sato. T. Yamamoto, Y. Shimizu, and M. Tamura, *J. Chromatogr. 498*: 25 (1990).
149. C. Ostman, A. Bemgard, and A. Colmsjo. *J. High Resolut. Chromatogr. 15*: 437 (1992).

150. J. J. Vreuls, G. J. de Jong, and U. A. Th. Brinkman, *Chromatographia 31*: 113 (1991).
151. C. Ostman and U. Nilsson. *J. High Resolut. Chromatogr. 15*: 745 (1992).
152. C. Grobe-Rhode, H. G. Kicinski, and A. Kettrup, *Chromatographia 29*: 489 (1990).
153. W. Pfau and D. H. Phillips, *J. Chromatogr. 570*: 65 (1991).
154. K. Jinno, K. Yamamoto, H. Nagashima, T. Ueda, and K. Itoh, *J. Chromatogr. 517*: 193 (1990).
155. C. E. Kibbey and M. E. Meyerhoff, *J. Chromatogr. 641*: 49 (1993).
156. S. A. Wise, L. C. Sander, R. Lapouyade, and P. Garrigues, *J. Chromatogr. 514*: 111 (1990).
157. R. Ohmacht, M. Kele, and Z. Matus. *Chromatographia. 39*: 668 (1994).
158. S.-H. Chen, C. E. Evans, and V. L. McGuffin, *Anal. Chim. Acta 246*: 65 (1991).
159. W. Maher, F. Pellegrino, and J. Furlonger, *Microchem. J. 39*: 160 (1989).
160. H. J. Gotze, J. Schneider, and H.-H. Herzog, *Fresenius' J. Anal. Chem. 340*: 27 (1991).
161. A. N. Gachanja and P. J. Worsfold, *Anal. Proc. 29*: 61 (1992).
162. Ya. I. Korenman, V. A. Vorob'ev, and Zhilinskaya, *J. Appl. Chem. USSR 63*: 850 (1990).
163. O. G. Mironov, N. A. Pisareve, T. L. Shchekaturina, and B. P. Lapin, *Hydrobiological J. 26*: 68 (1990).
164. D. Thompson, D. Jolley, and W. Maher, *Microchem. J. 47*: 351 (1993).
165. Z.-H. Zhao, W.-Y. Quan, and D.-H. Tian, *Sci. Total Environ. 92*: 145 (1990).
166. Z.-H. Zhao, W.-Y. Quan, and D.-H. Tian, *Sci. Total Environ. 113*: 197 (1992).
167. F. Martin, I. Hoepfner, and G. Scherer, *Environ. International 15*: 41 (1989).
168. P. P. Fu, *Drug Metabolism Reviews 22*: 209 (1990).
169. L. S. Von Tungelin and P. P. Fu, *J. Chromatogr. 461*: 315 (1989).
170. A. D. Deshpande, *Arch. Environ. Contam. Toxicol. 18*: 900 (1989).
171. D. J. Freeman and F. C. R. Cattell, *Environ. Sci. Technol. 24*: 1581 (1990).
172. S. R. Wild, K. S. Waterhouse, S. P. McGrath, and K. C. Jones, *Environ. Sci. Technol. 24*: 1706 (1990).
173. J. R. Mazzeo, I. S. Krull, and P. T. Kissinger, *J. Chromatogr. 550*: 585 (1990).
174. A. I. Krylov, I. O. Kostyuk, and N. F. Volynets. *J. Anal. Chem. 50*: 495 (1995).
175. J. Lintelmann, K. Wadt, R. Sauerbrey, and A. Kettrup, *Fresenius' J. Anal. Chem 352*: 735 (1995).
176. H. Li and R. Westerholm, *J. Chromatogr. A. 664*: 177 (1994).
177. M. T. Galceran and E. Moyano, *J. Chromatogr. A. 683*: 9 (1994).
178. C. M. Pace and L. D. Betowski, *J. Am. Soc. Mass Spectrom.* 6:597 (1995).
179. D. R. Doerge, J. Clayton, P. P. Fu, and D. A. Wolfe. *Bio. Mass Spec. 22*: 654 (1993).
180. T. J. Gremm and F. H. Frimmel, *Chromatographia 38*: 781 (1994).
181. M. A. Rogriquez-Delgado, M. J. Sanchez, V. Gonzalez, and F. Garcia-Montelongo, *Fresenius' J. Anal. Chem. 345*: 748 (1993).
182. M. A. Garcia, O. Jimenez, and M. L. Marina, *J. Chromatogr. A. 675*: 1 (1994).
183. M. A. Garcia and M. L. Marina, *J. Chromatogr. A. 687*: 233 (1994).
184. D. Lopez-Lopez, S. Rubio-Barroso, and M. Polo-Diez, *J. Liq. Chromatogr. 18*: 2397 (1995).
185. M. A. Rodriguez-Delgado, M. J. Sanchez, V. Gonzalez, and F. Garcia-Montelongo, *J. Chromatogr. A. 697*: 71 (1995).

186. M. R. Hadjmohammadi and M. H. Fatemi, *J. Liq. Chromatogr. 18*: 2569 (1995).
187. K. Jinno, T. Ibuki, N. Tanaka, M. Okamoto, J. C. Fetzer, W. R. Biggs, P. R. Griffiths, and J. M. Olinger, *J. Chromatogr. 461*: 209 (1989).
188. L. C. Sander and S. A. Wise, *Anal. Chem. 61*: 1749 (1989).
189. W. Hesselink, R. H. N. A. Schiffer, and P. R. Kootstra, *J. Chromatogr. A. 697*: 165 (1995).
190. M. Makela and L. Pyy, *J. Chromatogr. A. 699*: 57 (1995).
191. J. Chen, D. Steenackers, A. Medvedovici, and P. Sandra, *J. High Resolut. Chromatogr. 16*: 605 (1993).
192. A. Robbat and T.-Y. Liu, *J. Chromatogr. 513*: 117 (1990).
193. J. C. Fetzer and W. R. Biggs, *Chromatographia 27*: 118 (1989).
194. S. F. Y. Li, H. K. Lee, and C. P. Ong, *J. Liq. Chromatogr. 12*(16):3251 (1989).
195. K. Jinno, Y. Miyashita, S.-I. Sasaki, J. C. Fetzer, and W. R. Biggs, *Environ. Monitor. Assessment 19*: 13 (1991).
196. P. P. Fu, Y. Zhang, Y.-L. Mao, L. S. Von Tungeln, Y. Kim, H. Jung, and M.-J. Jung. *J. Chromatogr. 642*: 107 (1993).
197. J.-S. Shiow, S. S. Hung, L. E. Unruh, H. Jung, and P. P. Fu, *J. Chromatogr. 461*: 327 (1989).
198. M. Funk, H. Frank, F. Oesch, and K. L Platt. *J. Chromatogr. A. 659*: 57 (1994).
199. S. K. Yang, M. Mushtaq, Z. Bao, H. B. Weems, M. Shou, and X.-L. Lu, *J. Chromatogr. 461*: 377 (1989).
200. K. Jinno, Y. Saito, R. Malhan née Chopra, J. J. Pesek, J. C. Fetzer, and W. B. Biggs, *J. Chromatogr. 557*: 459 (1991).
201. U. L. Nilsson and A. L. Colmsjo, *Chromatographia 32*: 334 (1991).
202. B. E. Il, S. N. anin, Yu. S. Nikitin, *Russian J. Phys. Chem. 65*: 1399 (1991).
203. A. L. Lafleur and E. F. Plummer, *J. Chromatogr. Sci. 29*: 532 (1991).
204. M. Olsson, L. C. Sander, and S. A. Wise, *J. Chromatogr. 477*: 277 (1989).
205. A. Malik and K. Jinno, *J. High Resolut. Chomatogr. 14*: 117 (1991).
206. H. Lamparczyk, P. Zarzycki, R. J. Ochocka, and D. Sybilska, *Chromatographia 30*: 91 (1990).
207. H. Lamparczyk, P. Zarzycki, R. J. Ochocka, M. Asztemborska, and D. Sybilska, *Chromatographia 31*: 157 (1991).
208. V. M. Pozhidaev, K. A. Pozhidaeva and L. K. Belokopytova, *Chem. Technol. Fuels Oils 26*: 158 (1990).
209. E. Menichini, A. D. Domenico, L. Bonanni, E. Corradetti, L. Mazzanti, and G. Zucchetti, *J. Chromatogr. 555*: 211 (1991).
210. W. L. Hinze, D. Y. Pharr, Z. S. Fu, and W. G. Burkert, *Anal. Chem. 61*: 422 (1989).
211. T. Babelek and W. Ciezkowski, *Environ. Geol. Water Sci. 14*: 93 (1989).
212. A. J. Kubis, K. V. Somayajula, A. G. Sharkey, and D. M. Hercules, *Anal. Chem. 61*: 2516 (1989).
213. R. J. Van de Nesse, G. J. M. Hoogland, J. J. M. De Moel, C. Gooijer, U. A. Th. Brinkman, and N. H. Velthorst, *J. Chromatogr. 552*: 613 (1991).
214. C. T. Jafvert, *Environ. Sci. Technol. 25*: 1039 (1991).
215. D. A. Edwards, R. G. Luthy, and Z. Liu, *Environ. Sci. Technol. 25*: 127 (1991).
216. Y. F. Yik, C. P. Ong, S. B. Khoo, H. K. Lee, and S. F. Y. Li, *Environ. Monitoring Assess. 19*: 73 (1991).

217. Y. F. Yik, C. P. Ong, S. B. Khoo, H. K. Lee, and S. F. Y. Li, *J. Chromatogr. 589*: 333 (1992).

218. A. Bockelen and R. Niessner, *Fresenius' J. Anal. Chem. 346*: 435 (1993).

219. E. R. Brouwere, A. N. J. Hermans, H. Lingeman, and U. A. Th. Brinkman. *J. Chromatogr. A. 669*: 45 (1994).

220. N. J. Fendinger and D. Glotfelty, *Environ. Toxicol. Chem. 9*: 731 (1990).

221. H. Gusten, D. Horvatic, and A. Sablijic, *Chemosphere 23*: 199 (1991).

222. C. J. Moretti and R. D. Neufeld, *Wat. Res. 23*: 93 (1989).

223. W. E. Acree and A. I. Zvaigzne, *Physics Chem. Liquids 23*: 225 (1991).

224. S. Johnsen, I. S. Gribbestad, and S. Johansen, *Sci. Total Environ. 81/82*: 231 (1989).

225. S. Onodera, K. Igarashi, A. Fukuda, J. Ouchi, and S. Suzuki, *J. Chromatogr. 466*: 233 (1989).

2

Polychlorinated Biphenyls

JUAN-CARLOS MOLTÓ, GUILLERMINA FONT,
YOLANDA PICÓ, AND JORDI MAÑES
University of Valencia, Valencia, Spain

I. INTRODUCTION

Polychlorinated biphenyls (PCBs) are various commercial mixtures of chloro-phenyls produced by chlorination of a biphenyl in the presence of a catalyst. There are 209 theoretically possible PCB congeners. The commercial products (Aroclor [USA], Chlorphen [Germany], and Kanechlor [Japan]) are complex mixtures containing as many as 132 congeners, and each mixture consists of different but overlapping assemblages of PCBs [1]. Because of their widespread usage from 1930 until the late 1970s as dielectric fluids in transformers, capacitors, hydraulic fluids, fire retardants, paint, pigments, and in the paper and cardboard industries, together with their high hydrophobicity, lipid solubility, and persistence, they have permeated practically every environmental medium in the world [2–5] and accumulated in fatty material [6–10]. Besides a generalized PCB contamination, there have also been specific problems associated with accidental leakage and fires resulting in extreme incidents of food contamination [7].

The toxicity of PCBs differs for each congener and ranges from the highly toxic ones, which are potent inducers of enzymes (P-450 and glucuronyl transferases), to the moderately toxic, which are more potent inhibitors of dopamine and other neurotransmitters [11]. PCBs have immunosuppressive activity, are tumor promoters, and interfere with calcium utilization (producing their well-known negative effect on eggshell formation in birds) [12]. They are classified as carcinogens by both the U.S. Environmental Protection Agency (EPA) and the International Agency for Research on Cancer (IARC). Because of their persistence and ubiquity, and their potential for bioaccumulation and biomagnification, PCB monitoring is important for the conservation of the environment and biota [13], and PCBs are included in the priority pollutant lists published by the EPA and by the European Union (EU) [14].

The diverse methods used to determine PCBs in environmental and food samples have been well documented over the last few years. Wells has written two interesting chapters on organic contamination [15] and analysis of PCBs [16]. There is an excellent review by Wells et al. on analysis of non- and mono-*ortho* chlorobiphenyls [17]. Other reviews deal with PCBs in the environment [16,18], marine organic pollutants [19], analysis of organic micropollutants in the lipid fraction of foodstuffs [20], and analysis of water pollution by PCBs [21]. The chromatographic techniques for separation and determination have proven to be the analyst's most important tools for reaching the required analytical accuracy and precision. Extraction and cleanup are also considered relevant in residue analysis. Many of these determinations are performed at very low levels and require considerable skill and expertise if they are to be done correctly.

This chapter reviews the recent developments in the chromatographic analyses of PCBs in the environment and foods, including techniques used for the separation of non-*ortho* and mono-*ortho*-chlorobiphenyls. Sample treatment, complexity of the extraction and cleanup procedures for isolating analytes from matrix components and other contaminants, fractionation and group separation, chromatographic identification and determination, and various detection methods are covered to offer a broad view of current trends in the analytical field. Some analytical details about real samples are presented in the last section.

II. SAMPLE PRETREATMENT

Sampling is a time-consuming and error-prone step in the analytical procedure. The equipment used for sampling and storage must be treated according to the procedures recommended in the literature [18].

Movement through the atmosphere and air–surface exchange processes are very important in the global cycling and redistribution of PCBs and must be considered to explain their presence in remote areas of the globe [22]. Air samples for PCB determination are collected using high-volume samplers [23,24]. For sampling, polyurethane foam is most often used [23–25], but both XAD-2 [26] and silica [27] have also been employed. Air particles are usually captured using glass fiber filtration [22–26]. The volumes of air pumped through the sorbents are hundreds to thousands of cubic meters.

Water sampling is difficult because the hydrophobic nature of these compounds usually leads to very low concentrations of PCBs in water [19,21,28]. For example, a complete survey of coastal waters requires that the PCB concentration be examined in water, sediment, and related biotransformation indicators in target organisms (mussels, fish). In order to standardize an appropriate sampling program, various aspects have to be considered: (a) parameters for the choice of sample sites; (b) ways of collecting the sample; and (c) preparation of the sample for short-term storage on board ship and for further analysis in the ground laboratory [29].

The distribution of these contaminants in water is heterogenous. There is a concentration gradient from freshwater sources to the sea, and in seawater the PCB concentration is greater in the surface layers. The water sample taken must represent the water environment. At the same time, the organic contaminants in the water column are distributed between water, organic colloids, and the organic component of particulate material. This particulate material is removed from the water samples retaining them by using glass fiber filters [19,21,28,30] or tangential flow membrane filters [30]. Another technique for removing the particulate matter is continuous flow centrifugation [30,31].

A main problem in water sampling is contamination by the sampler, solvents, adsorbent material, chemicals, or sampler penetration through the surface layer of the sea, which may be high in pollutants [19,28].

Samples of sediment and tissues collected in the field are usually preserved by freezing, either on board ship or in the laboratory [32]. Rapid preservation is vital if the integrity of the sample is to be maintained.

Studies dealing with food contaminants are usually done either on total diet samples or on individual food products [20]. In dietary intake studies, the collected samples are normally pooled samples of food products (meat, fruit, vegetables, drink, etc.) that a person has consumed in one day (24 h). In most studies, the collected samples are initially freeze-dried to remove the water content [8,9].

A. Extraction

A schematic overview of the methods of PCB analysis is given in Figure 1, which shows the various extraction, cleanup, and separation techniques that have been used to prepare environmental samples for the measurement of PCBs by ECD or MS.

1. Liquid–Liquid Extraction (LLE)

LLE is a well-established technique used for isolating PCBs. These methods are the backbone of residue analysis protocols for governmental agencies that have been established for PCB determination alone or with other specific compounds,

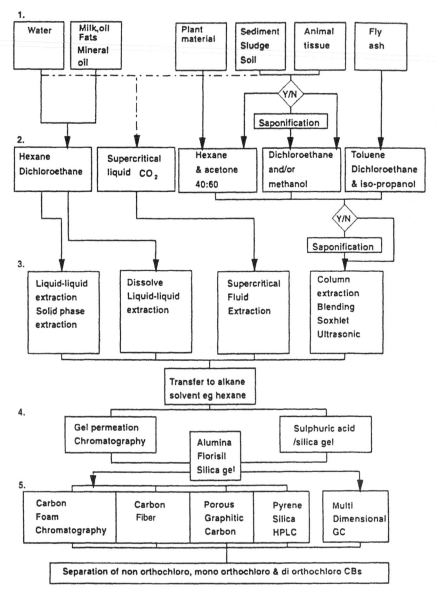

FIGURE 1 Schematic flow diagram for the determination of PCBs, including non-*ortho* and mono-*ortho*-chloro congeners in environmental matrices. (From Ref. 16).

and they are well documented in recent reference texts and have been extensively reviewed [17,20].

LLE can be performed using discontinuous extraction systems, in which aqueous samples are extracted by shaking with dichloromethane [33,34] or hexane [29,31,35–37], or by being stirred with pentane [28,38]. Sewage sludge or sediment samples are extracted by shaking with cyclohexane [39] or hexane [36] or by sonicating with mixtures of dichloromethane–hexane [40] or acetone–hexane [41]. Traditional extraction with acetone–water left standing overnight has also been reported [42].

The microwave-assisted extraction (MAE) of PCBs from soil, sand, and organic compost has been evaluated. Analytical scale MAE is gaining an important place among sample preparation techniques for solid matrices because it uses much less organic solvent for extracting PCBs, reduces extraction times, and increases sample throughput by using multivessel systems that allow simultaneous MAE of multiple samples [43,44].

Biological samples, such as tern eggs and young tissues, mussels, other molluscs, fat of poles, different fish species, and pine needles, are extracted with hexane [45–47], acetone–hexane [48], or dichloromethane [49]. Some methods employ a previous saponification with ethanolic KOH [46,47].

When analyzing food and milk, they are subjected to an LLE with acetone–petroleum ether [50], hexane–acetonitrile–ethanol [51], heptane [52], or hexane [53], sometimes helped by the addition of a strong acid that destroys the fats from the matrix [52,53].

Batch continuous extraction has been employed to extract large volumes of water samples. The sample is first stirred with pentane and then extracted for several hours, during which the flask with pentane is replaced at intervals [30].

Soxhlet extraction consists of continuous extraction of a sample with an appropriate solvent mixture at elevated temperatures for at least several hours. This method is recommended for isolating contaminants that are difficult to extract with conventional LLE.

Sewage sludge and sediment, food, and environmental samples have been widely analyzed using Soxhlet extraction [15,17,18,20,54]. Water must be eliminated prior to Soxhlet extraction by air drying [55,56], freeze drying [54,57–59], or mixing the sample with anhydrous sodium sulfate [54,60–70]. During the process a water separator could be applied for a time to remove the water [71].

The main advantage to the LLE procedure is that it is a well-established technique with a great deal of available data about accuracy and precision. Geissler and Schoeler [38] performed a statistical evaluation of conventional LLE, SPE, and steam distillation. The results obtained proved that PCBs in the ng/L concentration range are best extracted from water by LLE. In this study LLE showed higher recoveries and lower standard deviations than the use of commercial C18 phase cartridges or extractive steam distillation.

Soxhlet extraction and sonication have been compared [40]. Sonication was finally chosen because it gives the least amount of coextracted impurities. It is less time consuming, and good recoveries are obtained. Contrasting of LLE, Soxhlet, and saponification has been done. The repeatability of the LLE method compared favorably with the others [72].

The major disadvantage is the use of large volumes of expensive, high-purity organic solvent, which eventually must be disposed of. This technique is extremely time-consuming and not easily automated. Another drawback is the frequent emulsion formation.

2. Solid Phase Extraction (SPE)

In the last decade, SPE technology has been increasingly used to isolate PCBs from environmental matrices and foods [73,74]. It has been applied mainly in water and air, which are the principal environmental sinks of persistent, toxic anthropogenic chemicals such as PCBs. Analysis of the ultra-trace levels of contaminants present in them requires specialized extraction techniques.

One technique commonly used is adsorption onto a polyurethane foam [23–25,75] or a macroreticular resin such as XAD 2 or XAD 4 [21,26,76] for both water and air samples. These adsorbents have the disadvantage of requiring extensive cleanup and special storage conditions for the sorbent.

An alternate approach that overcomes some problems of XAD resins in water analysis is the use of reversed phase C18 cartridges [77,78], glass columns [79–81], or disks [82]. The sample capacity and breakthrough volumes of this reverse phase allow sampling volumes of approximately 10 L.

Milk is a particular fatty matrix where PCBs are usually present. This matrix is a colloidal dispersion formed by nonpolar lipids that are present in chilomicrons. Before the SPE they must be disrupted by addition of an organic solvent [83,84]. A technique for transferring of fat and PCBs from cod liver oil into the lipophilic gel Lipidex 5000 is also described [85]. Subsequent elution of the gel separated about 60% of the fat from the sample.

The advantages of SPE are that it is faster, easier, and cheaper than other methods and reduces solvent requirements considerably compared to LLE; but it does not eliminate them completely. Most studies also show that SPE provides recoveries similar to those obtained with LLE [80–82]. SPE cartridges and disks are susceptible to plugging and require sophisticated and expensive technology to be fully automated.

Matrices can be ground with the solid phase and analytes desorbed with small amounts of organic solvent. This process has been called matrix solid phase dispersion (MSPD). It has been tested on powdered milk employing Florisil and sodium sulfate as solid adsorbents and a mixture of acetone and hexane for eluting [86], and on fish tissue using C18 and acid silica [87] or Florisil [88] cocolumns for direct, on-line cleanup. The performance of the method compared favorably with the traditional approach, and the sample size, analysis time, and general cost are lower.

Solid phase microextraction (SPME) is a recent technique that can be considered a particular kind of SPE. With this technique, a fiber placed onto a special GC syringe is submerged in an aqueous matrix. After that, the analytes are allowed to diffuse to the fiber; then the fiber is inserted directly into the injector of a gas chromatograph. Finally, the analytes are thermally desorbed and analyzed. Until now, available fibers for SPME were poly(dimethylsiloxane) and polyacrylate. Other fibers are now under study and will probably be on the market in the near future. When performing SPME, some factors that must be considered are stirring rate, extracting temperature, absorption time, salinity of the aqueous matrix, presence of other contaminants, desorption temperature rate, and memory effects. This technique is solventless, sensitive at ng/L levels, and requires very low volumes of sample (ca 2 to 5 mL). Moreover the fiber is reutilized for many analyses, and the process can be completely automated. Because of these advantages, SPME of PCBs from waters should be a preferred technique in the near future [89].

3. Supercritical Fluid Extraction (SFE)

Today, SFE is an emerging technique for sample preparation and has a multitude of environmental and food applications. There are many reports on the use of SFE to extract PCBs and other organic compounds from particulates, soil, sediments, fish, tissue, and other solid matrices [90]. The supercritical fluids present unique physicochemical properties. They have a gaslike viscosity and zero surface pressure and allow efficient penetration into macromolecular materials, which makes the analytes more diffusive. However, they can become assimilated to a liquid because of their densities and solvating power. An alternate sample preparation technique utilizing SPE followed by SFE has attracted a great deal of attention [91–94]. It makes it possible to extract liquid matrices. The recent applications of SFE to PCB analysis are summarized in Table 1.

Carbon dioxide (CO_2), nitrous oxide (N_2O) and difluorochloromethane ($CHClF_2$) have been compared as supercritical fluids for PCB extraction [95,96]. While SFE with pure CO_2 yielded the lowest recoveries, $CHClF_2$ consistently gave the highest extraction efficiencies, most likely because of its high dipole moment. However, the main solvent used is CO_2 because it is nontoxic, available with a high degree of purity, and low in cost, and its solvation characteristics can be altered by changing either the pressure or the temperature of the fluid [90]. Yang et al. [97] proposed using subcritical water to extract PCBs from soil and sediment, since increasing the water temperature decreases its polarity enough to make it an efficient PCB extraction solvent.

There are many ways to optimize the extraction of analytes. Langenfeld et al. [98] proved the effect of using high temperatures and pressure on the extraction

Table 1 Methods for the Extraction of PCBs in Environmental and Food Samples by Supercritical Fluid Extraction

Matrix	Sample preparation/extractant	Operating conditions	Cleanup and trapping system	Ref.
Water	Extraction with C18 disk or cartridge/CO_2 modified with acetone or methanol	Various	2 mL of acetone or hexane	[91–93]
Air	Sample collected in denuders/CO_2	400 atm, 100°C extraction 30 min	hexane	[27]
Milk	Lyophilized sample added to Florisil/CO_2	120 Kg/cm², 65°C	1 mL heptane	[94]
Dust, street	/CO_2	200 bar 40°C, 3–5 hours	Trap on C18 columns Elution 1 mL toluene	[107]
Sewage sludge	Comparison of $CHClF_2$, N_2O, and CO_2	Various	2.5–5 mL dichloromethane	[95]
Sewage sludge	Sample added with Na_2SO_4 (anh.)/CO_2 Methanol, Ethanol	Various Methanol, Ethanol	Trap on C18, Florisil or silica Elution with 1.8 mL heptane	[101]
River sediment	/CO_2	650, 350 and 150 atm 50 and 200°C static extraction	5 mL of acetone	[98]
River sediment	Sample added with Na_2SO_4 (anh.)/CO_2 or NO_2	Various static extractions	Trap on silica, Florisil or graphitized carbon Elution with dichloromethane cleanup with copper for sulfur removal	[96]
River sediment	/CO_2 and CO_2 modified with toluene	Various	Trap on C18, C8, phenyl, diol, silica, cyano and amino elution with benzene or dichloromethane	[108]
Soil and sediment	/water	50 (or 100) atm 50 to 300°C 10–60 min	3 mL of hexane, toluene or cyclohexane	[97]

Matrix	Modifier/fluid	Conditions	Cleanup/Analysis	Ref.
Soil and sediment	/CO$_2$, 15% Cl$_2$CH$_2$	200 kg/cm^2, 80°C 60 min dynamic extraction	15 mL dichloromethane cleanup with two minicolumns packed with graphitized carbon black and Florisil Sulfur removal with copper granulates	[99]
Soil	/CO$_2$	15 to 50 MPa 40–150°C Static and dynamic conditions	1 mL of hexane	[102]
Sediment	/CO$_2$	35 MPa, 100°C 21 min, 20 dynamic extraction	Trap on C18 elution 1 mL isooctane hexane Cleanup with Florisil minicolumn Sulfur removal with mercury	[102]
Crab meat	Sample added with Na$_2$SO$_4$ anh. and basic alumina/CO$_2$	Various	On-line transfer into HPLC system to separate PCB congeners	[109]
Crab meat	Sample treatment with Na$_2$SO$_4$ anh. and basic alumina/CO$_2$	14.5 MPa, 60°C 35 min extraction	On-line transfer into the GC	[110]
Fish tissue	Sample treatment with Na$_2$SO$_4$ anh. and basic alumina/CO$_2$	5000 psi, 100°C 0.5 min static and 20 min dynamic range extraction	Trap on C18 Elution with isooctane Cleanup down size Florisil column	[104]
Fish tissue	Lyophilized and treatment with neutral alumina/CO$_2$	350 atm, 150°C 10 min static and 30 min dynamic	Trap on C18 Elution with 2 mL isooctane	[105]
Fish tissue	Lyophilized and added with Na$_2$SO$_4$(anh.)/CO$_2$ pure or modified with methanol	218 bar, 60°C 40 min and 30 min dynamic	Trap on Florisil Elution with heptane and dichloromethane acid silica cleanup	[106]

efficiencies for selected PCB congeners. Modifying CO_2 with an organic solvent such as dichloromethane [99] or acetone [91–93] has also been used to optimize extractions.

Apart from the fact that PCBs must be extracted from their matrix, the system used to trap the PCBs after the SFE must act efficiently. Until now, the most popular trapping technique was liquid collection [99]. Recently, SFE extracts have been collected on sorbents such as silicagel or bonded phase packing and then eluted with liquid solvents for subsequent analysis. Furton and Lin [100] showed that the PCB extraction efficiencies of SFE were highly dependent on the type of solid phase sorbent. Overall, the relative order of recoveries from highest to lowest is as follows: silica > phenyl > kaolin > Florisil > C8 > C18.

SFE has been compared with Soxhlet extraction [96,99,101–106] and with LLE [102]. Some advantages of SFE over other extraction techniques are that SFE is fast and can be used on-line, combined with gas, liquid, or supercritical fluid chromatography [90].

4. Other Techniques

Dialysis membranes filled with solvents can be used to monitor persistent lipophilic pollutants in seawater and freshwater. The membranes are filled with hexane and exposed for several weeks in the water and sediments [111]. When the dialytic procedure is optimized, satisfactory and highly reproducible analyte recoveries can be obtained in a few days while separating > 90% of the lipid material in a single operation [112].

Research has been done on the preparation of food and environmental samples for residue analysis of PCBs by continuous steam distillation [113,114]. Although some authors report good recoveries [113,114], others find unacceptably low recoveries [38]. For this reason, it is not very widely used.

Bleidner vapor phase extraction [115] has been applied to the trace analysis of PCB in water sediments. The method is reproducible but the recoveries obtained are low in all cases.

B. Cleanup

Cleanup is performed to eliminate the coextracted material. These procedures effectively remove the bulk of coextractants such as lipids, sulfur, carotenoids, and pigments and also other compounds that may interfere with the final determination.

1. Chemical Treatments

It is mainly a lipid removal treatment and the most important drawback is its destructive character. Chemical treatment can be applied directly over the sample prior to the LLE with an organic solvent. Adding concentrated hydrochloric [52] or sulfuric acid to the sample [53] destroys the lipids present by dehydration and oxidation reactions.

Different samples have also been saponified with potassium hydroxide in ethanol [46,47]. However, when the reaction occurs at high temperatures, highest chlorinated PCBs lose chlorine atoms [64].

A chromic acid digestion extraction technique was compared to conventional solvent extraction for recovery of a series of PCBs from centrifuged Niagara River water. The digestion technique was more efficient than conventional solvent extraction. The relative recovery (undigested/digested) decreased exponentially with increasing Kow. This implies that digestion and extraction recovers both the dissolved fraction and the fraction bound to dissolved organic matter (DOM), while conventional solvent extraction recovers only the dissolved fraction [31].

Chemical treatments can also be applied to the final extract prior to GC analysis. Concentrated extracts of organic solvents are often washed with sulfuric acid [35–37,46,47,53,58,71,86,116]. Stability of 84 pesticides and 12 PCBs was studied after treatment with sulfuric acid. The results were applied to the analysis of samples with different fact contents and compared with the results obtained using Florisil. The treatment with acid proved to have a narrower field of application than the treatment with Florisil column [117]. Alkali and oxidative treatments [84,118] were used to confirm the presence of PCBs in sample extracts. Chromium (VI) oxide provides clean chromatographic profiles, but this technique is not recommended for the determination of low chlorine containing Aroclors in environmental samples. Quantification errors can be diminished by carring out parallel runs with similar concentrations of suspected Aroclors or choosing individual PCBs that are not degraded by this method. Figure 2 shows the chromatograms obtained from a water sample containing Aroclor 1254 before and after treatment with chromium (VI) acid [118].

Elemental sulfur in sediments results from degradation of biological materials. The interference is quite serious when ECD is used due to the solubility in most common organic solvents. In traditional analytical schemes, the sulfur interference has been dealt with using metallic copper [28,40,54,55], potassium cyanide solution, or elemental mercury [57], thiosulfate, sodium hydroxide, alumina, and water [59].

2. Column Cleanup

The use of adsorption chromatography for cleanup of lipid samples is well established. Florisil is the most used normal phase to remove interfering compounds. Tanabe et al. [56,61–63,65] proposed a widely employed method in which the extract is transferred to 20 g Florisil, followed by elution of acetonitrile-hexane. Generally, the tendency of the cleanup methods is to reduce the size of the Florisil column (5 g) [45,50,51,54,60,65,70,119,120] or to purify on one [29,41] or two Florisil microcolumns arranged in series with a glass tube [121].

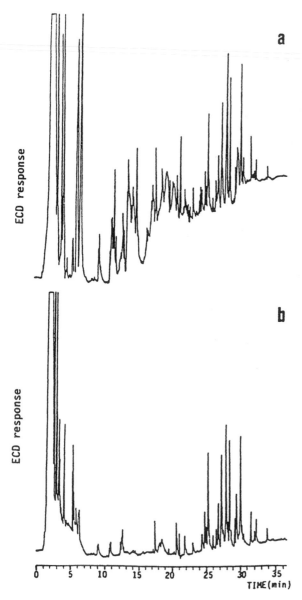

FIGURE 2   GC-ECD chromatograms obtained with a BP-5 column from a water sample containing Aroclor 1254, (a) before treatment and (b) after treatment with chromium (VI) oxide. (From Ref 118.)

Alumina and silica gel deactivated by the addition of a small percentage of water [64,67] or activated by water elimination are also employed [39,42,49,52,57,59, 69,85,122].

Sample cleanup using deactivated alumina has been compared with Florisil. Better results were obtained with the latter [40].

Multilayer columns containing different normal phases or the same phase subjected to different treatment, such as acid silica/basic silica/alumina column [34], deactivated basic alumina/modified silicagel column [46,48,123,124], silicagel/acid silicagel/sodium sulfate column [75], Florisil/silicagel column [55] or acid silicagel/Florisil [66], have been used effectively for cleanup.

3. Gel Permeation Chromatography (GPC)

GPC [68,69,125] was used to remove the majority of the lipid biogenic material. Many systems consisting of various types of columns, gels, and solvents have been utilized since the introduction of GPC as a cleanup technique into the field of PCB trace analysis. Poustka et al. [126] present an optimization of gel permeation chromatography for separation of PCBs from fats. The use of chloroform as a single mobile phase is generally preferred because PCBs elute at the lowest volume as a narrow fraction. Although the elution profiles for PCBs differed, various tested fats eluted with similar volumes.

C. Group Separation

The increased concern about the environmental consequences of widespread polychlorinated dibenzo-*p*-dioxins (PCDDs), polychlorinated dibenzofurans (PCDFs), and PCB contamination has focused attention on the separation and subsequent enrichment of these compounds from biological matrices and cocontaminants.

Adsorption chromatography using Florisil, silica, or alumina has been tested for separating non-*ortho*- and mono-*ortho*-substituted PCBs from the remaining ones. Three main problems were observed: lipid removal cannot be effected simultaneously on the same column, the volume of eluents required is considerable, and the planar congeners were only partially eluted [127,128]. Organochlorine pesticides and PCBs contained in the same extract are separated by passing the extract through silica gel. The first fraction eluted with hexane contains PCBs, *p,p'*-DDE, HCB, Aldrin, and Heptachlor [28,56,61–63,65,67,70,71, 116,120,129]. Rodriguez et al. [130] studied the simultaneous separation and determination of a mixture of hydrocarbons and organochlorine compounds by using a two-step microcolumn process based on the use of silica and alumina. The method offers good potential for the fractionation and determination of a large variety of organic compounds.

The possibility of separating PCBs according to planarity (and hence toxicity) is very useful. Certain activated carbons are well suited for separation of planar halogenated hydrocarbons on the bases of their planarities and degrees of chlorination. Non-*ortho* coplanar PCB members (IUPAC Nos. 77, 126, and 169) were separated from other PCB isomers and congeners using carbon column chromatography [47,68,85,131,132]. A mixture of activated carbon and Celite 545 as a carbon support packed in a disposable tube was used for the fractionation of PCB technical formulations and several samples [133]. Non-*ortho* PCBs were separated from others using a column with Florisil and carbon [134].

HPLC is a very convenient and efficient method for the separation of the coplanar PCBs and mono-*ortho* substituted congeners. With the HPLC carbon column these compounds were separated with an optimal gradient elution [42,52,135,136]. This method was applied for determining coplanar PCB congeners in food of animal origin and human milk [137]. Activated carbon dispersed by C-18 produces four discrete fractions containing bulk PCBs, mono-*ortho* PCBs, non-*ortho* PCBs and PCDDs and PCDFs [138].

Fractionation was also performed using an HPLC system with an aminopropyl silica column. A PCB fraction was obtained by heart-cutting the HPLC eluent [49]. With this column it is possible to isolate two fractions containing (1) the PCB congeners and lower polarity chlorinated pesticides and (2) the more polar chlorinated pesticides. For the normal-phase LC fractionation, hexane was used as a mobile phase for the isolation of PCBs and lower polarity pesticides, and 5% methylene chloride in hexane was used for the isolation of the more polar pesticides [125].

Normal phase HPLC on a pyrenylethylsilica (PYE) [139] or on a silica column [71] was utilized for separating PCBs and organochlorine pesticides from PCDDs and PCDFs. Nitrophenylpropylsilica was used alone [64] or combined with a pyrenylethyl column [140]. A two dimensional HPLC system was proposed to provide well-defined fractions of mono-tetra-*ortho*-PCBs substituted with one to four chlorine atoms in ortho position, non-*ortho*-PCBs, and non-substituted in *ortho* position PCDDs, PCDFs, and PAHs [140].

With the HPLC step the coplanar congeners present only in small amounts can be concentrated to a considerable extent. They are detected in the gas chromatograms without any interference by coextractives or other PCB congeners. The main advantages of HPLC are the higher efficiency, reproducibility, and speed of this method compared to open liquid chromatography. Other reasons for using HPLC are the lower solvent consumption, which reduces solvent-introduced contamination, and less manipulation of the sample by the operator. Porous graphitic carbon columns can be operated in reversed flow mode to recover adsorbed fractions like dioxins. This is not the case for the pyrenylsilica column. Extracts injected onto the HPLC columns must be free of lipids, especially the PYE column, whereas activated charcoal bears large amounts of lipids [17].

III. CHROMATOGRAPHIC TECHNIQUES

A. Gas Chromatography

Gas chromatography is the principal analysis technique in many laboratories, and other chromatographic techniques are rarely used. Most of the early analyses utilized low-resolution packed column gas chromatographic separation of PCB mixtures [141]. High-resolution analysis for PCB mixtures was reported in 1971 [142], and there have been continued improvements in both the resolution capabilities of capillary columns and detection methods.

Each process includes injection, separation, and detection. The complexity of PCB separation requires the use of capillary columns and either splitless injection [143] or on-column injection [144]. No significant difference could be found between splitless and on-column injection [145]. Injection of larger extract volumes into a programmed temperature vaporizer (PTV) was described for the determination of PCBs in water [79]. The carrier gas used is normally hydrogen [136] or helium as second choice [146]. To facilitate gas chromatographic separation, some advances in the use of electronic pressure programming have been made [147], but current use is limited.

1. Gas Chromatographic Column

The most important development in gas liquid chromatography was the introduction of capillary columns, which enabled the determination of PCB congeners. The introduction of fused silica capillary columns in 1979 initiated such advantages as convenience, high resolution, and short analysis times. Capillary columns can be characterized by their inside diameters. Currently there are three general types of capillary columns: (1) microbore (i.d. 0.10 mm), (2) conventional (0.25–0.32 mm i.d.), and (3) wide-bore (i.d. 0.53 mm). Microbore columns exhibit very high efficiency but require specialized instrumentation. Conventional capillary columns are not as efficient as microbore ones. They require dedicated capillary instrumentation and are used in most analytical laboratories that require high resolution of closely eluting mixtures of chemicals. Wide-bore columns offer less efficiency than the microbore or the conventional capillary columns but require no specialized instrumentation such as a special inlet system. The use of wide-bore capillary columns is a quick and fairly inexpensive means to achieve capillary column separation efficiencies without the expense of purchasing new gas chromatographs with capillary inlets [148].

Capillary columns are currently used for the separation and determination of the complex congener pattern. For routine analysis, nonpolar stationary phases based on methyl polysiloxane containing 5% phenyl groups are used in most cases, e.g., SE-54, CP Sil 8, CB, or DB-5. They provide a high separation efficiency, but there occur always some congener pairs that cannot be resolved. An excellent alternative is the HT5 column, which is coated with a carborane-methyl-polysiloxane phase. It exhibits much more efficient separation than the usual DB-5 column, especially for the higher chlorinated congeners [136]. A new apolar phase containing high-molecular-weight hydrocarbon similar to squalane has been applied in the quantification of PCB congeners in seal blubber extract showing better results than on columns coated with phases similar to SE-54. The PCB 138, a major PCB congener in both technical mixtures and environmental samples, was well separated [149].

The PCB elution patterns using capillary columns of several polarities are different. A dual column system, where the two columns are installed in parallel in a GC oven, offers a convenient route to improve the quality of PCB analysis by GC-ECD. Several pairs of capillary columns working in parallel have been reported to analyze environmental samples, such as DB-5 and DB-1701 [143], SE-54 and OV-1701 [76], DB-5 and DB-17 [77,150], Sil-8 and HT-5 [97], SE-54 and Sil-19 [30], and DB-17 and a series combination of a Sil-8 and an HT-5 [106].

Coupling series of two or more GC columns in a single gas chromatograph has been described [151]. This practice offers a multitude of possibilities for PCB separation and should be expected to increase in popularity in the future.

2. Multidimensional Gas Chromatography (MDGC)

MDGC is a separation technique employing two columns connected by a valve system, which makes it possible to transfer a fraction (heart cutting) of the eluent from the first to the second column. The very high resolution power of GC-GC and its ability to reduce total analysis time make it a suitable tool for the analysis of coplanar PCBs.

MDGC-MS, with a Sil-8 and an FFAP column, was used for the determination of coplanar PCB congeners in biological samples [152]. MDGC-ECD is recommended, using a combination of HP-Ultra 2 and an FFAP column, for determining mono-*ortho* PCBs in mixtures and aquatic organisms [153]. At least nine of the conformational stable chiral PCBs are present in commercial PCB formulations and are expected to accumulate in the environment. The use of MDGC with an achiral–chiral column combination was described for the enantiomer separation from technical mixtures [154]. The enantiomeric ratio of the chiral PCB 132 was determined by different achiral columns for preseparation and a Chirasil-Dex column and quantified by ECD and MS-SIM [155]. Figure 3 shows the MDGC-ECD chromatograms of PCB cut fraction from a human milk sample.

On the other hand the use of this kind of instrumentation is rather complicated because of the large number of parameters that have to be regulated. The need for demanding and expensive instrumentation and specialized operators can be considered the main disadvantage of this technique. MDGC is still not a routine technique, but it is a powerful research technique for isomer-specific PCB analysis at trace level [151].

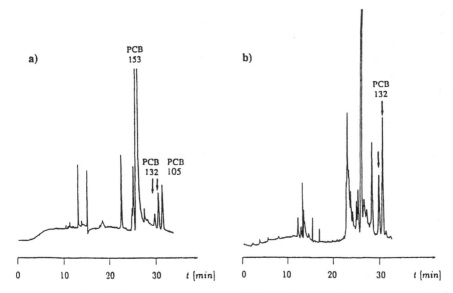

FIGURE 3 MDGC-ECD chromatograms of PCB cut fraction from a human milk sample, showing the separation of enantiomers of PCB 132. (a) DB-5 and (b) OV-1701 columns. (From Ref. 155.)

3. Coupling Chromatographic Separation Techniques

On-line liquid chromatography–gas chromatography has been applied to the determination of PCBs [156]. Multidimensional open tubular column SFC-SFC is a promising [157], but time consuming, technique for resolving complex mixtures. A two-dimensional packed capillary-to-open tubular column SFC-SFC system was reported. It uses a valve switching interface in line with a cryogenic trap for refocusing of analyte fraction cuts in the second dimension [158]. Few reports have utilized on-line SFE-GC for trace analysis of PCBs from complex matrices. SFE-GC has been utilized for selective extraction of PCBs from lipidic matrices in different biological species. The method reduces the potential for sample contamination and loss of analyte [110]. No application LC-GC or SFE-SFC has been described for the determination of non- and mono-*ortho*-PCBs [17].

4. Specific Detectors

Electron Capture Detector (ECD)

The ECD is an extremely sensitive tool for the analysis of organochlorine compounds, and it has been the most popular because of the high sensitivity and selectivity toward compounds such as PCBs. Selection of the ECD is normally based on the fact that this detector is most widely used for PCB quantitation and is generally present in the laboratory for routine analysis. Optimal performance was obtained by adjusting the four experimental parameters, pulse voltage, detector temperature, makeup gas flow rate, and the reference current using the signal ratio of the eluting peaks as the criterion [159]. The optimum flow for an ECD is much higher than carrier gas flow through the column. Thus an additional detector makeup gas flow is necessary, as argon–methane or nitrogen [160]. The response of a detector also varies with the purity of gas passing through it. Cleanliness and selectivity can be affected by many interfering compounds such as fatty substances or phthalate esters. The linearity of an ECD working in the pulse modulated mode is approximately of four orders of magnitude. The extreme sensitivity of ECD makes it vulnerable to dirt and overloading. Typically, between 80 and 150 congeners are present in complex environmental samples. The response of a chlorinated compound depends significantly on both the number and the position of chlorine atoms in the molecule, and consequently there is a wide range of response factors of PCB congeners. Because isomeric PCBs can exhibit variable ECD response, bias may arise when two or more congeners are present in a single chromatographic peak [161]. The response curves of the ECD to halocarbons have been fully reviewed [162]. The limitations of the linear range of ECD are well-known and documented. Most workers have attempted to calibrate the detector using the most linear portion of the response curve. Single point calibration does not have sufficient accuracy. Bracketing calibration points, covering the upper and lower end of the working range, are only valid if the measurement is made in the more linear portion of the ECD response. Establishing a multipoint calibration twice a week is recommended [17]. ECD has been used as final detection and MS to verify the peak identities [163,164]. The operation of a GC with ECD in parallel to GC-MS, and under identical conditions, can be extremely helpful with environmental samples, especially if the presence of matrix or other interfering compounds cannot be excluded [165].

Mass Spectrometry Detector

Conventional mass spectrometry is less sensitive than the ECD but can confirm the molecular weight of the different homologues and identify the presence of other environmental contaminants, especially when they coelute. MS offers various approaches at different levels of sophistication, from conventional GC-MS to the high-resolution version HRGC-HRMS through the MS-MS technique. A further advantage of the MS technique is the use of labelled internal standards, which increases accuracy and precision and, if properly applied, can perform reference measurements. Finally, identification of congeners is very important when changes in exposure are to be monitored or identification of the sources is required. GC-MS is a primary analytical technique for samples containing large amounts of other chlorinated compounds, because chlorine detectors such as ECD cannot discriminate between PCBs and interferences [166]. Electron impact (EI) is the most widely used ionization technique because of its universal response towards organic compounds. The EI mass spectra of PCBs are characterized by intense molecular ions and typical chlorine clusters. To gain sensitivity, PCB mixtures were separated by capillary gas chromatography (HRGC) and detected by MS in the selected ion monitoring (SIM) by acquiring several characteristic ions [94,98,133,134,140,167–169]. HRGC-MS with the mode SIM compares favorably with the scan mode (whole mass spectra) [170]. However, SIM mode can be unsatisfactory when interferences having identical masses to those of PCB congeners are present. To reduce incorrect identifications, very high mass resolving power is required.

Chemical ionization has additional discrimination capabilities and was used successfully to increase both specificity and sensitivity. While the positive ion chemical ionization (PICI) has a sensitivity similar to that of positive ion electron impact mass spectrometry, negative ion chemical ionization (NICI) produces abundant negative ions for many organohalogenated compounds [93,171].

Ion mobility spectrometry (IMS) is an atmospheric pressure ionization technique that is undergoing a considerable renaissance in mass spectrometry. The development of a high-temperature ion mobility spectrometer was described for interfacing to chromatographic techniques. Standard mixtures of PCBs were studied by GC-IMS as a quantitative technique showing good results [172].

The ion trap detector has been applied to analyze water samples [82,89,92].

MS-MS coupled with HRGC was demonstrated to be a highly sensitive, selective, and specific technique, particularly when used as multiple reaction monitoring (MRM exhibits many advantages in comparison with other scan modes used) [173]. The use of HRGC combined with MS is becoming more and more widely used.

Other Detectors

The atomic emission detector (AED) provides a specific detection for any element. Response factors are independent of the structure of a compound, but with the lower sensitivity of chlorine detection by AED, more efficient preconcentration of the extract has to be carried out prior to the analyses [174,175]. A detection technique for GC employing multichannel AED with a helium microwave induced plasma (MIP) has recently been introduced. Fragmentation of eluted substances into atoms occurs in the MIP, and the light emitted by excited atoms is then simultaneously detected by a photodiode array (PDA). The lower sensitivity of MIP-AED toward PCBs, compared with those of ECD or MS-SIM, requires more efficient preconcentration of analytes [175].

A technique combining both MS and Fourier transform infrared spectroscopy (FTIR) to analyze simultaneously the components from a single gas chromatographic injection has been applied to the quantitative and qualitative characterization of PCB isomers. The sensitivity of vibration spectroscopy to subtle differences in structure was shown to be highly complementary to GC-MS for qualitative identification of individual PCB isomers and congeners [176].

The electrolytic conductivity detector (ELCD) may be used as an alternative for the detection of organohalides. Because the ELCD can be configured in the halogen mode, halogens are selectively detected. Instrument specifications indicate that the ELCD is approximately 10 times less sensitive than the ECD. However, the higher specificity of the ELCD makes it particularly useful whenever the increased sensitivity of the ECD is not essential, as it is sufficient for most scientific studies of PCBs [105,177]. The response of the ELCD is directly proportional to the number of halogen atoms present in the molecule. A dual detector system with an ECD and an ELCD was evaluated on many environmental samples of different matrices, and it lessened the risk of false positives [150].

B. Liquid Chromatography

Few investigations have been reported on the determination of PCBs by high-performance liquid chromatography. Dioxetane chemiluminescence detection in liquid chromatography (LC) based on photosensitized on-line generation of singlet molecular oxygen can be applied as normal phase or reversed phase LC to determine PCBs, and it offers some inherent selectivity for planar PCBs [178]. A method has been described using HPLC with photodiode array detection for planar and certain other PCBs in marine biota samples. The HPLC system consisted of a guard column and two analytical PYE columns controlled by two electrical valves to backflush the guard column with hexane. This method agreed well with those determined by GC-ECD or GC-HRMS [179].

The advantage of supercritical fluid chromatography (SFC) is that supercritical fluid possesses solvating properties similar to those of a liquid, and the solute diffusion coefficients are more than two orders of magnitude greater than those found in liquids. Therefore comparable efficiencies to HPLC can be obtained with shorter analysis times. The retention characteristic of PCBs on cyanopropyl and octadecylsilane columns using carbon dioxide and nitrous oxide as eluents are shown, and a method is proposed for the determination of PCBs in sediments [180]. It has been demonstrated that the determination of PCBs by SFC using microbore columns and UV detection can be used at the ppm level [181].

IV. DETERMINATION AND VALIDATION METHODS
A. Determination Methods

The commercial products of PCBs are complex mixtures, and each mixture consists of different, but overlapping, assemblages of these compounds. Because the mixtures were used in such a wide range of industrial and domestic applications, the PCB composition of environmental samples is usually complex.

The most frequently used detection system for PCBs is ECD. With this detector, early methods used to measure PCBs relied upon comparison of sample gas chromatograms and those obtained for one or more commercial mixtures. The criteria for the selection of commercial PCB standards were variable. However, the choice was usually made according to the similarities between the chromatographic peak patterns observed for the standard mixture and the PCBs in the analyte. At present, many laboratories still use the same methodology. This approach is referred as the Aroclor method [161]. ECNICI coupled with a sequential classification method (SCM) has been employed to determine the total PCB content by using one of the Aroclor standards [171]. Aroclor 1260 was used for external standard quantifying by CG-ELCD and CG-ECD [177]. In other reports determination was done by measuring the peak heights of seven major peaks from the Aroclor 1254 [182] and Aroclor 1242 [183,184]. The PCB mixture Clorphen A 60 and a pattern-matching algorithm was used for determination [79]. Recently compiled data on the composition of commercial Aroclor mixtures and ECD response factors for all 209 PCB congeners are used to develop bias estimates associated with determination of PCBs. The standard Aroclor method resulted in 20% underestimation of total PCB concentration in a sediment sample compared to the congener specific detection method [164].

Identification and determination of individual PCB congeners can be based on relative retention times and on individual peak area responses [114]. To minimize the influence of experimental conditions and apparatus on the response it is preferable to use octachloronaphthalene as the external extandard [76,185]. However, considerable differences in the relative responses for PCBs are found in the published data [186].

The variable ECD response to PCB isomers introduces errors when multicomponent peaks are quantified by congener specific procedures, and the errors are greater when the highest and the lowest chlorinated congeners are quantified. Nevertheless, when the traditional Aroclor method is applied, the errors are even bigger [161].

Some methods for processing peak areas by least-squares procedures to determine the best value for the retention time factor have been reported [187].

The use of GC-NICI provides additional parameters for discriminating congeners from the interferences when an environmental sample is analyzed for total PCB or Aroclor content [171]. In the past, one of the drawbacks to applying congener specific procedures was the lack of calibration standards. Nowadays this situation has changed. Thus there seems to be little justification for using the traditional Aroclor method of PCB quantitation.

Individual congener standards are so costly that purchasing all of them is difficult. To reduce costs, PCB congeners have been identified by using Kanechlor mixtures as secondary reference standards for GC-ECD and MS analysis [169].

The most precise determination of PCBs is obtained by quantifying all the congeners separately. However, analytical procedures used for this purpose are relatively laborious and expensive. This makes them unsuitable for routine use. For practical reasons a limited number of representative congeners are employed [98,150,175].

An homologous series of eight trichlorophenyl alkyl ethers (TCPE) can be used as internal standards for the identification of PCBs in complex mixtures using the ECD [188].

When the MS is used, the ^{13}C labelled PCBs can be employed to study recoveries in the environmental samples. Many studies have used isotope dilution techniques in the determination of planar PCBs [127,140,189]. Wells et al. recommended their use as standards for the determination of the non-*ortho* PCBs that occur at much lower concentrations [17].

B. Validation Methods

One method for validating analytical procedures is by analyzing well-characterized reference materials. In recent years the National Institute of Standards and Technology (NIST) has developed several Standard Reference Materials (SRMs) to assist in validating measurements of PCB congeners. These SRMs represent three different levels of analytical difficulty: (1) simple calibration solutions containing a number of analytes, (2) complex natural mixtures (or a partially processed matrix), and (3) natural matrix materials. In 1985 and 1986 calibration solution SRMs were issued containing PCB congeners (SRM 1585, 1493, and 2262). Cod liver oil (SRM 1588) serves as an excellent surrogate for a tissue extract with a high lipid content. Several natural SRMs have been developed for determination of PCBs in marine and sediment samples such as SRM 1588 in cod liver oil, SRM 1939 in river sediment, SRM 1941 in marine sediment, SRM 1974 in mussel tissue [190], and recently SRM in fish homogenate [105].

The Community Bureau of Reference (BCR) offers also certified reference materials (CRM). There are two fish oil matrices available, CRM 349 in cod liver oil and CRM 350 in mackerel oil. Figure 4 shows the chromatogram from CRM 349 [146].

Six chlorinated biphenyls have been certified in a dried sewage sludge that is available as certified reference material (CRM 392) [163]. The PCB congeners were chosen because of their presence in industrial mixtures, wide occurrence in environmental samples, toxicity, scope for chromatographic techniques, and mention in legislation. Although certified PCB reference material is available from the Community Bureau of Reference (BCR), in practical applications a solution of mixed PCBs having certified contents considerably simplifies the calibration of the chromatographic systems used for the analysis. For this reason a CRM 365 has been developed consisting of a calibration solution with ten PCB congeners in isooctane solution at levels adequate for trace PCB analysis [191]. SRM 1945 made from whale blubber has been developed for the validation of methods used in the analyses of marine mammal tissues [192].

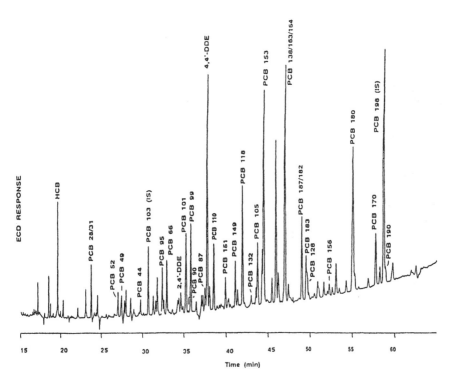

FIGURE 4 GC-ECD chromatogram of the PCB and lower polarity pesticide fraction isolated from CRM 349 on a C-18 column. (From Ref. 146.)

Several studies have shown a large variance and significant low bias in interlaboratory studies. Clearly there are some systematic errors in the analysis for PCBs in soils [193]. There is a long history of interlaboratory studies documenting large differences in results between laboratories when analyzing otherwise identical soils contaminated with PCBs. The sources of these differences are concentration dependent and due to widespread systematic (nonrandom) errors [194]. Various factors affecting the isolation of PCBs from soils for GC determination were compared using data from an interlaboratory study. Results from laboratories using Soxhlet extraction were significantly more accurate than those obtained using sonication, especially at higher concentrations, but with equal precision. The use of nonpolar solvents showed significantly lower accuracy than the use of more polar solvents; precision was similar. When using Florisil column cleanup, results were significantly more accurate and precise at lower concentrations than those obtained without cleanup procedures [195]. Some factors affecting the accuracy are the linear dynamic range, calibration range, and detector drift of the gas chromatographic instrument. Laboratories should be more critical in the examination of linearity of PCB calibrations. One easy solution is to use a "dilute to match" approach in which all samples are diluted or concentrated to match one standard [196]. Other interlaboratory studies have been done on other matrices like seal blubber and marine sediment extracts [197] or on fish oil [198]. Statistically significant differences within internal data of sample blocks can be detected by modelling data with multivariate methods, such as principal component analysis [199].

V. ENVIRONMENTAL AND FOOD ANALYSIS

The global distribution of a chemical is the result of the sum of all local and regional, horizontal and vertical, transportation processes in the atmosphere and hydrosphere, including the exchange between environmental compartments. These translocations are superimposed by abiotic and biotic transformation processes. PCBs are prominent environmental contaminants. Evaporation, particle binding, and deposition are the main routes for the long transport of PCBs. They can be found in diverse areas of the environment all over the world and have been shown to be present in almost every compartment of the global ecosystem, especially in most human and animal adipose samples, milk, and in all aquatic ecosystems.

A. Air Samples

The occurrence of PCBs in air is related to the characteristics of the locations and to the local industrial activities. There are remote, rural, urban areas and occupational environments where a wide range of concentrations can be monitored.

There is evidence that the occurrence of PCBs in clean marine air, which is not overlaid with continental air masses, is not a remainder of an incomplete air-to-sea deposition but reflects the PCB levels in the marine surface water and thus indicates an air/sea equilibrium. Fourteen PCBs have been quantified in air samples of the tropospheric boundary layer of the Atlantic Ocean by HRGC-ECD using several columns, and there was a correlation between the levels of the lower chlorinated congeners in air over the South Atlantic and the surface water temperature [129].

The atmospheric stability of PCB concentration is in direct contrast with the trends observed in other environmental compartments. Several air samples were collected at Bloomington, IN, analyzed by GC-ECD with a DB-5 column, and quantified by the Aroclor method. The concentration had not changed from the level detected six years earlier [24]. When the atmospheric concentrations of PCBs in Bermuda were analyzed as above, they showed very little change from twenty years before [23].

Detection by GC-ECD with confirmation by GC-MS has provided evidence that the PCB concentrations in the U.K. have declined since the 1960s, but the rate of decline in recent years has been much slower [22]. GC-MS analyses of air and inhalable particles from Rome indicate that lower chlorinated congeners are more abundant in the vapor phase than on particles [26]. Dust samples from German cities collected at places with high-speed traffic showed high levels of PCBs when analyzed by GC-MS. Tire abrasion and also diesel soot may be responsible for the high PCB contamination of some street dust samples [107].

B. Water Samples

Aquatic ecosystems have been contaminated by direct dumping of PCB and waste fluids containing PCBs. The range of PCB concentrations in water is large, from hundreds to several ng/L. PCB concentrations in sea water have decreased in recent years by a factor of 2–5 [200]. In saline waters, which present the lowest levels of all natural waters, it has been observed that concentrations of PCBs reported in the early studies sometimes were significantly higher than those identified in samples collected after 1985. These observations may reflect a decline in PCB levels in open ocean or an improvement in the analytical methods. Due to the gradient established from river water to the sea, there is a redistribution of PCBs in surface waters caused by turbulence and joining by the main current of the river [200,201].

For environmental samples determination if often done by GC-ECD with two columns of different polarity. Studies on contamination of the Antarctic environment began in the late 1960s, when PCBs and pesticides were detected in soil and snow samples. Those finding testified to the planetary diffusion of such compounds. From the data obtained during a campaign in Antarctica, an appreciable level of PCB contamination was confirmed in the entire Terra Nova Bay area [202]. Moreover, one can observe that PCB concentration and distribution follow preferential paths that could be related to water flow from glacier thaws [76]. Near Spain's industrial complexes, Aroclors were detected by GC-ECD [81]. Different PCB congener patterns were found in water and suspended particulate matter when GC-MSD was used. In the pattern of the dissolved PCBs, there was a decrease in the concentration from less chlorinated to more chlorinated PCBs. This is plausible because in this direction the aqueous solubility decreases and the hydrophobicity increases [75].

C. Soil, Sediment, and Sewage Sludge Samples

Of the total global commercial production of some 1.2 million tons of PCBs it is estimated that about 31% are still present in the environment and have become widely distributed throughout the biosphere. The low solubility of PCBs suggests that in soil, sediment, and sewage sludge samples the distribution should be strongly favored by organic matter.

Concentrations of the total PCBs and the most toxic non-*ortho* PCBs in soil samples collected from an industrial area of the island of Bahrain in Arabia were obtained by GC-ECD using a SPB capillary column [54], and with the same column MS-SIM was applied to the surface horizons of selected Scottish peat profiles analyses [203]. The increase in total PCB concentration from north to south demonstrates the influence of an industrial environment on background pollution. A thermal desorption gas chromatograph/mass spectrometer (TDGC-MS) was evaluated for on-site detection of PCBs in soil and sediments by the U.S. Environmental Protection Agency, and the degree of data quality was high [204].

Sediments collected from the Adriatic Sea's eastern coastal waters were monitored by CG-ECD [57], and a sample of Hudson River sediment was analyzed by CG-ECD with a DB-5 column [95]. Sediments from SW-Netherlands were analyzed by CG with two ECD and an SE-54 and a Sil 10 column. In general, concentrations found were high [59]. PCBs have also been found in the lagoon of Venice. As a consequence of rapid algal decomposition, sediment PCB concentrations were observed to increase. The analyses were done on an HRGC-ECD system and on an HRGC-MS-SIM equipped with SE-54 capillary columns [205]. Eight toxic PCBs studied were present in the sediments from Sheboygan River, a Wisconsin tributary to Lake Michigan. PCB congeners were quantified using MDGC-ECD with DB-5 and DB-1 columns [55]. PCB levels were determined by CG-MS-SIM with an HP5 column for Hamilton Bay sediments [103] or CG-MS with a DB-5 column for Dokay Bay sediments [99].

Recently there has been increased concern about the presence of potentially toxic organic contaminants in sewage sludges that are disposed of on agricultural land. Twelve sewage sludges from rural, urban, and industrial waste water treatment from northwest England were analyzed by capillary GC-ECD with a 5% phenyl methyl silicone column. Twenty-nine PCB congeners were routinely detected and quantified in all the samples. If these sludges were ploughed into arable land at normal rates, the PCB levels in soil would rise only slightly. However, if the same quantity of sludge were applied to the surface of pasture grassland, there would be significant elevations in the potential transfer of PCBs to grazing livestock [54]. With a Sil 8 column and the same system a municipal sewage system was examined for the sources of semivolatile organic compounds like PCBs that appeared as an industry-correlated parameter [71]. Also, GC-ECD with two columns was used to confirm the results obtained from sewage sludges [101,124]. The three most toxic coplanar PCBs, 77, 126 and 169, were identified and quantified at ultra trace levels in Swiss sewage sludge samples by applying HRGC-MS-SIM as well as NICI-MS using a DB-5 column. NICI-MS was the preferred detection method for the determination of non-*ortho* PCBs [206].

D. Aquatic Biota Samples

The increasing pollution of the marine environment is a threat to the health of organisms inhabiting the seas, as well as to marine mammals and human beings as consumers of such organisms. Some organochlorines found at the highest concentrations in Arctic fish are PCBs, and the implications of these contaminants for the northern ecosystems and the people who depend on them are still not clear [207]. Mussels are ideal biomonitors not only because of their well-known ability to bioconcentrate organic contaminants but also because they integrate contaminant uptake from the water, sediment, and suspended particles by virtue of their benthic, filter-feeding lifestyle. Mussels from the St. Lawrence River bioaccumulated the major organic industrial contaminants such as PCBs, which were identified and quantified by CG with dual capillary HP5 and HP17 columns using four mixture of Aroclors for instrument calibration [66]. CG-ECD was also used to study mussels in Dutch tidal waters [208]. Applying the same technique the PCB mussel contents from the estuary waters of New York State [209], the Western Mediterranean coast [58], Galicia (Spain) [121], and the Netherlands [72] are monitored. This information gives a general indication of their potential for food chain biomagnification and early warning of changes in environmental pollution, which will ultimately affect organisms at higher trophic levels.

Several researchers analyzed contaminants in tissues of mussels, fish, or marine mammals by CG-ECD using different columns and determination methods.

In order to study the levels of contaminants in flatfish in the Hvaler Archipelago (southeastern Norway), fifteen PCBs were selected, and the concentration levels were calculated by the sum of the selected congeners [210]. PCB contamination of the Hudson River estuary and bioaccumulation in species consumed by humans was studied. The samples were analyzed on a glass capillary column coated with Apiezon L. The total PCB was determined by adding up the individual congener concentrations of 74 compounds. Typical PCB residue contained smaller proportions of metabolically persistent PCB congeners than those found in common foodstuffs derived from warm-blooded animals [211]. To examine the PCB content of paddlefish roe in the Ohio River a glass megabore column SP-2401 was employed. The Aroclor 1260 was used as a standard. Findings of high PCB levels in all mature roe samples would certainly warrant a warning to the public [123].

According to studies recently carried out with a DB-1 column, concentrations of PCBs in fish from tropical Asian countries were low, except in some samples proceeding from Vietnam [56].

To assess the potential reduction in environmental contaminants from several species of Great Lakes fish, a comprehensive study was conducted on the level of contaminant consumption at the dinner table. PCBs were analyzed using a column SE-30, and Aroclors 1254 and 1260 were used for the determination. People should be encouraged to skin carp fillets before cooking [120].

Within the last decade, Alaska has experienced a tremendous increase in human industrial activities. Since most marine mammals are at or near the top of the food chain, selected tissues were collected from them and analyzed. Results were expressed as the sum of 15 measured PCB congeners. As a consequence of these studies, blubber was selected as the primary tissue for organic contamination analyses [125]. Other studies on tissues collected from dying and dead seals from the coast of Northern Ireland showed that the organochlorines found at the highest concentrations in blubber were PCBs. It was suggested that this pollution may exacerbate the effect of infectious diseases. Last analyses were done using an SPB 608 column to screen and an SPB 5 column to confirm positive samples; Aroclors 1260 and 1254 were used in the determination [122]. Various tissues of seals and a dolphin have been examined using an SE-54 column and a second DB-1701 column to confirm the results. The recovery rates were determined by analyzing standard reference material [116].

Some authors use CG-ECD followed by confirmation by GC-MS to determine the toxic non-*ortho*- and mono-*ortho* PCBs.

Cod liver oil was intended as a vitamin A and D supplement. The total PCB concentrations were analyzed by ECD with an SE-54 column and by EI-MS-SIM with an SE-30 column. There was found 1 μg/g of toxic PCBs [85]. GC-ECD with capillary columns of different polarities followed by confirmation by CG-EI-MS-SIM was applied to biota samples in Spain [212]. Figure 5 shows the GC-ECD chromatograms obtained from an extract of biota sample using a DB-5 and a DB-17 column. Similar analyses were realized in the District of Columbia (USA) [68], in the Buffalo River (New York) [213], in Taiwan [87], and in Canada [104].

Other researchers preferred gas chromatographic mass spectrometric methods to determine only coplanar PCBs.

The biological activity of PCB congeners depends on their structure. There are two classes of PCBs exhibiting immunotoxicity, the coplanar non-*ortho* and the mono-*ortho* isomers. They can be analyzed by CG-MS. An isotopic dilution HRGC-HRMS method of four PCBs was proposed to determine coplanar PCBs in fish from the Great Lakes, and in marine mammals from the Atlantic Ocean. The total PCB concentration in the Great Lakes fish samples is high when compared to the samples from the Atlantic Ocean [214]. In this way, these compounds were identified in Main River fish [135] and in Rhine fish [215] by MID with a DB-1 column. Concentrations of the coplanar congeners in fish expressed as TCDD equivalents (TE) were 300 and 80 times higher than those of the polychlorinated dibenzo-*p*-dioxins (PCDDs) and therefore more problematic. The use of NICI-SIM joined to a DB-5 column made it possible to monitor most of the highly chlorinated PCBs found in Great Lakes fish samples [216]. In fish

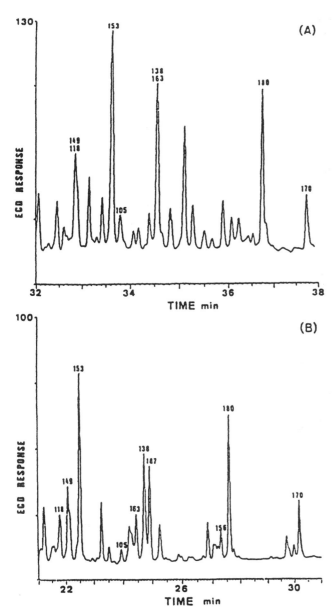

FIGURE 5 GC-ECD traces of an extract of biota sample of *Sardina pilchardus* after sulfuric acid cleanup with (a) DB-5 and (b) DB-17 columns. (From Ref. 212.)

samples from Sweden (Järnberg), the USA [69], and Taiwan [88], CG-MS was used to determine PCBs.

E. Food Samples

Both terrestrial and aquatic food chains can accumulate certain environmental contaminants to toxic concentrations. PCBs bioconcentrate in fat-rich samples, such as milk. Samples of cow milk from Spain were determined by GC-ECD like Aroclor 1260. Reported levels were higher than those found in Greece and Canada [60]. PCB congeners were analyzed in different milks, showing that the PCB levels are, in most instances, lower than those of organochlorine pesticides [83,84]. There are results demonstrating that the apparent dioxin-like toxicity of PCB residues in Aroclor and in human milk is mainly due to the PCBs 77, 105, and 126 [217]. PCB congeners have been determined in human milk samples by MDGC-ECD [155]. GC-MS analyses of milk samples showed the presence of PCBs in all samples [218], especially in those proceeding from industrial areas [189].

Dietary intakes of PCBs were obtained by multiplying the residue levels in the food by the amount of food consumed. The lack of knowledge on the toxicological impacts of organochlorine residues in foods is one of the most important causative factors of chemical intoxication of humans in developing countries.

Pollution of foodstuffs with PCBs is widespread in Vietnam. The daily dietary intake of PCBs was comparable to those in developed countries. Cereals and vegetables were the predominant sources of PCBs [65]. Food pollution and dietary intakes of PCBs were relatively low in India [63]. In Australia the dietary intake of PCBs was higher than those observed in most developed and developing countries. Analysis of some foodstuffs collected from Papua New Guinea and the Solomon Islands revealed that the contamination patterns and residue levels of organochlorines were similar to those observed in Australia [62].

The GC-ECD was used to analyze raw foodstuffs collected from Bangkok. Residues of PCBs were detected in all samples and quantified with Kanechlor mixtures. These results suggest that PCB contamination is slowly appearing in Thailand and other developing countries [61]. The content of six selected PCB congeners in total diet has been determined by GC-ECD. Congeners most frequently found were PCB 138 and 153. The mean daily intake of the sum of the three congeners 138, 153, and 180 was 2.3 µg/person. This value compares favorably with a value calculated from the PCB content of selected foods and consumption data [219].

PCB increased 12.9 times in average concentration from plankton to fish. The percentage use of higher chlorinated PCB homologues increased from 54 to 56% of the total PCB. These compounds continue to enter into the Great Lakes, where they are bioconcentrated by the biota and biomagnified through successively higher trophic levels [67].

VI. CONCLUSIONS

A variety of PCB analytical methods have been developed for different matrices. PCB analysis is complex enough to require several steps. Most of them include an extraction and a cleanup step. In the last decade SPE technology has been increasingly used to isolate PCBs. SPE, MSPD, and SPME present advantages over other methods when analyzing air and water samples. SFE is an emerging technique reaching high extraction efficiencies from solid matrices in short times. All these techniques will lead to future restriction of the use of halogenated organic solvents; moreover they can be used on-line, or combined with gas, liquid, or SFC.

An effective cleanup procedure is essential to eliminate coextractants as lipophilic materials. Chemical treatments, adsorption chromatography, and GPC are employed to remove the majority of the lipid biogenic material. GPC is the only technique for which automated systems are available. The possibility of separation of PCBs according to planarity is very useful. HPLC supplies convenient and efficient method for the separation and concentration of coplanar and mono-*ortho* substituted congeners.

GC-ECD and GC-MS are the most accepted techniques in routine laboratories. The use of multicolumn GC allows the determination of nearly all the congeners. MDGC is still not a common technique, but it is a powerful tool for PCB isomer separation. The most accurate approach is to quantify on the basis of individual congeners. A multilevel calibration of both the ECD and MS should be used. A good choice is to use the ECD to quantify and MS to verify peak identities. Intercomparison exercises or use of reference materials is recommended.

There have been numerous indications of PCB pollution in various parts of the world. The most disquieting fact about PCBs is their presence in practically all environmental samples analyzed. Toxicological studies on individual PCBs and commercial formulations indicate that these compounds are highly toxic and have both carcinogenic and teratogenic properties in both aquatic and terrestrial animals. Due to their persistence and tendency to accumulate in sediments and biota samples, PCB pollution will remain an environmental concern for many years to come, representing a considerable risk of chronic ecotoxicity. Interest in individual congener concentrations has also increased because recent toxicological data indicate that the potency of congeners varies widely. Unfortunately, most of the congeners are difficult to analyze.

There is still much work to be done in the field of real samples.

REFERENCES

1. G. Bommanna, G. Loganathan, and K. Kannan, Global organochlorine contamination trends: an overview, *AMBIO 23*(3): 187 (1994).
2. S. Safe, Development of bioassays and approaches for the risk assessment of 2,3,7,8-tetrachlorodibenzo-*p*-dioxin and related compounds, *Environ. Health Perspect. 101* (Suppl. 3): 317 (1993).
3. T. Watson, US rules hinder research on disposal of PCBs (news), *Nature 359*: 6 (1992).
4. B. Weiss, Environmental hazards: real or exaggerated? (letter; comment), *Science 262*: 638 (1993).
5. M. Oehme, Dispersion and transport paths of toxic persistent organochlorines to the Arctic-levels and consequences, *Sci. Total. Environ. 106*: 43 (1991).
6. B. T. Hargrave, G. C. Harding, W. P. Vass, P. E.Erickson, B. R. Fowler, and V. Scott, Organochlorine pesticides and polychlorinated biphenyls in the Arctic Ocean food web, *Arch. Environ. Contam. Toxicol. 22*: 41 (1992).
7. B. G. Loganathan and K. Kannan, Time perspectives or organochlorine contamination in the global environment, *Marine Pollut. Bull. 22*(12): 582 (1991).
8. D. Kinloch, H. Kuhnlein, and D. C. Muir, Inuit foods and diet: a preliminary assessment of benefits and risks, *Sci. Total. Environ. 122*: 247 (1992).
9. J. Gilbert, The fate of environmental contaminants in the food chain, *Sci. Total. Environ. 143*: 103 (1994).
10. V. Gajduskova and R. Ulrich, Use of polychlorinated biphenyl congener analysis for monitoring food and raw material of animal origin, *Vet. Med. Praha. 37*: 471 (1992).
11. R. J. Fielder and D. Martin, *General and Applied Toxicology* (B. Ballantyne, T. Marrs, and P. Turner, eds.), Macmillan, Wimbledon, 1994, p. 1133.
12. J. H. Dean, J. B. Cornacoff, G. J. Rosenthal, and M. I. Luster, *Principles and Methods of Toxicology* (A. W. Hayes, ed.), Raven Press, New York, 1994, p. 1065.
13. S. Safe, Toxicology, structure-function relationship, and human and environmental health impacts of polychlorinated biphenyls: progress and problems, *Environ. Health Perspect. 100*: 259 (1992).
14. F. Bro-Rasmussen, EEC water quality objectives for chemical dangerous to aquatic environments (List 1), *Rev. Environ. Contam. Toxicol. 137*: 83 (1994).
15. D. E. Wells, *Environmental Analysis, Techniques, Applications and Quality Assurance* (D. Barceló, ed.), Elsevier, Amsterdam, 1993, p. 80.
16. D. E. Wells, *Environmental Analysis Techniques, Applications and Quality Assurance* (D. Barceló ed.), Elsevier, Amsterdam, 1993, p. 113.
17. P. Hess, J. de Boer, W. P. Cofino, P. E. G. Leonards, and D. E. Wells, Critical review of the analysis of non- and mono-*ortho*-chlorobiphenyls, *Sci. Total Environ. 703*: 417 (1995).
18. V. Lang, Polychlorinated biphenyls in the environment, *J. Chromatogr. 595*: 1 (1992).
19. H. Huhnerfuss and R. Kallenborn, Chromatographic separation of marine organic pollutants, *J. Chromatogr. 580*: 191 (1992).
20. A. K. Liem, R. A. Baumann, A. P. J. M. de Jong, E. G. Van der Velde, and P. van Zoonen, Analysis of organic micropollutants in the lipid fraction of foodstuffs, *J. Chromatogr. A 624*: 317 (1992).

21. G. Font, J. Mañes, J. C. Moltó, and Y. Picó, Current developments in the analysis of water pollution by polychlorinated biphenyls, *J. Chromatogr. A. 733*: 449 (1996).
22. K. C. Jones, R. Duarte-Davison, and P. A. Cawse, Changes in the PCB concentration of United Kingdom air between 1972 and 1992, *Environ. Sci. Technol. 29*: 272 (1995).
23. S. Y. Panshin and R. A. Hites, Atmospheric concentration of polychlorinated biphenyls at Bermuda, *Environ. Sci. Technol. 28*: 2001 (1994).
24. S. Y. Panshin and R. A. Hites, Atmospheric concentrations of polychlorinated biphenyls at Bloomington, Indiana, *Environ. Sci. Technol. 28*: 2008 (1994).
25. W. E. Cotham and T. F. Bidleman, Polycyclic aromatic hydrocarbons and polychlorinated biphenyls in air at an urban and rural site near Lake Michigan, *Environ. Sci. Technol 29*: 2782 (1995).
26. L. Turrio-Baldassarri, A, Carere, A. di Domenico, S. Fuselli, N. Iacovella, and F. Rodriguez, PCDD, PCDF, and PCB contamination of air and inhalable particulates in Rome, *Fresenius' J. Anal. Chem. 348*: 144 (1994).
27. M. S. Krieger and R. A. Hites, Measurement of polychlorinated biphenyls and polycyclic aromatic hydrocarbons in air with a diffusion denuder, *Environ. Sci. Technol. 28*: 1129 (1994).
28. A. G. Kelly, I. Cruz, and D. E. Wells, Polychlorobiphenyls and persistent organochlorine pesticides in seawater at the pg/L level. Sampling apparatus and analytical methodology, *Anal. Chim. Acta 276*: 3 (1993).
29. P. Garrigues, J. F. Narbonne, M. Lafaurie, D. Ribera, P. Lemaire, C. Raoux, X. Michel, J. P. Salaun, J. L. Monod, and M. Romeo, Banking of environmental samples of short-term biochemical and chemical monitoring of organic contamination in coastal marine environments: the GICBEM experience (1986–1990), Groupe Interface Chimie Biologie des Ecosystèmes Marins, *Sci. Total. Environ. 139–140*: 225 (1993).
30. J. H. Hermans, F. Smedes, J. W. Hofstraat, and W. P. Cofino, Method for estimation of chlorinated biphenyls in surface waters: influence of sampling method on analytical results, *Environ. Sci. Technol. 26*: 2028 (1992).
31. M. S. Driscoll, J. P. Hasset, C. L. Fish, and S. Litten, Extraction efficiencies of organochlorine compounds from Niagara river water, *Environ. Sci. Technol. 25*: 1432 (1991).
32. G. T. Ankley, K. Lodge, D. J. Call, M. D. Balcer, L. T. Brooke, P. M. Cook, R. G. Kreis, Jr., A. R. Carlson, R. D. Johnson, and G. J. Niemi, Integrated assessment of contaminated sediments in the lower Fox River and Green Bay, Wisconsin, *Ecotoxicol. Environ. Safety 23*: 46 (1992).
33. R. Frank, L. Logan, and B. S. Clegg, Pesticide and polychlorinated biphenyl residues in waters at the mouth of the Grand, Saugeen, and Thames Rivers, Ontario, Canada, 1986–1990, *Arch. Environ. Contam. Toxicol. 21*: 585 (1991).
34. H. Miyata, O. Aozasa, S. Ohta, T. Chang, and Y. Yasuda, Estimated daily intakes of PCDDS (polychlorinated dibenzo-*p*-dioxins), PCDFs (polychlorinated dibenzofurans) and non-*ortho* coplanar PCBs via drinking water in Japan, *Chemosphere 26*: 1527 (1993).
35. S. Fingler, V. Drevenkar, B. Tkalcevic, and Z. Smit, Levels of polychlorinated biphenyls, organochlorine pesticides and chlorophenols in the Kupa river water and in drinking waters from different areas in Croatia, *Bull. Environ. Contam. Toxicol. 49*: 805 (1992).

36. V. W. Chui, S. Y. Lam Leung, and T. C. Chan, Residues of polychlorinated biphenyls (PCBs) in fish, water and sediment from Shing Mun River, *Biomed. Environ. Sci. 4*: 399 (1991).

37. S. Fingler, B. Tkalcevic, Z. Frobe, and V. Drevenkar, Analysis of polychlorinated biphenyls, organochlorine pesticides and chlorophenols in rain and snow, *Analyst 119*: 1135 (1994).

38. A. Geissler and H. F. Schoeler, Analysis of chloropesticides and PCB in water. Statistical evaluation of four enrichment methods, *Chemosphere 23*: 1029 (1991).

39. P. Frost, R. Camenzind, A. Magert, R. Bonjour, and G. Karlaganis, Organic micropollutants in Swiss sewage sludge, *J. Chromatogr. 643*: 379 (1993).

40. F. Paune, J. Rivera, I. Espadaler, and J. Caixach, Determination of polychlorinated biphenyls in sewage sludges from Catalonia (N. E. Spain) by high-resolution gas chromatography with electron-capture detection, *J. Chromatogr. A 684*: 289 (1994).

41. B. Bush, S. Dzurica, K. Wood, and E. C. Madrigal, Sampling the Hudson River estuary for PCBs using multiplate artificial substrate samplers and congener-specific gas chromatography in 1991, *Environ. Toxicol. Chem. 13(8)*: 1259 (1994).

42. H. Steinwandter, Research in environmental pollution. VIII. Identification of non-o,o-Cl and mono-o,o-Cl substituted PCB congeners in Hessian sewage sludges. *Fresenius' J. Anal. Chem. 344*: 66 (1992).

43. V. Lopez-Avila, R. Young, R. Kim, and W. F. Beckert, Accelerated extraction of organic pollutants using microwave energy, *J. Chromatogr. Sci. 33*:481 (1995).

44. V. Lopez-Avila, J. Benedicto, C. Charan, R. Young, and W. F. Beckert, Determination of PCBs in soil sediments by microwave-assisted extraction and GC/ECD or ELISA, *Environ. Sci. Technol. 29*: 2709 (1995).

45. L. Castillo, E. Thybaud, T. Caquet, and F. Ramade, Organochlorine contaminants in common tern (*Sterna hirundo*) eggs and young from the river Rhine area (France), *Bull. Environ. Contam. Toxicol. 53*: 759 (1994).

46. S.-G. Chu, Z.-Q. Xi and X.-B. Xu, Polychlorinated biphenyl (PCB) congeners in mussel and other molluscs from Da Chem island, East China Sea, *Bull. Environ. Contam. Toxicol. 55*: 682 (1995).

47. J. Falandysz, K. Kannan, S. Tanabe, and R. Tatsukawa, Concentration and 2,3,7,8-Tetrachlorodibenzo-p-dioxin toxic equivalents of non-ortho coplanar PCBs in adipose fat of Poles, *Bull. Environ. Contam. Toxicol. 53*: 267 (1994).

48. C. F. Mason and J. R. Ratford, PCB congeners in tissues of European otter (*Lutra lutra*), *Bull. Environ. Contam. Toxicol. 53*: 548 (1994).

49. H. Kylin, E. Grimvall, and C. Ostman, Environmental monitoring in polychlorinated biphenyls using pine needles as passive samplers, *Environ. Sci. Technol. 28*: 1320 (1994).

50. V. Prachar, M. Veningerova, J. Uhnak, and J. Kovacicova, Polychlorinated biphenyls in mother's milk and adapted cow's milk, *Chemosphere 29*: 13 (1994).

51. S. M. Dogheim, M. el Shafeey, A. M. Afifi, and F. E. Abdel Aleem, Levels of pesticide residues in Egyptian human milk samples and infant dietary intake, *J. Assoc. Off. Anal. Chem. 74*: 89 (1991).

52. V. Bohm, E. Schulte, and H. P. Thier, Polychlorinated biphenyl residues in food and human milk: determination of co-planar and mono-ortho substituted congeners, *Z. Lebensm. Unters. Forsch. 196*: 435 (1993).

53. V. Hietaniemi and J. Kumpulainen, Isomer specific analysis of PCBs and organo-chlorine pesticides in Finnish diet samples and selected individual foodstuffs, *Food Addit. Contam. 11*: 685 (1994).

54. R. E. Alcock and K. C. Jones, Polychlorinated biphenyls in digested UK sewage sludge, *Chemosphere 26*: 2199 (1993).

55. W. Sonzogni, L. Maack, T. Gibson, and J. Lawrence, Toxic polychlorinated biphenyl congeners in Sheboygan River (USA) sediments, *Bull. Environ. Contam. Toxicol. 47*: 398 (1991).

56. K. Kannan, S. Tanabe, and R. Tatsukawa, Geographical distribution and accumula-tion features of organochlorine residues in fish in Tropical Asia and Oceania, *Environ. Sci. Technol. 29*: 2673 (1995).

57. M. Picer and N. Picer, Long-term trends of DDTs and PCBs in sediment samples collected from the eastern Adriatic coastal waters, *Bull. Environ. Contam. Toxicol. 47*: 864 (1991).

58. M. Sole, C. Porte, D. Pastor, and J. Albaiges, Long-term trends of polychlorinated biphenyls and organochlorinated pesticides in mussels from the Western Mediterra-nean coast, *Chemosphere 28*: 897 (1994).

59. J. Stronkhorst, P. C. Vos, and R. Misdorp, Trace metals, PCBs, and PAHs in benthic (epipelic) diatoms from intertidal sediments: a pilot study, *Bull. Environ. Contam. Toxicol. 52*: 818 (1994).

60. L. M. Hernández, M. A. Fernández, B. Jiménez, M. J. González, and J. F. García, Organochlorine pollutants in meat and cow's milk from Madrid (Spain), *Bull. Envi-ron. Contam. Toxicol. 52*: 246 (1994).

61. S. Tanabe, K. Kannan, M. S. Tabucanon, C. Siriwong, Y. Ambe, and R. Tatsukawa, Organochlorine pesticides and polychlorinated biphrenyl residues in foodstuffs from Bangkok, Thailand, *Environ. Pollut. 72*: 191 (1991).

62. K. Kannan, S. Tanabe, R. J. Williams, and R. Tatsukawa, Persistant organochlorine residues in foodstuffs from Australia, Papua New Guinea and the Solomon Islands: contamination levels and human dietary exposure, *Sci. Total. Environ. 153*: 29 (1994).

63. K. Kannan, S. Tanabe, A. Ramesh, A. Subramanian, and R. Tatsukawa, Persistent organochlorine residues in foodstuffs from India and their implication on human dietary exposure, *J. Agric. Food Chem. 40*: 518 (1992).

64. N. Kannan, G. Petrick, D. E. Schultz-Bull, and J. C. Duinker, Chromatographic techniques in accurate analysis of chlorobiphenyls, *J. Chromatogr. 642*: 425 (1993).

65. K. Kannan, S. Tanabe, H. T. Quynh, N. D. Hue, and R. Tatsukawa, Residue pattern and dietary intake of persistent organochlorine compounds in foodstuffs from Viet-nam, *Arch. Environ. Contam. Toxicol. 22*: 367 (1992).

66. J. L. Metcalfe and M. N. Charlton, Freshwater mussels as biomonitors for organic industrial contaminants and pesticides in the St. Lawrence River, *Sci. Total. Environ. 97–98*: 595 (1990).

67. M. S. Evans, G. E. Noguchi, and C. P. Rice, The biomagnification of polychlorinated biphenyls, toxaphene, and DDT compounds in a Lake Michigan offshore food web, *Arch. Environ. Contam. Toxicol. 20*: 87 (1991).

68. T. R. Schwartz, D. E. Tillitt, K. P. Feltz, and P. H. Peterman, Determination of mono- and non-*o,o'*-chlorine substituted polychlorinated biphenyls in Aroclors and envi-ronmental samples, *Chemosphere 26*: 1443 (1993).

69. P. J. Marquis, R. L. Hanson, M. L. Larsen, W. M. de Vita, B. C. Butterworth, and D. W. Kuehl, Analytical methods for a national study of chemical residues in fish. II: Pesticides and polychlorinated biphenyls, *Chemosphere 29*: 509 (1994).

70. A. S. Perry, I. Sidis, and A. Zemach, Organochlorine insecticide residues in birds and bird eggs in the coastal plain of Israel, *Bull. Environ. Contam. Toxicol. 45*: 523 (1990).

71. R. Rieger and K. Ballschmiter, Semivolatile organic compounds—polychlorinated dibenzo-*p*-dioxins (PCDD), dibenzofurans (PCDF), biphenyls (PCB), hexachlorobenzene (HCB), 4,4'-DDE and chlorinated paraffins (CP)—as markers in sewer films, *Fresenius' J. Anal. Chem. 352*: 715 (1995).

72. K. Booij and C. van der Berg, Comparison of techniques for the extraction of lipids and PCBs from benthic invertebrates, *Bull. Environ. Contam. Toxicol. 53*: 71 (1994).

73. G. Font, J. Mañes, J. C. Moltó, and Y. Picó, Solid phase extraction in multiresidue pesticide analysis of water, *J. Chromatogr. 642*: 135 (1993).

74. Y. Picó, J. C. Moltó, J. Mañes, and G. Font, Solid phase techniques in the extraction of pesticides and related compounds from foods and soils, *J. Microcol. Sep. 6*: 331 (1994).

75. R. Götz, P. Enge, P. Friesel, K. Roch, B. Schilling, B. Hartmann, and H. Wunsch, Quantification of polychlorinated biphenyls (PCBs) and hexachlorobenzene (HCB) in the picogram/liter range in water of the river Elbe, *Fresenius' J. Anal. Chem. 348*: 694 (1994).

76. L. Morselli, S. Zappoli, and A. Donati, Identification, quantification and distribution of polychlorinated biphenyls (PCB) in an Antarctic marine environment: Terranova Bay, Ross Sea, *Ann. Chim. 79*: 677 (1989).

77. J. L. Bernal, M. J. Del Nozal, J. Atienza, and J. J. Jimenez, Multi-determination of PCBs and pesticides by use of a dual GC column–dual detector system, *Chromatographia 33*: 67 (1992).

78. J. Lenicek, J. Holoubek, M. Sekyra, and S. Kocianova, Determination of polychlorinated biphenyls, chlorinated pesticides, chlorinated phenols, polycyclic aromatic hydrocarbons and chlorophenoxyalkane acids in water after concentration on Separon SGX C18, *Chem. Listy 87*: 852 (1993).

79. E. Stottmeister, H. Hermenau, P. Hendel, T. Welsch, and W. Engewald, Solid-phase extraction–programmed-temperature vaporizer (PTV) injection in GC analysis of toxaphene (camphechlor) and PCB in aqueous samples, *Fresenius' J. Anal. Chem. 340*: 31 (1991).

80. J. C. Moltó, Y. Picó, J. Mañes, and G. Font, Analysis of polychlorinated biphenyls in aqueous samples using C18 glass column extraction, *J. AOAC Int. 75*: 714 (1992).

81. Y. Picó, J. C. Moltó, M. J. Redondo, E. Viana, J. Mañes, and G. Font, Monitoring of the pesticide levels in natural waters of the Valencia Community (Spain), *Bull. Environ. Contam. Toxicol. 53*: 230 (1994).

82. A. Kraut-Vass and J. Thoma, Performance of an extraction disk in synthetic organic chemical analysis using gas chromatography–mass spectrometry, *J. Chromatogr. 538*: 233 (1991).

83. M. J. Redondo, Y. Picó, J. Server-Carrió, J. Mañes, and G. Font, Organochlorine residue analysis of commercial milks by capillary gas chromatography, *J. High Resolut. Chromatogr. 14*: 597 (1991).

84. Y. Picó, M. J. Redondo, G. Font, and J. Mañes, Solid-phase extraction on C18 in the trace determination of selected polychlorinated biphenyls in milk, *J. Chromatogr. A. 693*: 339 (1995).

85. C. Weistrand and K. Noren, Liquid–gel partitioning using Lipidex in the determination of polychlorinated biphenyls in cod liver oil, *J. Chromatogr. 630*: 179 (1993).

86. M. T. Galceran, F. J. Santos, D. Barceló, and J. Sanchez, Improvements in the separation of polychlorinated biphenyl congeners by high-resolution gas chromatography. Application to the analysis of two mineral oils and powdered milk, *J. Chromatogr. 655*: 275 (1993).

87. Y. C. Ling, M. Y. Chang, and I. P. Huang, Matrix solid-phase dispersion extraction and gas-chromatographic screening of polychlorinated biphenyls in fish, *J. Chromatogr. A 669*: 119 (1994).

88. Y. C. Ling and I. P. Huang, Multiresidue-matrix solid-phase dispersion method for determining 16 organochlorine pesticides and polychlorinated biphenyls in fish, *Chromatographia 40*: 259 (1995).

89. D. W. Potter and J. Pawliszyn, Rapid determination of polyaromatic hydrocarbons and polychlorinated biphenyls in water using solid-phase microextraction and GC-MS, *Environ. Sci. Technol. 28*: 298 (1994).

90. I. J. Barnabas and J. R. Dean, Supercritical fluid extraction of analytes from environmental samples, *Analyst 119*: 2381 (1994).

91. P. H. Tang, J. S. Ho, and J. W. Eichelberger, Determination of organic pollutants in reagent water by liquid–solid extraction followed by supercritical fluid elution, *J. AOAC Int. 76*: 72 (1993).

92. J. S. Ho, P. H. Tang, J. W. Eichelberger, and W. L. Budde, Liquid–liquid disk extraction followed by SFE and GC-Ion-Trap MS for the determination of traces of organic pollutants in water, *J. Chromatogr. Sci. 33*: 1 (1995).

93. V. S. Ong and R. A. Hites, Determination of pesticides and polychlorinated biphenyls in water: a low-solvent method, *Environ. Sci. Technol. 29*: 1259 (1995).

94. A. G. Mills and T. M. Jefferies, Rapid isolation of polychlorinated biphenyls from milk by a combination of supercritical-fluid extraction and supercritical-fluid chromatography, *J. Chromatogr. 643*: 409 (1993).

95. S. B. Hawthorne, J. J. Langenfeld, D. J. Miller, and M. D. Burford, Comparison of supercritical chlorodifluoromethane, nitric oxide, and carbon dioxide for the extraction of polychlorinated bipheyls and polycyclic aromatic hydrocarbons, *Anal. Chem. 64*: 1614 (1992).

96. R. Tilio, S. Kapila, K. S. Nam, R. Bossi, and S. Facchetti, Reduction-elimination of sulfur interference in organochlorine residue determination by supercritical-fluid extraction, *J. Chromatogr. A 662*: 191 (1994).

97. Y. Yang, S. Bowadt, S. B. Hawthorne, and D. J. Miller, Subcritical water extraction of polychlorinated biphenyls from soil and sediment, *Anal. Chem. 67*: 4571 (1995).

98. J. J. Langenfeld, S. B. Hawthorne, D. Miller, and J. Pawliszyn, Effects of temperature and pressure on supercritical-fluid extraction efficiencies of polycyclic aromatic hydrocarbons and polychlorinated biphenyls, *Anal. Chem. 65*: 338 (1993).

99. P. Tong and T. Imagawa, Optimization of supercritical fluid extraction for polychlorinated biphenyls from sediments, *Anal. Chim. Acta 310*: 93 (1995).

100. K. G. Furton and Q. Lin, Variation in the supercritical-fluid extraction of polychlori-

nated bipheyls as a function of sorbent type, extraction cell dimensions, and fluid flow rate, *J. Chromatogr. Sci. 31*: 201 (1993).

101. S. Bowadt, B. Johansson, F. Pelusio, B. R. Larsen, and C. Rovida, Solid-phase trapping of polychlorinated biphenyls in supercritical-fluid extraction, *J. Chromatogr. A 662*: 424 (1994).

102. E. G. Van der Velde, W. De Haan, and A. K. Liem, Supercritical fluid extraction of polychlorinated biphenyls and pesticides from soil. Comparison with other extraction methods, *J. Chromatogr. 626*: 135 (1992).

103. H. B. Lee and T. E. Peart, Optimization of supercritical carbon dioxide extraction for polychlorinated biphenyls and chlorinated benzenes from sediments, *J. Chromatogr. A 663*: 87 (1994).

104. H. B. Lee, T. E. Peart, A. J. Nimi, and C.R. Knipe, Rapid supercritical carbon dioxide extraction method for determination of polychlorinated biphenyls in fish, *J. AOAC Int. 78*: 437 (1995).

105. R. C. Hale and M. O. Gaylor, Determination of PCBs in fish tissues using supercritical fluid extraction, *Environ. Sci. Technol. 29*: 1043 (1995).

106. S. Bowadt, B. Johansson, P. Fruekilde, M. Hansen, D. Zilli, B. Larsen, and J. de Boer, Supercritical fluid extraction of polychlorinated biphenyls from lyophilized fish tissue, *J. Chromatogr. A. 675*: 189 (1994).

107. Y. Yang and W. Baumann, Study of polychlorinated biphenyls in street dust by supercritical fluid extraction-gas chromatography/mass spectrometry, *Fresenius' J. Anal. Chem. 354*: 56 (1996).

108. L. J. Mulcahey, J. L. Hedrick, and L. T. Taylor, Collection efficiency of various solid-phase traps for off-line supercritical fluid extraction, *Anal. Chem. 63*: 2225 (1991).

109. H. R. Johansen, G. Becher, and T. Greibrokk, Determination of planar PCBs by combining on-line SFE-HPLC and GC-ECD or GC/MS, *Anal. Chem. 66*: 4068 (1994).

110. H. R. Johansen, G. Becher, and T. Greibrokk, Determination of PCBs in biological samples using on-line SFE-GC, *Fresenius' J. Anal. Chem. 344*: 486 (1992).

111. A. Sodergren, Monitoring of persistent lipophilic pollutants in water and sediment by solvent-filled dialysis membranes, *Ecotoxicol. Environ. Saf. 19*: 143 (1990).

112. J. Meadows, D. Tillitt, J. Huckins, and D. Schroeder, Large-scale dialysis of sample lipids using a semipermeable membrane device, *Chemosphere 26*: 1993 (1993).

113. C. Hemmerling, C. Risto, B. Augustyniak, and K. Jenner, The preparation of food and environmental samples for residue analysis of pesticides and PCBs by continuous steam distillation, *Nahrung 35*: 711 (1991).

114. L. Ramos, G. P. Blanch, L. Hernandez, and M. J. Gonzalez, Recoveries of organochlorine compounds (polychlorinated biphenyls, polychlorinated dibenzo-*p*-dioxins and polychlorinated dibenzofurans) in water using steam-distillation–solvent extraction at normal pressure, *J. Chromatogr. A 690*: 243 (1995).

115. I. Schuphan, W. Ebing, J. Holthoefer, R. Krempler, E. Lanka, M. Ricking, and H. J. Pachur, Bleidner vapour-phase extraction technique for the determination of organochlorine compounds in lake sediments, *Fresenius' J. Anal. Chem. 336*: 564 (1990).

116. S. Mossner, I. Barudio, T. S. Spraker, G. Antonelis, G. Early, J. R. Geraci, P. R. Becker, and K. Ballschmiter, Determination of HCHs, PCBs, and DDTs in brain

tissues of marine mammals of different age, *Fresenius' J. Anal. Chem. 349*: 708 (1994).

117. J. L. Bernal, J. L. Del Nozal, and J. J. Jimenez, Some observations on cleanup procedures using sulphuric acid and Florisil, *J. Chromatogr. 607*: 303 (1992).

118. E. Viana, J. C. Moltó, J. Mañes, and G. Font, Clean-up and confirmatory procedures for gas chromatographic analysis of pesticide residues. Part II, *J. Chromatogr. A. 678*: 109 (1994).

119. S. M. Dogheim, E. N. Nasr, M. M. Almaz, and M. M. el Tohamy, Pesticide residues in milk and fish samples collected from two Egyptian Governorates, *J. Assoc. Off. Anal. Chem. 73*: 19 (1990).

120. M. E. Zabic, M. J. Zabic, A. M. Booren, M. Nettles, J. H. Song, R. Welch, and H. Humphrey, Pesticides and total polychlorinated biphenyls in chinook salmon and carp harvested from the Great Lakes: effects of skin-on and skin-off processing and selected cooking methods, *J. Agric. Food Chem. 43*: 993 (1995).

121. E. Alvarez Piñeiro, J. Simal Lozano, and A. Lage Yusty, Gas chromatographic determination of polychlorinated biphenyls in mussels from Galicia, Spain, *J. AOAC Inc. 77*: 985 (1994).

122. S. H. Mitchell and S. Kennedy, Tissue concentrations of organochlorine compounds in common seals from the coast of Northern Ireland, *Sci. Total. Environ. 115*: 163 (1992).

123. D. T. Gundersen and W. D. Pearson, Partitioning of PCBs in the muscle and reproductive tissues of paddlefish, Polyodon spathula, at the falls of the Ohio River, *Bull. Environ. Contam. Toxicol. 49*: 455 (1992).

124. H. Steinwandter, Research in environmental pollution. IX. A new micromethod for the extraction of polychlorinated biphenyls (PCBs) from sewage sludges, *Fresenius' J. Anal. Chem. 347*: 436 (1993).

125. M. M. Schantz, B. J. Koster, S. A. Wise, and P. R. Becker, Determination of PCBs and chlorinated hydrocarbons in marine mammal tissues, *Sci. Total. Environ. 139–140*: 323 (1993).

126. J. Poustka, J. Hajslova, and K. Holadova, The comparison of elution profiles of polychlorinated biphenyls and fats in various gel-permeation chromatographic systems, *Int. J. Environ. Anal. Chem. 57*: 83 (1994).

127. E. Storr-Hansen and T. Cederberg, Determination of coplanar polychlorinated biphenyl (PCB) congeners in seal tissues by chromatography on active carbon, dual-column high resolution GC-ECD and high resolution GC-high resolution MS, *Chemosphere 24*: 1181 (1992).

128. S. J. Harrad, A. S. Sewart, R. Boumphrey, R. Duarte-Davison, and K. C. Jones, Method for the determination of PCB congeners 77,126 and 169 in biotic and abiotic matrices, *Chemosphere 24*: 1147 (1992).

129. J. Schreitmueller and K. Ballschmiter, Levels of polychlorinated biphenyls in the lower troposphere of the North- and South-Atlantic Ocean. Studies of global baseline pollution. XVII, *Fresenius' J. Anal. Chem. 348*: 226 (1994).

130. O. M. Rodriguez, P. G. Desideri, L. Lepri, and L. Checchini, Simultaneous separation and determination of hydrocarbons and organochlorine compounds by using a two-step microcolumn, *J. Chromatogr. 555*: 221 (1991).

131. B. G. Loganathan, K. Kannan, I. Watanabe, M. Kawano, K. Irvine, S. Kumar, and H. C. Sikka, Isomer-specific determination and toxic evaluation of polychlorinated biphenyls, polychlorinated/brominated dibenzo-*p*-dioxins and dibenzofurans, poly-brominated biphenyl ethers, and extractable organic halogen in carp from the Buffalo River, New York, *Environ. Sci. Technol. 29*: 1832 (1995).

132. J. Falandysz, N. Yamashita, S. Tanabe and R. Tatsukawa, Isomer-specific analysis of PCBs including toxic coplanar isomers in canned cod livers commercially processed in Poland, *Z. Lebensm. Unters. Forsch. 194*: 120 (1992).

133. A. Kocan, J. Petrik, J. Chovancova, and B. Drobna, Method for the group separation of non-ortho-, mono-ortho- and multi-ortho-substituted polychlorinated biphenyls and polychlorinated dibenzo-*p*-dioxins/polychlorinated dibenzofurans using activated carbon chromatography, *J. Chromatogr. A. 665*: 139 (1994).

134. T. L. King, J. F. Uthe, and C. J. Musial, Rapid semi-micro method for separating non-ortho chlorinated chlorobiphenyls from other chlorobiphenyls, *Analyst 120*: 1917 (1995).

135. H. Steinwandter, Research in environmental pollution. IV. Identification of non-*o,o'*-Cl and mono-*o,o'*-Cl substituted PCB congeners in Main river fish, *Fresenius' J. Anal. Chem. 342*: 416 (1992).

136. V. Bohm, E. Schulte, and H.-P. Thier, Carbon column HPLC and carbone-phase capillary GLC for the separation of polychlorinated biphenyl congeners, *Fresenius' J. Anal. Chem. 348*: 297 (1994).

137. V. Bohm, E. Schulte, and H. P. Thier, Determination of coplanar polychlorinated biphenyl congener residues in food using HPLC fractionation, *Z. Lebensm. Unters. Forsch. 192*: 548 (1991).

138. K. P. Feltz, D. E. Tillitt, R. W. Gale, and P. H. Peterman, Automated HPLC fractionation of PCDDs and PCDFs and planar and nonplanar PCBs on C_{18}-dispersed PX-21 carbon, *Environ. Sci. Technol. 29*(3): 709 (1995).

139. U. Pyell and P. Garrigues, Clean-up by high-performance liquid chromatography of polychlorodibenzo-*p*-dioxins and polychlorodibenzofurans on pyrenylethylsilica gel column, *J. Chromatogr. A 660*: 223 (1994).

140. C. Bandh, R. Ishaq, D. Broman, C. Naf, Y. Ronquist-Nii, and Y. Zebuhr, Separation for subsequent analysis of PCBs, PCDD/Fs, and PAHs according to aromaticity and planarity using a two-dimensional HPLC system, *Environ. Sci. Technol. 30*: 214 (1996).

141. S. H. Safe, Polychlorinated biphenyls (PCBs): environmental impact, biochemical and toxic responses, and implications for risk assessment, *Crit. Rev. Toxicol. 24*: 87 (1994).

142. D. Sissons and D. Welti, Structural identification of polychlorinated biphenyls in commercial mixtures by gas liquid chromatography, nuclear magnetic resonance and mass spectrometry, *J. Chromatogr. 60*: 15 (1971).

143. E. Storr-Hansen, Comparative analysis of thirty polychlorinated biphenyl congeners on two capillary columns of different polarity with non-linear multi-level calibration, *J. Chromatogr. 558*: 375 (1991).

144. C. Y. Chen and Y. C. Ling, Optimizing the gas-chromatographic separation and detection of polychlorinated biphenyls by use of electronic pressure programming and experimental design, *J. High Resolut. Chromatogr. 17*: 784 (1994).

145. L. G. M. T. Tuinstra, A. H. Roos, B. Griepink, and D. E. Wells, Inter-laboratory studies of the determination of selected chlorobiphenyl congeners with capillary gas-chromatography using splitless and on-column injection techniques, *J. High Resolut. Chromatogr. 8*: 475 (1985).

146. M. M. Schantz, R. M. Parris, J. Kurz, K. Ballschmiter, and S. A. Wise, Comparison of methods for the gas chromatographic determination of PCB congeners and chlorinated pesticides in marine reference materials, *Fresenius' J. Anal. Chem. 346*: 766 (1993).

147. B. W. Hermann, L. M. Freed, M. Q. Thompson, R. J. Phillips, K. J. Klein, and W. D. Snyder, CGC (capillary GC) using a programmable electronic pressure controller, *J. High Resolut. Chromatogr. 13*: 361 (1990).

148. J. D. Tessari and D. T. Winn, Relative retention ratios of 38 chlorinated pesticides and polychlorinated biphenyls on five gas–liquid chromatographic columns, *J. Chromatogr. Sci. 29*: 1 (1991).

149. W. Vetter, B. Luckas, F. Biermans, M. Mohnke, and H. Rotzsche, Gas-chromatographic separation of polychlorinated biphenyls on a new apolar capillary column, *J. High Resolut. Chromatogr. 17*: 851 (1994).

150. G. S. Durell and T. C. Sauer, Simultaneous dual-column, dual-detector gas-chromatographic determination of chlorinated pesticides and polychlorinated biphenyls in environmental samples, *Anal. Chem. 62*: 1867 (1990).

151. B. Larsen and S. Bowadt, *HRGC Separation of PCB Congeners. The State of the Art*, Proc. 15th Int. Symp. on Capil. Chromatogr., Vol. 1, 1993, p. 503.

152. E. Sippola and K. Himberg, Determination of toxic PCB congeners in biological samples by multi-dimensional gas chromatography–mass spectrometry (GC–GC–MS), *Fresenius' J. Anal. Chem. 339*: 510 (1991).

153. J. de Boer, Q. T. Dao, P. G. Wester, S. Bowadt, and U. A. T. Brinkman, Determination of mono-ortho sustituted chlorobiphenyls by multidimensional gas chromatography and their contribution to TCDD equivalents, *Anal. Chim. Acta 300*: 155 (1995).

154. A. Glausch, G. J. Nicholson, M. Fluck, and V. Schurig, Separation of the enantiomers of stable atropisomeric polychlorinated biphenyls (PCBs) by multi-dimensional gas chromatography on Chirasil-Dex, *J. High Resolut. Chromatogr. 17*: 347 (1994).

155. A. Glausch, J. Hahn, and V. Schurig, Enantioselective determination of chiral 2,2',3,3',4,6'-hexachlorobiphenyl (PCB 132) in human milk samples by multi-dimensional gas chromatography/electron capture detection and by mass spectrometry, *Chemosphere 30*: 2079 (1995).

156. I. L. Davies, K. E. Markides, M. L. Lee, M. W. Raynor, and K. D. Bartle, Applications of coupled LC-GC: review, *J. High Resolut. Chromatogr. 12*: 193 (1989).

157. H. J. Cortes, C. D. Pfeiffer, G. L. Jewett, and B. E. Richter, Direct introduction of aqueous eluents for on-line coupled liquid chromatography-capillary gas chromatography, *J. Microcol. Sep. 1*: 28 (1989).

158. Z. Juvancz, K. M. Payne, K. E. Markides, and M. L. Lee, Multi-dimensional packed capillary coupled to open tubular column supercritical-fluid chromatography using a valve-switching interface, *Anal. Chem. 62*: 1384 (1990).

159. C. Y. Chen and Y. C. Ling, Optimizing electron-capture detector performance for

gas-chromatographic analysis of polychlorinated biphenyls, *Chromatographia 33*: 272 (1992).

160. Anonymous, *Determination of DDTs and PCBs by Capillary Gas Chromatography and ECD*, United Nations Environment Programme, 1988.

161. R. P. Eganhouse and R. W. Gossett, Sources and magnitude of bias associated with determination of polychlorinated biphenyls in environmental samples, *Anal. Chem. 63*: 2130 (1991).

162. M. Dressler, *Selective Gas Chromatographic Detectors* (J. Chromatogr. Libr., Vol. 36), Elsevier, Amsterdam, 1986.

163. B. Griepink, E. A. Maier, H. Muntau, and D. E. Wells, Certified reference materials chlorobiphenyls (IUPAC no. 28, 52, 101, 118, 153 and 180) in dried sewage sludge—CRM 392, *Fresenius' J. Anal. Chem. 339*: 173 (1991).

164. Y. V. Gankin, A. E. Gorrshteyn, and A. Robbat, Identification of PCB congeners by gas chromatography electron capture detection employing a quantitative structure-retention model, *Anal. Chem. 67*: 2548 (1995).

165. K. Ballschmiter, R. Bacher, A. Mennel, R. Fischer, U. Riehle, and M. Swerev, The determination of chlorinated biphenyls, chlorinated dibenzodioxins, and chlorinated dibenzofurans by GC-MS, *J. High Resolut. Chromatogr. 15*(4): 260 (1992).

166. S. Facchetti, Mass spectrometry in the analysis of polychlorinated biphenyls, *Mass Spectrom. Rev. 12*: 173 (1993).

167. K. Oxynos, K. W. Schramm, P. Marth, J. Schmitzer, and A. Kettrup, Chlorinated hydrocarbons -(CHs) and PCDD/F-levels in sediments and breams (*Abramis brama*) from river Elbe (contribution to the German Environmental Specimen Banking), *Fresenius' J. Anal. Chem. 353*: 98 (1995).

168. J. Petrik and A. Kocan, Isomer-specific analysis of technical mixtures of polychlorinated biphenyls by high-resolution GC-MS, *Chem. Listy 86*: 694 (1992).

169. R. Boonyathumanondh, S. Watanabe, W. Laovakul, and M. Tabucanon, Development of a quantification methodology for polychlorinated biphenyls by using Kanechlor products as the secondary reference standard, *Fresenius' J. Anal. Chem. 352*: 261 (1995).

170. S. Musil and J. Lesko, Isotopic ratio of molecular patterns via gas chromatography–mass spectrometry with selected-ion monitoring as a chemometric tool, *J. Chromatogr. 664*: 253 (1994).

171. C. Y. Ma and C. K. Bayne, Differentiation of Aroclors using linear discrimination for environmental samples analysed by electron-capture negative-ion chemical ionization mass spectrometry, *Anal. Chem. 65*: 772 (1993).

172. D. Jones, A. G. Brenton, D. E. Games, A. H. Brittain, S. Taylor, D. Kennedy, and P. Smith, Ion mobility spectrometry as a detection technique for the separation sciences, *Rapid Commun. Mass Spectrom. 7*: 561 (1993).

173. L. Zupancic-Kralj, J. Marsel, B. Kralj, and D. Zigon, Application of tandem mass spectrometry to the analysis of chlorinated compounds, *Analyst 119*: 1129 (1994).

174. Y. Horimoto, S. Nishi, K. Yamaguchi, and R. Oguchi, Method for quantitative analysis of total polychlorinated biphenyls by capillary GC, *Bunseki Kagaku 41*: 95 (1992).

175. J. Hajslova, P. Cuhra, M. Kempny, J. Poustka, K. Holadova, and V. Kocourek,

Determination of polychlorinated biphenyls in biotic matrices using gas chromatography-microwave-induced plasma atomic emission spectrometry, *J. Chromatogr. A. 699*: 231 (1995).

176. D. M. Hembree, N. R. Smyrl, W. E. Davis, and D. M. Williams, Isomeric characterization of polychlorinated biphenyls using gas chromatography–Fourier-transform infrared and gas chromatography–mass spectrometry, *Analyst 118*: 249 (1993).

177. J. Greaves, E. Harvey, and R. J. Huggett, Evaluation of gas chromatography with electrolytic conductivity detection and electron capture detection and use of negative chemical ionization GC–MS for the analysis of PCBs in effluents, *Environ. Toxicol. Chem. 10*: 1391 (1991).

178. H. A. G. Neiderlaender, M. J. Nuijens, E. M. Dozy, C. Gooijer, and N. H. Velthorst, Dioxetane chemiluminescence detection in liquid chromatography based on photosensitized on line generation of singlet molecular oxygen; a thorough examination of experimental parameters and application to polychlorinated biphenyls, *Anal. Chim. Acta 297*: 349 (1994).

179. M. M. Krahn, G. M. Ylitalo, J. Buzitis, C. A. Sloan, D. T. Boyd, S. L. Chan, and U. Varanasi, Screening for planar chlorobiphenyl congeners in tissues of marine biota by high-performance liquid chromatography with photodiode array detection, *Chemosphere. 29*: 117 (1994).

180. K. Cammann and W. Kleiboehmer, Supercritical-fluid chromatography of polychlorinated biphenyls on packed columns, *J. Chromatogr. 522*: 267 (1990).

181. F. I. Onuska, K. A. Terry, S. Rukushika, and H. Hatano, Microbore columns vs. open tubular columns for supercritical fluid chromatography in environmental analysis: separation of polychlorinated biphenyls and terphenyls, *J. High Resolut. Chromatogr. 13*: 317 (1990).

182. V. W. Burse, M. P. Korver, P. C. McClure, J. S. Holler, D. M. Fast, S. L. Head, D. T. Miller, D. J. Buckley, J. Nassif, and R. J. Timperi, Problems associated with interferences in the analysis of serum for polychlorinated biphenyls, *J. Chromatogr. 104*: 117 (1991).

183. S. R. Finch, D. A. Lavigne, and R. P. W. Scott, One example where chromatography may not necessarily be the best analytical method, *J. Chromatogr. Sci. 28*: 351 (1990).

184. K. M. Carroll, M. R. Harkness, A. A. Bracco, and R. R. Balcarcel, Application of a permeant/polymer diffusional model to the desorption of polychlorinated biphenyls form Hudson river sediments, *Environ. Sci. Technol. 28*: 253 (1994).

185. S. E. McGroddy, J. W. Farrington, and P. M. Gschwend, Comparison of the in situ and desorption sediment–water partitioning of polycyclic aromatic hydrocarbons and polychlorinated biphenyls, *Environ. Sci. Technol. 30*: 172 (1996).

186. M. Cigánek, M. Dressler, and V. Lang, Relative electron-capture detector response of selected polychlorinated biphenyl congeners. Influence of detector temperature and design, *J. Chromatogr. A. 668*: 441 (1994).

187. J. L. Spencer, J. P. Hendricks, and D. Kerr, Least-squares analysis of gas chromatographic data for polychlorinated biphenyl mixtures, *J. Chromatogr. A 654*: 143 (1993).

188. M. Morosini and K. Ballschmiter, Retention indices of 28 polychlorinated biphenyls

in capillary gas chromatography referred to 2,4,6-trichlorophenyl alkyl ethers as RI-standards, *Fresenius' J. Anal. Chem. 348*: 595 (1994).

189. J. A. van Rhijn, W. A. Traag, P. F. van de Spreng, and L. G. Tuinstra, Simultaneous determination of planar chlorobiphenyls and polychlorinated dibenzo-*p*-dioxins and -furans in Dutch milk using isotope dilution and gas chromatography–high-resolution mass spectrometry, *J. Chromatogr. 630*: 297 (1993).

190. S. A. Wise, M. M. Schantz, R. M. Parris, R. E. Rebbert, B. A. Benner, and T. E. Gills, Standard reference materials for trace organic contaminants in the marine environment, *Analusis 20*: 57 (1992).

191. T. Rymen, S. Clark, A. Boenke, P. J. Wagstaffe and A. S. Lindsey, Reference materials for PCB analysis: production and certification of ten polychlorinated biphenyls in an iso-octane reference solution, *Fresenius' J. Anal. Chem. 343*: 553 (1992).

192. M. M. Schantz, B. J. Koster, L. M. Oakley, S. B. Schiller, and S. A. Wise, Certification of polychlorinated biphenyl congeners and chlorinated pesticides in a whale blubber standard Reference Material, *Anal. Chem. 67*: 901 (1995).

193. D. E. Kimbrough, R. Chin, and J. Wakakuwa, Industry-wide performance in a pilot performance evaluation sample programme for hazardous materials laboratories. II. Precision and accuracy of [determination of] polychlorinated biphenyls, *Environ. Sci. Technol. 26*: 2101 (1992).

194. D. E. Kimbrough, R. Chin, and J. Wakakuwa, Wide-spread and systematic errors in the analysis of soils for polychlorinated biphenyls, I. A. review of inter-laboratory studies, *Analyst 119*: 1277 (1994).

195. D. E. Kimbrough, R. Chin, and J. Wakakuwa, Wide-spread and systematic errors in the analysis of soils for polychlorinated biphenyls. II. Comparison of extraction systems, *Analyst 119*: 1283 (1994).

196. D. E. Kimbrough, R. Chin, and J. Wakakuwa, Wide-spread and systematic errors in the analysis of soils for polychlorinated biphenyls. III. Gas chromatography, *Analyst 119*: 1293 (1994).

197. J. de Boer, J. van der Meer, L. Reutergardh, and J. A. Calder, Determination of chlorobiphenyls in cleaned-up seal blubber and marine sediment extracts: inter-laboratory study, *J. AOAC. Int. 77*: 1411 (1994).

198. M. T. Galceran, G. Rauret, F. J. Santos, and D. Wells, Interlaboratory exercises in quality assurance. II. Analysis of chlorinated biphenyls (CBs) and organochlorine pesticides (OCPs), *Quim. Anal. 14*: 177 (1995).

199. W. J. Dunn, D. L. Stalling, T. R. Schwartz, D. De Vault, S. Wold, P. A. Bergqvist, K. Wiberg, and C. Rappe, A chemometric study of polychlorinated dibenzo-*p*-dioxins and polychlorinated dibenzofurans in Great Lakes fish, *IARC. Sci. Publ.* 175 (1991).

200. E. K. Duursma, J. Nieuwenhuize, and J. M. Van Liere, Polychlorinated biphenyl equilibria in an estuarine system, *Sci. Total. Environ. 79*: 141 (1989).

201. S. Raccanelli, P. Pavoni, A. Marcomini, and A. A. Orio, Polychlorinated biphenyl pollution caused by resuspension of surface sediments in the lagoon of Venice, *Sci. Total. Environ. 79*: 111 (1989).

202. L. Morselli, S. Zappoli, and A. Donati, Evaluation of the occurrence of PCBs in Terra Nova Bay, Ross sea. Results of the second campaign, *Ann. Chim. 81*: 503 (1991).

203. J. M. Bracewell, A. Hepburn, and C. Thomson, Levels and distribution of poly-chlorinated biphenyls on the Scottish land mass, *Chemosphere 27*: 1657 (1993).

204. A. Robbat, T. Y. Liu, and B. M. Abraham, Evaluation of a thermal-desorption gas chromatograph–mass spectrometer: on site detection of polychlorinated biphenyls at a hazardous waste site, *Anal. Chem. 64*: 358 (1992).

205. B. Pavoni, C. Calvo, A. Sfriso, and A. A. Orio, Time trend of PCB concentrations in surface sediments from a hypertrophic, macroalgae populated area of the lagoon of Venice, *Sci. Total. Environ. 91*: 13 (1990).

206. V. Raverdino, R. Holzer, and J. D. Berset, Comparison of high resolution gas chromatography with electron impact and negative ion mass spectrometry detection for the determination of coplanar polychlorobiphenyl congeners in sewage sludges, *Fresenius' J. Anal. Chem. 354*: 477 (1996).

207. W. L. Lockhart, R. Wagemann, B. Tracey, D. Sutherland, and D. J. Thomas, Presence and implications of chemical contaminants in the freshwaters of the Canadian Arctic, *Sci. Total. Environ. 122*: 165 (1992).

208. A. C. Smaal, A. Wagenvoort, J. Hemelraad, and I. Akkerman, Response to stress of mussels (Mytilus edulis) exposed in Dutch tidal waters, *Comp. Biochem. Physiol. C. 100*: 197 (1991).

209. C. S. Hong, B. Bush, and J. Xiao, Coplanar PCBs in fish and mussels from marine and estuarine waters of New York State, *Ecotoxicol. Environ. Safety. 23*: 118 (1992).

210. A. Goksoyr, A. M. Husoy, H. E. Larsen, J. Klungsoyr, S. Wilhelmsen, A. Maage, E. M. Brevik, T. Andersson, M. Celander, M. Pesonen, et al., Environmental contami-nants and biochemical responses in flatfish from the Hvaler Archipelago in Norway, *Arch. Environ. Contam. Toxicol. 21*: 486 (1991).

211. B. Bush, R. W. Streeter, and R. J. Sloan, Polychlorobiphenyl (PCB) congeners in striped bass (Morone saxatilis) from marine and estuarine waters of New York State determined by capillary gas chromatography, *Arch. Environ. Contam. Toxicol. 19*: 49 (1990).

212. M. D. Pastor, J. Sanchez, D. Barceló, and J. Albaigés, Determination of coplanar polychlorobiphenyl congeners in biota samples, *J. Chromatogr. 629*: 329 (1993).

213. B. Jimenez, J. Tabera, L. M. Hernandez, and M. J. Gonzalez, Simplex optimization of the analysis of polychlorinated biphenyls. Application to the resolution of a complex mixture of congeners of interest on a single gas-chromatographic column, *J. Chromatogr. 607*: 271 (1992).

214. D. W. Kuehl, B. C. Butterworth, J. Libal, and P. Marquis, Isotope-dilution high-resolution gas-chromatographic-high-resolution mass-spectrometric method for the determination of coplanar polychlorinated biphenyls: application to fish and marine mammals, *Chemosphere 22*: 849 (1991).

215. H. Steinwandter, Research in environmental pollution. V. Identification of non-*o,o'*-Cl and mono-*o,o'*-Cl substituted PCB congeners in Rhine river fish. *Fresenius' J. Anal. Chem. 342*: 467 (1992).

216. L. J. Schmidt and R. J. Hesselberg, A mass spectroscopic method for analysis of AHH-inducing and other polychlorinated biphenyl congeners and selected pesti-cides in fish, *Arch. Environ. Contam. Toxicol. 23*: 37 (1992).

217. C. S. Hong, B. Bush, J. Xiao, and H. Qiao, Toxic potential of non-ortho and mono-

ortho coplanar polychlorinated biphenyls in Aroclors, seals, and humans, *Arch. Environ. Contam. Toxicol. 25*: 118 (1993).

218. M. R. Driss, S. Sabbah, and M. . Bouguerra, High resolution gas phase chromatography-mass spectrometry of polychlorinated biphenyl congener residues in samples of biologic origin, *J. Chromatogr. 552*: 213 (1991).

219. R. Kibler and J. Lepschy von Gleissenthall, Intake of polychlorinated biphenyls (PCB) in the total diet, *Z. Lebensm. Unters. Forsch. 191*: 214 (1990).

3

Analysis of Nitrosamines in Foods and Beverages

Geoffrey Yeh, John D. Ebeler,
and Susan E. Ebeler
University of California, Davis, California

I. INTRODUCTION

The N-nitroso compounds occur in foods and biological systems as the products of the reactions between secondary amines and nitrite. The discovery in the early 1960s that highly carcinogenic N-nitroso compounds occur in foods necessitated the development of highly specific and sensitive methods for their analysis. The development of accurate analytical methods was also essential in the elucidation of chemical and biochemical mechanisms for the formation, metabolism, and toxicity of these compounds.

Nitrosamines can be categorized as volatile or nonvolatile depending on their relative vapor pressures. Volatile nitrosamines are nonpolar, low molecular weight compounds that possess sufficient vapor pressure to allow their removal from a food matrix by distillation. These compounds can be easily separated by gas chromatography (GC) without additional derivatization. Examples of volatile N-nitrosamines include N-nitrosodimethylamine (NDMA), N-nitrosodiethyl-

Dialkylnitrosamines

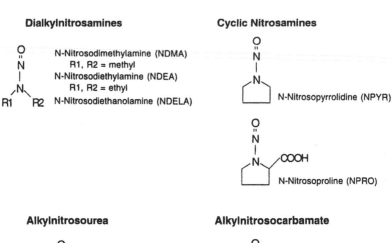

N-Nitrosodimethylamine (NDMA)
    R1, R2 = methyl
N-Nitrosodiethylamine (NDEA)
    R1, R2 = ethyl
N-Nitrosodiethanolamine (NDELA)

Cyclic Nitrosamines

N-Nitrosopyrrolidine (NPYR)

N-Nitrosoproline (NPRO)

Alkylnitrosourea

Alkylnitrosocarbamate

Alkylnitrosoguanidine

Alkylnitrosamide

FIGURE 1 Common types of volatile and nonvolatile *N*-nitrosamines.

amine (NDEA), and *N*-nitrosopyrrolidine (NPYR) (Fig. 1). Nonvolatile nitrosamines cannot be isolated by distillation techniques, require derivatization prior to analysis by GC, or are analyzed by high performance liquid chromatography (HPLC). Nonvolatile nitrosamines include *N*-nitrosoproline (NPRO) and *N*-nitrosodiethanolamine (NDELA) (Fig. 1).

This review provides a brief overview of the toxicity, formation, and occurrence of *N*-nitroso compounds and focuses on the analysis of volatile nitrosamines in foods and beverages. Several methods for the analysis of nonvolatile nitrosamines have recently been described and will be very briefly discussed.

II. *N*-NITROSO TOXICITY

Nitrosamines are known carcinogens that can also cause acute poisoning and death at doses that are well above normal exposure levels. The toxic effects almost always include centritubular necrosis of the liver, a pattern that appears to have no relation to species [1]. Hemorrhaging is also common at high doses. In addition, cyclic nitrosamines can interact with the central nervous system, causing convulsions or brain edema [2].

Potency varies greatly among the individual nitrosamines, with NDMA the most potent acute toxin in most species. Oxidation or hydroxylation of the alkyl groups causes a decrease in toxicity, while halogens and cyano groups cause a marked increase in potency [2]. In mice, NDMA tends to accumulate in the kidney and liver, and radioactive tracers have shown that some constituents of NDMA can bind to tissue, suggesting that nitrosamines can lead to alkylation of cellular components [3].

Nitrosamines are relatively stable compounds that require metabolic activation to become capable of alkylating nucleophilic macromolecules [4]. Nitrosamides, nitrosoureas, and nitrosocarbamates can alkylate macromolecules, including nucleic acids, directly. They are also less stable and can cause acute damage at the site of application. The proposed mechanism of nitrosamine activation involves α-hydroxylation by a cytochrome P-450 dependent mixed function oxidase. The resulting hydroxylated nitrosamine splits into a primary alkyldiazohydroxide and an aldehyde. The alkyldiazohydroxide splits to yield an alkyldiazonium ion which further decomposes to a carbonium ion and nitrogen. It is widely believed that the carbonium ion is the probable alkylating agent. This mechanism is poorly understood and new evidence suggests that other enzymes may be involved [3].

Nitrosamines have no apparent carcinogenic threshold [3] and their structure determines the organ specificity and potency. In most species, carcinogenicity of aliphatic nitrosamines seems to be inversely related to the size of the substituent groups, and carcinogenicity of cyclic nitrosamines is proportional to the size of the ring [3]. Asymmetrical nitrosamines target the esophagus, while symmetrical ones target the liver. Route of administration is independent of target organ with few exceptions [5]. Oxygen substitution greatly decreases potency, and double α-hydroxylated nitrosamines have not yet shown carcinogenicity [2].

Mutagenicity assays are frequently used to evaluate the potential for a compound to be carcinogenic, but bacterial mutagenicity assays can give misleading results in terms of the true carcinogenic potential of nitrosamines in animals. In general, nitrosamines are less potent as mutagens than as carcinogens. Development of in vitro mammalian cell cultures has shown that nitrosamines have similar activity in humans and animals, causing gene mutation, chromosome aberrations, and sister chromatid exchanges [4].

III. *N*-NITROSO FORMATION AND OCCURRENCE

N-Nitrosamines arise from the reaction of a secondary or tertiary amine with a nitrosating agent. The reaction usually takes place in acidic solution, the optimum pH being dependent upon the pK_a of the secondary amine. The electrophilic nitrosating agents arise from nitrite or nitrous acid and include dinitrogen trioxide, dinitrogen tetraoxide, and the nitrous acidium ion, which is the main nitrosating species for amides and carbamates [6]. Nitrate is abundant in food and can be reduced to nitrite in vivo by microorganisms, possibly leading to nitrosamine formation. Because of its importance as a precursor to nitrite and nitrosamine formation, rapid, sensitive methods for the estimation of nitrate levels in food and biological samples have been proposed [7]. Primary amines are generally not considered to be nitrosamine precursors because their *N*-nitrosation generally results in a highly unstable diazo compound [8]. Tertiary amines react slowly, although some complex tertiary amines with unique substituents can be readily nitrosated. There is no systematic approach to determining which substituent group will be replaced by the nitroso group [6].

Catalysts of the nitrosation reaction include metal ions, carbonyl compounds [4], halide ions, especially Br^- and I^-, thiocyanate anion, acetate, phthalate, and weak acids [6]. Certain phenols can undergo *C*-nitrosation to form nitrosophenols, which are nitrosation catalysts [9]. These phenols include resorcinol, *p*-nitrosophenol, 2,4-dinitrosorcinol, and those complex phenols containing a resorcinol moiety capable of *C*-nitrosation [10].

Nitrosation inhibitors include ascorbate [6], vitamin E, sulfhydryl compounds, many phenolic compounds, and tannins [4]. Hydroquinone and catechol are very good nitrosation inhibitors [10]. One proposed explanation for this is a competitive reaction with the phenol resulting in a lower nitrite concentration. Walker and coworkers [10] have reviewed both the catalytic and the inhibitory effects of phenolic compounds on *N*-nitrosation.

Nitrosamine content in many foods has been reviewed [6]. Some common exogenous food sources of *N*-nitroso compounds are cured meats (hot dogs, fried bacon, and ham), fish, and cheese. Alcoholic malt beverages such as beer and whiskey have been extensively tested for nitrosamines. Malt kilning with natural gas burners supplies nitrogen oxides which can react with naturally occurring amines to form nitrosamines [11]. However, incorporation of sulfur dioxide as a nitrosation inhibitor effectively reduces the formation of NDMA [9,12]. There is no relationship between application of heat alone and NDMA formation during the malt drying process [13]. Nitrosamines were not detected in wine, ciders, and some hard liquors [14].

Cosmetics, toiletries (containing triethanolamine as an emulsifier), pharmaceuticals, tobacco products, pesticides, herbicides, rubber products, water, and air all offer exposure to *N*-nitroso compounds with tobacco providing the highest potential human exposure [4]. Endogenous *N*-nitroso formation can occur following exposure to nitrate in saliva, gastric juice, urine, feces, vaginal exudate, and blood [15].

IV. GENERAL ANALYTICAL CONSIDERATIONS

As reviewed by Hill [3], numerous methods have been developed for nitrosamine analysis in foods, consumer products, and biological tissues, including bacon, cooking oils, margarine, butter, milk and milk powder, cheese, fish, malt, beer, blood, feces, saliva, gastric juice, urine, cosmetics, and rubber products. Exact sample preparation steps will vary depending on the sample, but general steps for the analysis of volatile nitrosamines include

> Addition of an internal standard (IS) to the sample to ensure and to monitor analyte recovery and reproducibility
> Isolation of the nitrosamines via distillation and/or extraction into a relatively nonpolar solvent
> Column chromatography or other sample clean up steps used in addition to or as an alternative to distillation/extraction in order to remove interferences
> Sample concentration to improve sensitivity
> Chromatographic separation with a sensitive detector

A. Safety Considerations for the Analyst

As discussed previously, nitrosamines are potent carcinogens, so the analyst should take great care to avoid exposures of any type. Safety precautions include using a ventilated fume hood, protective gloves, and safety glasses/goggles whenever nitrosamine containing standards or samples will be handled. Mechanical pipetting aids should be used for all solutions, and any pipettes used for nitrosamines should be kept separate in order to avoid contamination of other reagents. After completing an analysis, destroy all nitrosamine containing standards or samples by boiling with HCl, KI and sulfamic acid before disposal. Finally, the Thermal Energy Analyzer (TEA) detector vacuum pump should be vented to a fume hood to remove ozone that might be out-gassed into the atmosphere.

Personnel exposure to volatile nitrosamines can be monitored by passing laboratory air through sampling tubes filled with Tenax or other suitable adsorbents [16]. Using this procedure, Issenberg and Swanson [16] identified several areas that could result in the potential for personnel exposure, including during animal treatment, during mutagenesis assays, during in vitro metabolism studies, and in nonventilated walk-in refrigerators where nitrosamines may be stored. Appropriate corrective measures, including use of biological safety hoods and proper ventilation, were sufficient to reduce all nitrosamine levels to below the analytical detection limit of 0.3 ppb NDMA.

B. Contamination and Artifact Formation

Nitrosamines are photolabile; cover samples with aluminum foil or other protective material or use special nontransparent glassware, and work under subdued lighting whenever possible. Goodhead and Gough [17] observed minimal degradation of aqueous NDMA solutions for periods up to 8 months when properly stored.

To avoid artifactual formation of nitrosamines during the analytical procedure, add sodium or ammonium sulfamate, sodium azide, or hydrazine sulfate to the sample early in the procedure. These compounds destroy any nitrosating agents by reacting with nitrite to form nitrogen gas. Alternatively, nitrite mediated N-nitrosation reactions are effectively blocked by making the solution alkaline. This prevents conversion of nitrite to nitrous acid and to active N-nitrosating agents such as dinitrogen trioxide. However, Sen et al. [18] found sulfamic acid (1%) to be a better nitrosation inhibitor than 3N KOH during the distillation and analysis of nitrosamines in beer and ale. Addition of sulfamic acid gave an improved and more consistent recovery of NDMA from beer and ale and reduced the chance of artifactual formation of nitrosamine during distillation.

Artifact formation can be monitored by adding a standard amine such as di-n-butylamine to the sample at the beginning of the analysis and measuring the amount of corresponding nitrosamine, i.e., dibutylnitrosamine, that is formed during the analysis [18]. Hotchkiss et al. [19] investigated the potential for artifact formation by adding dimethylamine, pyrrolidine, and sodium nitrite to a non-nitrite containing bacon and analyzing for NDMA and NPYR. Without the addition of ammonium sulfamate and sulfuric acid during distillation, up to 1000 μg/kg of the corresponding N-nitrosamine formed. Addition of ammonium sulfamate alone reduced artifact production to 1–100 μg/kg.

Finally, organic solvents, rubber tubing, rubber stoppers, and solid phase extraction devices can all contain trace levels of nitrosamines or nitrosamine precursors (e.g., nitrogen oxides, secondary amines, etc.), so a reagent blank should always be performed in order to detect possible exogenous contamination [20,21].

V. ANALYSIS OF VOLATILE NITROSAMINES

A. Isolation, Sample Cleanup, and Concentration

The *N*-nitrosamines generally occur at trace levels requiring sample cleanup and concentration for accurate and sensitive quantitation. Collaborative studies have shown that a number of methods can give reliable results with detection limits as low as 0.1 ppb. Official AOAC International or International Agency for Research on Cancer [22] methods should be consulted for details before performing a specific analysis; general isolation procedures are described below.

1. Vacuum Distillation

Vacuum distillation was one of the early sample preparation techniques used for nitrosamine analysis. This technique exploits the volatility of *N*-nitrosamines, allowing them to be separated form the less volatile components of complex matrices. Because of the possibility for artifact formation, nitrosation inhibitors must be added prior to the distillation [18,19]. The distillates are trapped in an aqueous medium or cryogenically, extracted into an organic solvent such as dichloromethane, dried with Na_2SO_4, and concentrated using a Kuderna-Danish evaporator or a gentle nitrogen stream. Losses of the volatile nitrosamines during the concentration step can occur if the evaporation is allowed to proceed too vigorously [17].

2. Atmospheric Distillation

Atmospheric distillation employs the same basic isolation principle as vacuum distillation, but because atmospheric pressures are used the sample temperature is elevated. Since reduced pressures are not utilized, atmospheric distillations can be simpler than vacuum distillations, requiring less elaborate equipment.

Similar precautions for artifact formation and sample losses are necessary as described for vacuum distillation procedures. The temperature of the water used for cooling the condenser may affect recoveries, but Sen and Seaman [23] found little effect of water temperatures as high as 20°C. According to Sen and Seaman, atmospheric distillation may give slightly better reproducibility than vacuum distillation while giving comparable recovery and sensitivity. These authors reported 75–115% recovery of 0.08 to 10 ppb NDMA, NDEA, and *N*-nitroso-dipropylamine (NDPA) from beer. The minimum detection limit of the method was ~0.1 ppb using a GC-Thermal energy analyzer for quantitation. Atmospheric distillation procedures have been extensively used for food and beverage analysis [17,19,23–27].

3. Liquid-Liquid Extraction

Liquid-liquid extraction into methylene chloride has been used for the isolation of nitrosamines from a number of matrices including wastewater, beer, and other fermented beverages and foods [12,28–30]. Following extraction with methylene chloride, samples are dried with anhydrous Na_2SO_4 and concentrated using a Kuderna-Danish evaporator or gentle nitrogen stream.

Liquid-liquid extraction procedures can have complications from foam and emulsion formation [30]. The use of saturated NaCl and centrifugation may aid in breaking up the emulsion and separating the organic and aqueous phases [30]. The evaporation step can again lead to losses of the volatile nitrosamine analytes [30]. Recoveries using liquid-liquid extraction seem to be quite variable depending on the matrix, the analyte, and the analyte concentration. Weston [12] reported 55–67% recovery of 10–100 ng NDMA from water-alcohol mixtures. Recovery of NDMA from malt-water suspensions was 79–92%. In beer samples [30], recoveries of 70–80% were observed for NDMA, NDEA, NDPA, *N*-nitrosodibutyl-amine (NDBA), *N*-nitrosopiperidine (NPIP), NPYR, and *N*-nitrosomorpholine (NMOR) spiked at 1 ppb. Higher recoveries, 80–102%, were observed for samples spiked at higher levels, 5 and 25 ppb.

4. Column Chromatography with Celite

Column chromatography with Celite is a simple, rapid alternative to the distillation procedures. Samples are mixed with Celite (diatomaceous earth) and the mixture is packed as a dry powder or slurry into a liquid chromatographic column. Alternatively, the column is first packed with Celite and samples are transferred to the top of the column. The nitrosamines are then eluted with methylene chloride, concentrated, and analyzed by GC. Lipid materials can be removed by washing the columns with hexane prior to methylene chloride elution. This solid phase extraction method gives similar results to those obtained by atmospheric and vacuum distillation techniques for analyzing NDMA in malt beverages [18,31]. These authors observed no artifact formation. The technique is very effective in removing water and polar components which may interfere with the analysis and avoids many of the foaming and emulsion problems encountered with other techniques. Variations of this procedure are extensively used [9,11,27,32–34].

Takatsuki and Kikuchi [35] proposed a method for analysis of NDMA in fish products using silica gel and alumina solid phase extraction cartridges. The dried sample is extracted with methylene chloride, mixed with hexane, and placed on a silica gel column. The NDMA is adsorbed on the silica and subsequently eluted with methylene chloride–diethyl ether. This eluate is then placed on an alumina A cartridge, and the NDMA is isolated with a small volume of diethyl ether–methanol. The method does not require solvent evaporation and concentration steps which can result in the loss of volatile NDMA and the formation of artifactual NDMA. The authors report excellent NDMA recoveries (96.7%), but an internal standard was not used for quantitation, and artifact formation using a standard amine was not reported.

Airborne nitrosamines may also be trapped on cartridges packed with various adsorbents, including activated charcoal, alumina, Florisil®, Silica, and Tenax®. The trapped nitrosamines are eluted with methylene chloride, concentrated, and analyzed by GC. Artifact formation on these sorbents is possible and appears to be dependent on the concentration of amines and airborne nitrogen oxides which may be present [16,20].

B. Chromatography

Gas chromatographic separations on polar stationary phases are relatively straightforward. While many analyses rely on packed column separations, megabore and capillary column methods are also frequently used [30,35].

C. Detection

Utilization of sensitive and highly specific detectors has made it possible to quantitate trace levels of nitrosamines in complex mixtures with minimal sample preparation. While the nitrogen-phosphorus detector is specific for nitrogen containing compounds it has largely been replaced by the chemiluminescent detectors, which are very sensitive and highly specific for nitroso compounds. Conversion of nitrosamines to derivatives that can be sensitively measured has also been explored. While these techniques are excellent for quantitation, absolute confirmation of nitrosamine identity requires GC-mass spectrometric analysis.

1. Chemiluminescent Detectors

Most analytical procedures for the analysis of nitrosamines involve the use of a chemiluminescent detector such as the Thermal Energy Analyzer™ or TEA (Thermo Electron Corp., Waltham, Mass.). These detectors rely on a chemiluminescent reaction in which nitric oxide is oxidized by ozone to give excited NO_2^*. As reviewed by McWeeney [9], analytes from the GC enter into a catalytic pyrolysis tube where the N-NO bond is selectivity cleaved. The NO radical is converted to NO_2^* in a stream of ozone, and upon return to the ground state a photon is released, which is detected by a photomultiplier (Fig. 2). Because the TEA is very specific for nitrosamines, minimal sample cleanup steps are necessary, but some non-*N*-nitroso compounds can respond to the TEA (Table 1). Because of its excellent sensitivity, the TEA is highly suited for quantitative analyses. However, the TEA cannot give absolute identification, and positive confirmation by MS is required.

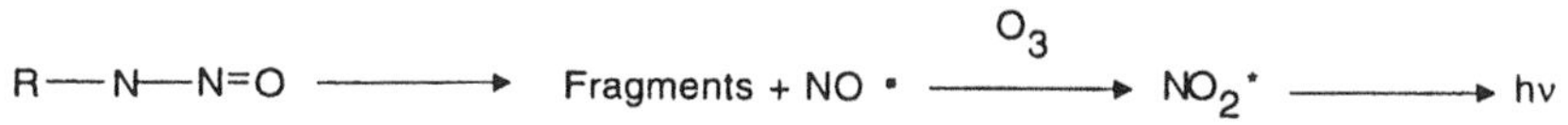

$$R\text{---}N\text{---}N{=}O \longrightarrow \text{Fragments} + NO^{\bullet} \xrightarrow{\;O_3\;} NO_2^{\bullet} \longrightarrow h\nu$$

FIGURE 2 Principle of chemiluminescence detection of nitrosamines.

Table 1 Thermal Energy Analyzer Response to Some Organic
Compounds (Molar Response Relative to NDMA)

Class	Compound	Response
N-nitrosamine	N-Nitrosodimethylamine	1.00
N-nitrosamide	N-Ethyl N-nitrosourea	0.02
N-nitrososulfonamide	N-Methyl N-nitroso toluenesulphonamide	0.17
N-nitramine	N-Nitrodimethylamine	0.40
S-nitroso	S-Nitrosotriphenylmethanethiol	0.53
C-nitro	Nitrobenzene	0.01
	2-Nitropropane	0.41
O-nitro	Ethyl nitrate	1.10
O-nitroso	Ethyl nitrate	0.70

Source: Ref. 9. (Reprinted with permission of Elsevier Science, Ltd.)

2. Mass Spectrometry

Nitrosamines give typical electron impact MS fragmentation patterns as reviewed
by Lijinsky [2]. The molecular ion for most low molecular weight nitrosamines is
generally prominent; the alkylnitrosamides are highly unstable and require soft
ionization techniques to provide molecular ions. Loss of OH (M-17) is common
and results in the base peak for the high molecular weight nitrosamines (Fig. 3).
Loss of NO (M-30) and HNO (M-31) are also common fragmentations of cyclic
and acyclic nitrosamines respectively (Fig. 3). Cleavage of the alkyl chain generally follows the α-cleavage patterns established for aliphatic amines. Using stable
isotopes Lijinsky and coworkers have evaluated the spectra of >130 compounds

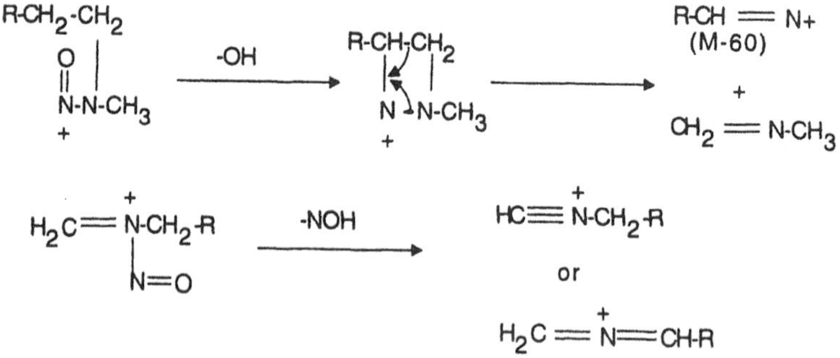

FIGURE 3 Common mass spectral fragmentation mechanisms of N-nitro-
samines. (From Ref. 2. Reprinted with permission of Cambridge University
Press.)

and found the fragmentation patterns to be quite regular, making this a highly useful tool for confirmation of nitrosamine structure.

3. Derivatizations

Numerous derivatives have been proposed including preparation of methyl esters, trimethylsilyl derivatives, and *N*-diethylthiophosphoryl derivatives [32,36,37]. The derivatives are analyzed by GC with a variety of detectors (e.g., flame ionization, nitrogen-phosphorus, and flame photometric detectors) allowing for selective and sensitive analyses. The techniques allow nonvolatile nitrosamines to be analyzed by gas chromatography, minimize sample cleanup steps, and offer significant cost savings by shifting from TEA detectors to other, less expensive detectors.

VI. ANALYSIS OF NONVOLATILE NITROSAMINES

A. Extraction and Chromatography

Nonvolatile nitrosamines present extraction and cleanup difficulties because of their wide polarity range. Common procedures employ extractions into relatively more polar solvents such as acetonitrile or acetone [38,39] and/or solid phase extraction using polar phases such as cyanopropyl [26,40]. Following the extraction and concentration steps, HPLC is the most common method for the determination of nitrosamines in food and beverages. Stationary phases usually consist of cyanoalkyl and saturated alkyl groups. Water is the main mobile phase with acetone, acetonitrile, or methanol as organic modifiers. Methanol and especially acetonitrile have been shown to decrease greatly the sensitivity of HPLC coupled to a TEA detector [26]. A problem with the use of reverse phase chromatography is that some *N*-nitroso compounds are sufficiently water soluble to elute with the solvent. Hotchkiss [5] reviews some proposed solutions to this problem.

The liquid mobile phases of HPLC are inherently incompatible with TEA, so several methods to turn the mobile phase into an aerosol have been introduced. The first is the coupling to TEA by a series of cold traps that strip the eluant of solvent [41]. Another effective coupling method is the use of a particle beam electron impact interface like that used for mass spectrometry. This method is particularly good because it provides a simultaneous coupling to the MS for structural confirmation [39]. These methods commonly have detection limits in the nanogram range.

B. Detection

1. Derivatization

A sensitive method of detection of nonvolatile nitrosamines involves the pre-chromatographic denitrosation by hydrobromic acid to a secondary amine that is derivatized with dansyl chloride. After separation by HPLC, the derivatives are detected by chemiluminescence [42]. Addition of peroxalate as a chemilumeno-genic reagent greatly increases the sensitivity of this method with a limit of detection in the picogram range [38].

2. Colorimetric Detection

Efficient colorimetric detection of volatile and nonvolatile nitrosamines involves reverse phase HPLC followed by post-column photohydrolysis in a UV photo-reactor. The liberated nitrite is then quantified by colorimetry with Griess reagent [43]. This method has sensitivity similar to that of HPLC-TEA and uses simple UV detection.

3. Amperometric Detection

Recently, a method using post-column oxidation of nitrite in HPLC eluants with voltammetric detection was developed with detection limits comparable to those for HPLC-TEA [44,45]. The method involves a post column redox reaction with iodide or Ce(IV). Iodide has been shown to provide greater sensitivity and a lower limit of detection, but the method also requires complete deoxygenation of the solution to avoid regeneration of nitric oxide from the reaction [45]. These techniques may be impractical for food analysis.

VII. CONCLUSIONS

A number of analytical methods for the analysis of volatile nitrosamines in foods, consumer products, and biological samples are available. The choice of method is dependent on the analytes to be measured and the sample matrix. Careful attention to the potential for analyte loss and artifact formation is essential; when adapting a method from one matrix to another, accuracy and precision should be confirmed using appropriate standards and reagent blanks.

The development of sensitive and selective analytical methods has provided extensive information on the occurrence and formation of volatile nitrosamines in foods and provided the information needed to change processing techniques to reduce or eliminate formation of and exposure to these potent carcinogens (e.g., malt and cured meats). However, as stated by Hotchkiss [5] ten years ago, there remains a critical need to develop reliable methods for the routine analysis of nonvolatile nitrosamines. Although much of the current work has been focused in this area and some promising methods have emerged, there is still limited informa-tion on the levels of nonvolatile nitrosamines in foods, beverages, and personal care products. This information will be essential in order to monitor the effects of processing on total nitrosamine formation, to assess the total human exposure to preformed nitrosamines, and to determine the overall health risks associated with this exposure.

REFERENCES

1. I. R. Rowland, The toxicology of *N*-nitroso compounds, *Nitrosamines: Toxicology and Microbiology* (M. J. Hill, ed.), Ellis Horwood, Chichester, 1988, pp. 117–141.
2. W. Lijinsky, *Chemistry and Biology of N-Nitroso Compounds*, Cambridge University Press, New York, 1992.
3. M. J. Hill, *Nitrosamines: Toxicology and Microbiology*, Ellis Horwood, Chichester, 1988.
4. A. R. Tricker and R. Preussman, Carcinogenic *N*-Nitrosamines in the diet: occurrence, formation, mechanisms, and carcinogenic potential, *Mutation Research 259*: 277–289 (1991).
5. J. H. Hotchkiss, A review of current literature on *N*-nitroso compounds in foods, *Advances in Food Res. 31*: 55–115 (1987).
6. A. R. Tricker and S. S. Kubaki, Review of the occurrence and formation of non-volatile *N*-nitroso compounds in food, *Food Additives Contam. 9*(1): 39–69 (1992).
7. B. J. Dull and J. H. Hotchkiss, A rapid method for estimating nitrate in biological samples using gas chromatography with a flame ionization detector or a thermal energy analyzer, *Food Chem. Toxic. 22*: 105–108 (1983).
8. E. Boyland, The effect of some ions of physiological interest on nitrosamine synthesis, *N-Nitroso Compounds Analysis and Formation* (P. Bogovski, R. Preussmann, and E. A. Walker, eds.), IARC, Lyon, 1972, pp. 124–126.
9. D. J. McWeeney, Nitrosamines in beverages, *Food Chemistry 11*:273–287 (1983).
10. E. A. Walker, B. Pignatelli, and M. Friesen, The role of phenols in the catalysis of nitrosamine formation, *J. Sci. Food Agric. 33*: 81–88 (1982).
11. R. Frommberger, NDMA in German beer, *Vol. Chem. Tox. 27*: 27–29 (1989).
12. R. J. Weston, N-nitrosamine content of New Zealand beer and malted barley, *J. Sci. Food Agric. 34*: 1005–1010 (1983).
13. M. M. Mangino and R. A. Scanlan, Formation of N-nitrosodimethylamine in direct-fire dried malt, *Am. Soc. of Brewing Chemists J.* January, pp. 55–57 (1981).
14. N. P. Sen, S. Seaman, and M. McPherson, Nitrosamines in alcoholic beverages, *J. Food Safety 2*: 13–18 (1980).
15. S. Leach, Mechanisms of endogenous N-nitrosation, *Nitrosamines, Toxicology and Microbiology* (M. J. Hill, ed.), Ellis Horwood, Chichester, 1988, pp. 69–87.
16. P. Issenberg and S. E. Swanson, Monitoring exposure of personnel to volatile nitrosamines in the laboratory environment, *N-Nitroso Compounds: Analysis, Formation and Occurrence* (E. A. Walker, L. Griciute, M. Castegnaro, and M. Börzsönyi, eds.), IARC, Lyon, 1980, pp. 531–540.
17. K. Goodhead and T. A. Gough, The reliability of a procedure for the determination of nitrosamines in food, *Food Cosmet. Toxicol. 13*: 307–312 (1975).
18. N. P. Sen, S. Seaman, and L. Tessier, Comparison of two analytical methods for the determination of dimethylnitrosamine in beer and ale, and some recent results, *J. Food Safety 4*: 243–250 (1982).
19. J. H. Hotchkiss, L. M. Libbey, J. F. Barbour, and R. A. Scanlan, Combination of a GC-TEA and a GC-MS-data system for the μg/kg estimation and confirmation of volatile *N*-nitrosamines in foods, *N-Nitroso Compounds: Analysis, Formation and Occurrence* (E. A. Walker, L. Griciute, M. Castegnaro, and M. Börzsönyi, eds.), IARC, Lyon, 1980, pp. 361–373.

20. D. P. Rounbehler, J. Reisch, and D. H. Fine, Some recent advances in the analysis of volatile *N*-nitrosamines, *N-Nitroso Compounds: Analysis, Formation and Occurrence* (E. A. Walker, L. Griciute, M. Castegnaro, and M. Börzsönyi, eds.), IARC, Lyon, 1980, pp. 403–416.

21. R. C. Massey, Analysis of *N*-nitroso compounds in foods and human body fluids, *Nitrosamines, Toxicology and Microbiology* (M. J. Hill, ed.), Ellis Horwood, Chichester, 1988, pp. 16–47.

22. H. Egan, *Environmental Carcinogens: Selected Methods of Analysis*, Vol. 6, IARC, Lyon, 1983.

23. N. P. Sen and S. Seaman, Gas-liquid chromatographic-thermal energy analyzer determination of *N*-nitrosodimethylamine in beer at a low parts per billion level, *J. Assoc. Off. Anal. Chem.* 64(4): 933–938 (1981).

24. K. S. Webb, T. A. Gough, A. Carrick, and D. Hazelby, Mass spectrometric and chemiluminescent detection of picogram amounts of *N*-nitrosodimethylamine, *Anal. Chem.* 51(7): 989–991 (1979).

25. N.-J. Sung, K. A. Klausner, and J. H. Hotchkiss, Incidence of nitrate, ascorbic acid, and nitrate reductase microorganisms on *N*-nitrosamine formation during Korean-style soysauce fermentation, *Food Additives and Contaminants* 8(3): 291–298 (1991).

26. R. C. Massey, P. E. Key, R. A. Jones, and G. L. Logan, Volatile, non-volatile and total *N*-nitroso compounds in bacon, *Food Additives and Contaminants* 8(5): 585–598 (1991).

27. N. P. Sen, S. W. Seaman, C. Bergeron, and R. Brousseau, Trends in the level of *N*-nitrosodimethylamine in Canadian and imported beers, *J. Agric. Food Chem.* 44: 1498–1501 (1996).

28. T. Kawabata, J. Uibu, H. Ohshima, M. Matsui, M. Hamano, and H. Tokiwa, Occurrence, formation and precursors on *N*-nitroso compounds in the Japanese diet, *N-Nitroso Compounds: Analysis, Formation and Occurrence* (E. A. Walker, L. Griciute, M. Castegnaro, and M. Börzsönyi, eds.), IARC, Lyon, 1980, pp. 481–490.

29. J. W. Rhoades, J. M. Hosenfeld, J. M. Taylor, and D. E. Johnson, Comparison of analyses of wastewaters for *N*-nitrosamines using various detectors, *N-Nitroso Compounds: Analysis, Formation and Occurrence* (E. A. Walker, L. Griciute, M. Castegnaro, and M. Börzsönyi, eds.), IARC, Lyon, 1980, pp. 377–385.

30. S. M. Billedeau, B. J. Miller, and H. C. Thompson, *N*-Nitrosamine analysis in beer using thermal desorption injection coupled with GC-TEA, *J. Food Sci.* 53(6): 1696–1698 (1988).

31. J. H. Hotchkiss, D. C. Havery, and T. Fazio, Rapid method for estimation of *N*-nitrosodimethylamine in malt beverages, *J. Assoc. Off. Anal. Chem* 64(4): 929–932 (1981).

32. J. Pensabene and W. Fiddler, Determination of ten nitrosamino acids in cured meat products, *J. Assoc. Off. Anal. Chem.* 73: 226–230 (1990a).

33. J. W. Pensabene, W. Fiddler, and J. G. Philips, Gas chromatographic-chemiluminescence method for the determination of volatile *N*-nitrosamines in minced fish meat and surimi meat frankfurters: collaborative study, *J. Assoc. Off. Anal. Chem.* 73(6): 947–952 (1990b).

34. J. Pensabene, W. Fiddler, and R. A. Gates, Solid phase extraction method for volatile

N-nitrosamines in hams processed with elastic rubber netting, *J. Assoc. Off. Anal. Chem. 75*: 438–442 (1992).

35. K. Takatsuki and T. Kikuchi, Determination of *N*-nitrosodimethylamine in fish products using gas chromatography with nitrogen-phosphorus detection, *J. Chromatogr. 508*: 357–362 (1990).

36. R. A. Scanlan and F. G. Reyes, An update on analytical techniques for *N*-nitrosamines, *Food Technology*, January, pp. 95–98 (1985).

37. H. Kataoka, S. Shindoh, and M. Makita, Selective determination of volatile *N*-nitrosamines by derivatization with diethyl chlorothiophosphate and gas chromatography with flame photometric detection, *J. Chromatogr. A. 723*: 93–99 (1996).

38. C. Fu, H. Xu, and Z. Wang, Sensitive assay system for nitrosamines utilizing high-performance liquid chromatography with peroxyoxalate chemiluminescence detection, *J. Chromatogr. 634*: 221–227 (1993).

39. S. M. Billedeau, T. M. Heinze, J. G. Wilkes, and H. C. Thompson, Jr., Application of the particle beam interface to high-performance liquid chromatography-thermal energy analysis and electron impact mass spectrometry for detection of non-volatile *N*-nitrosamines, *J. Chromatogr. A. 688*: 55–65 (1994).

40. Y. Y. Wigfield and C. C. McLenaghan, Liquid chromatographic and thermal energy analyzer detection of *N*-nitrosodiphenylamine in formulations of diphenylamine, *Bull. Environ. Contam. Toxicol. 45*: 853–857 (1990).

41. D. C. Havery, Determination of *N*-nitroso compounds by high-performance liquid chromatography with postcolumn reaction and a thermal energy analyzer, *J. Analytical Toxicol. 14*: 181–184 (1990).

42. Z. Wang, H. Xu, and C. Fu, Sensitive fluorescence detection of some nitrosamines by precolumn derivatization with dansyl chloride and high-performance liquid chromatography, *J. Chromatogr. 589*: 349–352 (1992).

43. G. Bellec, J. M. Cauvin, M. C. Salaun, K. Le Calvé, Y. Dréano, H. Gouérou, J. F. Ménez, and F. Berthou, Analysis of *N*-nitrosamines by high-performance liquid chromatography with post-column photohydrolysis and colorimetric detection, *J. Chromatogr. A. 727*: 83–92 (1996).

44. G. Favaro, G. A. Sacchetto, P. Pastore, and M. Fiorani, Liquid chromatographic determination of non-volatile nitrosamines by post-column redox reactions and voltammetric detection at solid electrodes. Study of a flow reactor system based on Ce(IV) reagent, *Anal. Chim. Acta 273*: 457–467 (1993).

45. G. A. Sacchetto, G. Favaro, P. Pastore, and M. Fiorani, Optimization of the amperometric detection of nitrite by reaction with iodide in a post-column reactor for liquid chromatography of non-volatile nitrosamines, *Anal. Chim. Acta 294*: 251–260 (1994).

4

Analysis of Foods and Related Samples for Heterocyclic Amine Mutagens/Carcinogens

MARK G. KNIZE AND JAMES S. FELTON
Lawrence Livermore National Laboratory, Livermore, California

I. INTRODUCTION

Diet has been associated with varying cancer rates in human populations for many years, yet the causes of the observed variation in cancer patterns have not been adequately explained [1]. Since 1980, a series of potent heterocyclic amines that are mutagenic and carcinogenic have been identified. These substances are formed in some heated foods, most notably muscle meats. The observed higher cancer rates at organ sites such as breast and colon in those populations consuming a "Western diet" high in meat have been attributed to high fat or low fiber. Heterocyclic amines are also associated with this diet and the meat it contains, and unlike fat or fiber, heterocyclic amines have clear genotoxic reactivity leading to the initiating events in the cancer process.

The heterocyclic amines are multisite carcinogens in mice and rats [2,3]. In addition, several of these compounds are among the most potent mutagenic substances ever tested in the Ames/*Salmonella* mutagenicity test [4].

A number of studies have shown the precursors for the formation of these mutagenic and carcinogenic compounds to be amino acids, such as phenylalanine, threonine, and alanine; creatine or creatinine; and sugars [5]. Cooking temperature and time are also important determinants in both the qualitative and the quantitative formation of these compounds in foods [6,7]. These variables influencing the formation process (presumably condensation reactions) create a large range of possible heterocyclic amine concentrations in foods, requiring the analysis of a large number of food samples to determine the sources of heterocyclic amines in the human diet. The health risk to the human population consuming these heterocyclic amines has been recently discussed [8–11] and has significant implications.

Now that many of the heterocyclic amine mutagens have been identified and synthesized, and shown to be animal carcinogens, the next research goals are to estimate human exposure by quantifying the amounts present in foods and to some extent in the smoke from cooking. To determine the importance of the heterocyclic amines in human health requires accurate dietary intake data on the amounts and types of heterocyclic amines to which humans are exposed. This exposure information must then be combined with the rodent carcinogenic potency assessment for calculation of estimates for human cancer risk.

Thus far, heterocyclic amines in foods are not regulated by government agencies, although their carcinogenic potencies, or the amounts detected, in many cases exceed those of many regulated compounds such as chlorinated compounds, pesticides, and mycotoxins [11]. Besides the usefulness for determining the human dose of these heterocyclic amines in foods, the quantitative chemical analysis can also be used to help devise cooking methods and food preparation strategies to reduce the formation of these compounds in foods.

The heterocyclic amines found in foods have stable multiple-ring aromatic structures and all have an exocyclic amino group. Structures of these compounds commonly detected in foods are shown in Figure 1, although additional mutagenic heterocyclic amines have been sporadically isolated from foods. All of the heterocyclic amines have characteristic UV spectra and high extinction coefficients, some of the compounds fluoresce, and all can be electrochemically oxidized, making UV absorbance, fluorescence, or electrochemical detection suitable methods. The aromatic structures of these heterocyclic amines give little fragmentation and therefore show large base peaks, making mass spectrometry a good detection method [12].

There are several factors that make the analysis of heterocyclic amines from foods a difficult problem. Heterocyclic amines are present in foods at very low part-per-billion levels (ppb). The low levels require that chromatographic efficiency, and both detector sensitivity and selectivity be optimized. Several of the heterocyclic amines are formed under the same reaction conditions, so the number of compounds to be quantified requires that the extraction, chromatographic separation, and detection be general enough to detect several of the heterocyclic

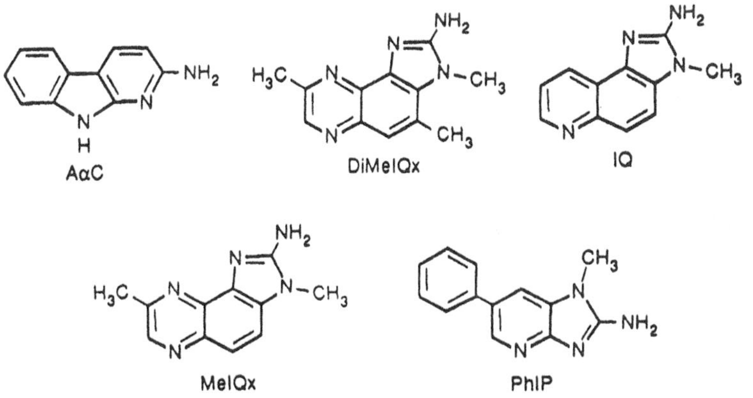

FIGURE 1 Structures and common names of five heterocyclic amines commonly found in cooked meats: AαC (2-amino-9H-pyrido[2,3-b]indole); DiMeIQx (2-amino-3,4,8-trimethylimidazo[4,5-f]quinoxaline); IQ (2-amino-3-methylimidazo[4,5-f]quinoline); MeIQx (2-amino-3,8-dimethylimidazo[4,5-f]quinoxaline); PhIP (2-amino-1-methyl-6-phenylimidazo[4,5-b]pyridine).

amines per chromatographic separation. The complexity and diversity of food samples needing to be analyzed require a rugged method not affected by the sample matrix.

II. SAMPLE EXTRACTION

A. Liquid/Liquid Extraction

The detection of bacterial mutagens in cooked meats led to efforts to isolate the mutagenic chemicals using a variety of extraction methods. Samples were extracted with acid or organic solvents [13,14]. The purity of the extracts was generally not sufficient to measure specific heterocyclic amines, and typically mutagenic potency was measured, an assay that performed well with crude mixtures.

Murray et al. [15] devised an analysis method specific enough to be used on samples after a simple extraction procedure. Food samples were dissolved in dilute HCl, washed with dichloromethane to remove oils and fats, and then extracted into ethyl acetate after pH adjustment to an alkaline condition. These were then derivatized and analyzed by GC/MS as described below.

In another laboratory, a solid-phase extraction method that uses liquid/liquid extraction as the first step was developed. The extraction was done on a solid support of diatomaceous earth (Keiselgur or Extrelut®), a sand-like porous material [16,17]. The inert diatomaceous earth carrier allowed efficient and rapid organic solvent extraction without any risk of emulsion formation and the require-ment of subsequent centrifugation, a commonly encountered problem when ex-tracting food. The solid support also minimized the amount of methylene chloride needed for the extraction because of the high surface area at organic/aqueous interface available in this procedure.

B. Extraction with Blue Cotton

A solid support containing a blue copper phthalocyanine trisulfonate linked to cotton (or rayon) was developed for its ability to adsorb aromatic compounds having three or more fused rings. Mutagens can be adsorbed to the blue cotton from saline solutions and eluted with a methanol/ammonia solution [18]. This material also adsorbs other contaminants from foods, such as aflatoxin B1 and polycyclic aromatic hydrocarbons [19]. The material is commercially available but its high cost may be one reason the material is used in relatively few published papers.

C. Solid-Phase Extraction with Disposable Columns

Solid-phase extraction (SPE) refers to procedures using disposable cartridges typically containing 100 mg to 500 mg of a solid, often silica-based, sorbent. Gross [17] developed a method to purify heterocyclic amines from foods and related products with the goal of high sample throughput and high analytical sensitivity. The key to this method is the coupling of liquid/liquid extraction to a cation exchange resin column (propylsulfonic acid silica), concentrating the sam-ple diluted from the large volume of organic solvent from the first extraction step. Using the cation exchange properties of the column allows selective washing and elution to purify the sample further.

Two simple elution steps subsequently separate heterocyclic amines into two groups, the carboline derivatives and the IQ-type compounds. Concentration of these eluates is conveniently achieved on a C_{18} silica cartridge. The IQ-type heterocyclic amines can be eluted and conveniently concentrated by passing over a C_{18} silica cartridge, rather than by rotary evaporation, and finally eluted with a methanol–ammonium hydroxide solution [16]. The multistep procedure does not include any evaporations requiring rotary evaporator concentration and is suited for routine quality control. Using disposable extraction materials in this method is desirable in the extraction of trace quantities of analytes to prevent cross-contamination from the glassware. The adsorption behavior of the heterocyclic amines varies, and careful optimization for maximum recovery and cleanup is the subject of ongoing research. Details of the extraction steps and recoveries of specific heterocyclic amines at various steps were reported [20–23].

Some food samples, flame-grilled meat and fish, and some industrially produced flavorings (so called "process flavors") show increased levels of chromatographic interferences and therefore require more sample cleanup. An additional extraction step was optimized using a weak cation exchange material with different selectivity than that of materials used in the previous purification steps. Interfering peaks in extracts prepared by diatomaceous earth-PRS extraction were significantly lowered using this treatment [23]. Another approach to the problem of sample peaks interfering with the chromatographic analysis of process flavors used a copper phthalocyanine affinity chromatography step [24] although experimental details were not given.

A single solid-phase extraction scheme was used to isolate three classes of genotoxic compounds from charcoal-grilled meat: heterocyclic amines, polycyclic aromatic hydrocarbons, and nitrogen-containing polycyclic aromatic hydrocarbons [25]. The method gave detection limits of 0.3 to 8.4 ng/g in a charcoal-grilled meat sample for the 12 compounds used and appears to be an important new idea in the analysis of mutagenic/carcinogenic compounds in food.

D. Antibodies Immunoaffinity Chromatography

Immunoaffinity chromatography was used for the purification of IQ and MeIQx from heated beef products [26]. Following homogenization, samples were adsorbed on XAD-2 resin, eluted, and applied to affinity columns that had antibodies immobilized on them to IQ or MeIQx. Resulting eluents were separated using HPLC and photodiode-array detection. One drawback is that specific antibodies are needed for each heterocyclic amine, but to determine only one marker compound, immunoaffinity column chromatography offers impressive specificity.

III. CHROMATOGRAPHY AND DETECTION

A. HPLC/Bacterial Mutagenic Activity Testing

Because the heterocyclic amines are found as natural products, frequently their biological activity is detected before their chemical structure has been determined. Thus, for unknown mutagens, the number of mutagenic compounds separable by HPLC can be determined with the relative contribution of each peak's mutagenic activity. In this analysis, a crude extract is separated by HPLC, eluent fractions collected by time, usually 1 or 2 minutes for each, and the Ames/*Salmonella* test can be used to test each collected fraction for mutagenic potency [27]. Specific strains of bacteria can be used to differentiate chemical classes to some degree.

Many foods and related samples have been analyzed by HPLC and mutagenic activity tested in collected fractions. In many cases this method is used as a preparative step to purify samples for later analysis by analytical HPLC and UV detection, mass spectrometry, or NMR spectrometry. The potency of heterocyclic amines enables a mutagenic response to be determined on subnanogram amounts for most compounds, thus preserving most of the sample, and this method can be used for peak confirmation if other methods are not available.

Some extracts of foods containing unknown mutagens were characterized by HPLC. This was done for meats [28,29], coffee substitutes [30], and grain foods [31]. While this method is very useful for characterizing and comparing the unknown mutagens, the more difficult task of mutagen identification, leading to chemical synthesis, is essential for biological experiments. Combined with dietary dose information, potency measurements enable the risk of the consumption of mutagenic aromatic amines to be estimated.

B.   HPLC-UV, Fluorescence or Electrochemical Detection

Baseline separation of all heterocyclic amines is a prerequisite for the multicompound analysis method. For the TSK gel ODS80 TM (TosoHaas, Montgomeryville, PA) column, chromatographic conditions have been optimized for the separation of 12 heterocyclic amines [22]. Further work examining the effect of pH and acetonitrile concentration on capacity factors was reported for this column [32]. A 2.1 mm internal diameter column was used successfully for heterocyclic amines and showed a 3-fold increase in sensitivity and reduced analysis times compared to the standard 4.6 mm internal diameter column; it also generated less HPLC waste [33].

Analyzing heterocyclic aromatic amines at nanogram levels with HPLC requires chromatograms free from interfering peaks. Quantitative determinations can be done by HPLC with a variety of detectors, but coextracted matrix components influence analyte detection limits more than does the absolute detector sensitivity. Thus effective sample preparation is the most critical part of heterocyclic amine analysis from foods using HPLC.

The detection limits of heterocyclic amines purified through solid-phase extraction depend on the food sample investigated. With meat fried at temperatures below about 230°C, detection limits may be as low as 0.1 nanogram per gram using UV detection [6,34]. Samples processed at higher temperatures, e.g., by grilling or flame-broiling, usually show higher levels of interfering peaks. The detection limits then worsen to about 2–5 nanograms per gram of food [20].

Figure 2 shows HPLC chromatograms for chicken breast meat grilled for 10, 20, 30, or 40 minutes at equal attenuation and amount of meat extracted. The figure shows the problems of interfering peaks in some samples grilled at longer times that greatly increase the number and size of interfering peaks. Caffeine (our internal standard) and any heterocyclic amines are much more difficult to integrate and confirm by UV spectra at the longer grilling times because of this problem.

Peak confirmation is a crucial problem when working with such low levels of heterocyclic amines, since coelution with other coextracted compounds can

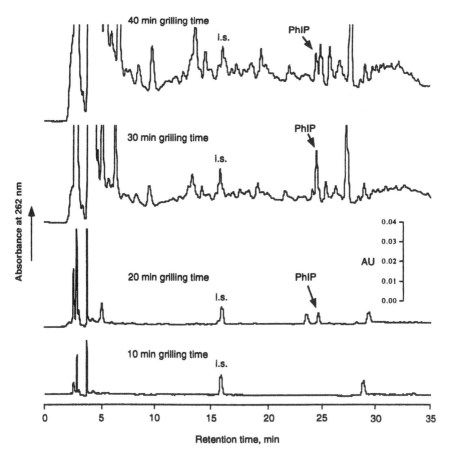

FIGURE 2 HPLC chromatograms (262 nm) of chicken breast meat grilled for 10, 20, 30, or 40 min, run on a TSK gel ODS80-TM column as optimized by Gross and Grüter [22]. This figure shows increased detection limits caused by interfering peaks at longer cooking times due to coextracted compounds. "i.s." is the internal standard.

occur. The most convenient and accessible instrument to identify heterocyclic amines on-line during an HPLC separation is the UV photodiode array detector. Most instruments allow the recording of UV spectra even at a 0.1 nanogram level. A photodiode array detection system efficiently prevents most false peak identifications and is essential to prevent the interpretation of false positive results. In our experience, even with modern photodiode array detection with spectral library matching by computer, human interpretation is needed for confirmation in many cases. Other peak confirmation methods like off-line mass spectrometry or muta-genicity testing have been used successfully in our laboratory.

Fluorescence detection is typically used in-line as a complement to photo-diode array detection. Not all heterocyclic amines fluoresce, but PhIP is com-monly detected in foods and shows about a ninefold greater peak signal than our photodiode array detection method, and the fluorescence signal typically shows fewer interfering peaks. In our laboratory we still require confirmation of fluores-cence peaks, so the sensitivity of the photodiode array detector limits the overall sensitivity of the detection scheme.

Reproducibility of the method over time was determined in our laboratory as part of a blind study. Two hamburger samples, one cooked rare and one cooked well-done, were repeatedly received in separate aliquots for quality control deter-mination. For the two samples analyzed repeatedly during the 24 months, MeIQx, PhIP, and DiMeIQx were consistently detected in the sample cooked well-done, but not in the sample cooked rare. Relative standard deviations (coefficients of variation) for the amounts determined were 34 for MeIQx, 22 for PhIP, and 38 for DiMeIQx for the well-done cooked samples. No sample degradation was seen in the samples stored frozen at $-4°C$. Average recoveries of PhIP, MeIQx, Di-MeIQx, IQ, and MeIQ spiked into a range of food samples were 34, 62, 74, 53, and 65%, respectively for 230 samples analyzed in our laboratory.

Electrochemical detection of heterocyclic amines was reported for beef extracts following extraction [24,32,35,36]. Electrochemical detection offers in-creased detector specificity over UV absorbance detection, although the sensi-tivities of the two detector types are about the same. Hydrodynamic voltograms for 7 heterocyclic amines showed that an electrode potential of $+1000$ mV gave the best detector response [32]. Importantly, a means of on-line peak confirmation is not possible with currently available detectors, greatly limiting their usefulness compared to the in-line UV photodiode array detectors currently available.

C. LC/MS

Mass spectrometry has many desirable features as a detector of heterocyclic amines. Mass spectrometry offers the selectivity of mass detection with the possi-bility of adding heavy-isotope-labeled internal standards to determine extraction recovery and act as chromatographic standards simultaneously. Pioneering work used laborious sample extraction methods and thermospray liquid interfaces, yet measurements of heterocyclic amines as low as 0.3 ng per gram of food were reported [37,38].

Recent work in this area benefited from improved extraction methods and electrospray-liquid-chromatography interfaces. Gross et al. showed LC/MS single ion plots for 11 heterocyclic amines and showed extracts of bacon to be free of interfering peaks at the masses monitored for the heterocyclic amines [33].

Galceran et al. used electrospray LC/MS with a 1 mm diameter column to analyze beef extracts for 5 heterocyclic amines [39].

Disadvantages of mass spectrometry include the expensive instrumentation required and the need for heavy-isotope-labeled internal standards. Since the heterocyclic amines do not fragment well, there are no abundant secondary ions for peak confirmation. Peak confirmation may be important if heavy-isotope internal standards are not available for an accurate retention-time check and if ion plots are not free from nearby peaks. A triple stage quadrupole mass spectrometer, however, offers peak confirmation by selection of the target ion in the first quadrupole and then monitoring for confirmatory fragments (MS/MS technique). The expense and complexity of operating such an instrument will limit its use as a routine method.

D. GC/MS

The most sensitive approach for heterocyclic amine analysis is that devised by Murray et al. [15]. Most of the heterocyclic amines exhibit poor chromatographic behavior for gas chromatography and thus require derivatization prior to injection. For this method extracts of food from a liquid/liquid extraction scheme were derivatized with 3,5-bistrifluoromethylbenzyl bromide at room temperature, washed with hexane, and extracted with ethyl acetate. The total yield for these extraction steps was reported to be about 40% as determined by radioactivity measurements with [^{14}C]MeIQx [40]. PhIP can give multiple products upon derivatization with this method, and another derivatizing agent, pentafluorobenzyl bromide, was reported to give apparently a single product [41].

The high chromatographic efficiency of capillary gas chromatography and the detection of negative ions using chemical ionization mass spectrometry are the advantages of this chromatography system. The specificity of single-ion monitoring does not require the degree of sample purification that is needed for the liquid chromatography detectors, so a much simpler extraction can be used.

Recovery for extraction and derivatization can be calculated from the use of heavy-isotope-labeled internal standards. The chemical ionization and negative-ion detection give a reported 1 pg detection limit using selected-ion monitoring [15]. The GC/MS method has not been optimized for as many different heterocyclic amines as the SPE/HPLC method, and IQ reportedly cannot be detected by this method [42]. Analysis of foods with this method give limits of detection at least 20-fold lower than HPLC methods [42,43].

Like the LC/MS technique, this method also requires expensive mass spectrometry instrumentation and the need for heavy-isotope-labeled internal standards, although some are available commercially. Column lifetime with gas chromatography is not as great as with HPLC, especially for the analysis of complex food extracts.

IV. RESULTS

A. Comparison of SPE-HPLC and Liquid/Liquid Extraction-Derivatization GC/MS

We examined two published methods for their suitability for heterocyclic amine analysis in foods, the SPE-HPLC method and the derivatization and GC/MS method. These differ in nearly every respect from sample preparation to the instrumentation required. We repeatedly analyzed two cooked meat samples for four heterocyclic amines commonly found in cooked meats, MeIQx, PhIP, Di-MeIQx, and AαC.

Ground beef patties (100 g, approximately 1.5×9 cm, 12% fat) were either fried on a steel griddle (temperature of the griddle varied from 173°C to 200°C during cooking) for 10 min per side or cooked over charcoal at a meat surface temperature of 350°C to 400°C for 6 min per side.

For the GC/MS method, samples were prepared by the method of Murray et al. [15,43] with the modifications noted. Cooked beef (10 g) was homogenized with 0.25 M hydrochloric acid (50 g) with a tissue homogenizer. This was the minimum amount of sample needed for complete homogenization with our equipment. Aliquots (12 g) were weighed into tubes spiked with a methanolic solution containing 25 ng [^{13}C,^{15}N$_2$]MeIQx (a kind gift of Dr. Nigel Gooderham), 25 ng [^{2}H$_3$] DiMeIQx (a kind gift of Dr. Robert Turesky), and 50 ng [^{2}H$_5$]PhIP). Samples were extracted according to the published method except that the specified 2×5 ml of ethyl acetate extraction solvent was increased to 2×10 ml due to poor separation of the phases. The combined organic extract was evaporated to dryness.

To each vial was added a 5% solution of 3,5-bistrifluoromethylbenzyl bromide in acetonitrile, and 20 μl of diisopropylethylamine. The reaction mixture was left for two days at room temperature, which increased derivative formation, and evaporated.

A Varian gas chromatograph with a DB-5 column coupled to an Extrel ELQ 400 mass spectrometer (Extrel Corp, Pittsburgh, PA) was used for analysis. The mass spectrometer was operated in the negative-ion-chemical-ionization-single-ion monitoring (SIM) mode. The mass spectrometer was tuned and programmed to monitor the derivatized compounds as negative ions at m/z 408 (AαC), m/z 438 (MeIQx), m/z 441 ([^{13}C,^{15}N$_2$]MeIQx), m/z 449 (PhIP), m/z 452 (DiMeIQx), m/z 454 ([^{2}H$_5$ PhIP]), and m/z 455 ([^{2}H$_3$ DiMeIQx]). Each ion used for quantification represents the fragment ion from the loss of one derivatizing group from the diderivatized parent molecule.

Samples were also extracted and analyzed by HPLC according to the procedures of Gross and Grüter [22]. Additional cleanup of the polar fraction for the charcoal-cooked sample was done as described by Gross et al. [23]. Amounts determined by both methods are corrected for sample losses using peak areas from spiked samples.

The samples were chosen to represent either high or low heterocyclic amine content, representing the extremes in sample complexity that we have encountered in previous work. AαC appears to be generated only in samples cooked at the higher temperature associated with grilling over charcoal or a gas flame, but when it is found, it can be present in relatively high amounts [22,33].

Figure 3 shows the results from a GC/MS analysis of a charcoal-cooked beef sample. The GC tracings show the major ions of the derivatized heterocyclic

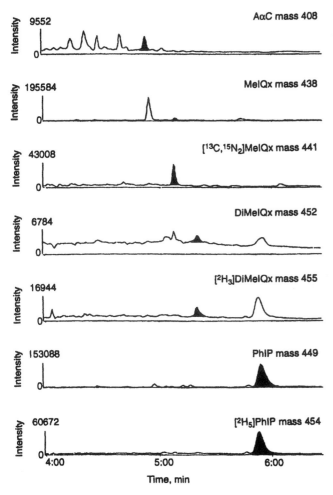

FIGURE 3 GC chromatograms of the extract of charcoal-cooked beef. Chromatographic conditions are given in the text.

amine and the heavy-isotope-labeled standards (when available) for the charcoal-cooked beef sample. The peaks of interest are shown as filled. Note that the chromatograms are not free of peaks at retention times other than those of the heterocyclic amines, even with the selectivity of the negative-ion-chemical-ionization and monitoring the relatively high masses of the fragments of the derivatized heterocyclic amine.

Figure 4 shows the HPLC chromatograms for the polar and apolar fraction from the charcoal-cooked beef sample. For PhIP and AαC the fluorescence chromatogram shows fewer interfering peaks and a larger signal-to-noise ratio. However, sample matching to reference spectra (shown at right) from the photodiode-array detector is essential for peak confirmation when using this method. All four compounds show a match to the spectrum of synthetic hetero-cyclic amines.

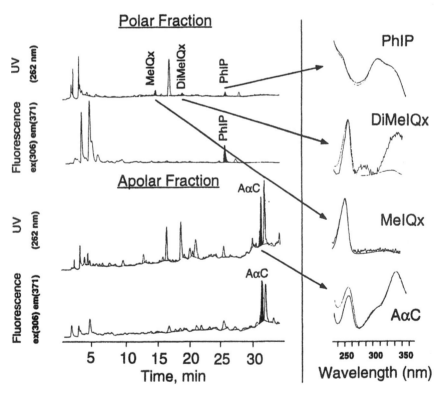

FIGURE 4 HPLC chromatograms of charcoal-cooked beef, showing polar and apolar extracts, fluorescence, and UV detection. Right: UV absorbance spectra comparing sample spectra (jagged lines) to reference spectra (smooth lines).

Table 1 Comparison of GC/MS and HPLC/UV Analysis Methods for Fried Hamburgers

Compound	Method	Trials, n	Amount mean, ng/g	Std. dev.
MeIQx	HPLC/UV	10	3.6	0.3
	GC/MS	19	4.6	1.3
DiMeIQx	HPLC/UV	10	0.8	0.08
	GC/MS	19	1.4	1.3
PhIP	HPLC/UV	10	8.6	1.2
	GC/MS	19	6.0	1.4
AαC	HPLC/UV	10	nd	—

nd, not detected (less than 0.5 ng/g).

Tables 1 and 2 show our results with repeated analysis of the two different meat samples. AαC is not detected in the fried meat sample (Table 1), and the AαC is not recovered when spiked into either sample when using the GC/MS method. Similar amounts of MeIQx, DiMeIQx, and PhIP are seen with each method. A t-test showed that the means for each analysis method were significantly different for quantification of both MeIQx and PhIP. When compared to the HPLC/UV method, the GC/MS method produced concentration levels that were 20% lower for PhIP and 25% higher for MeIQx. Although this difference may indicate method bias, it is more probably a reflection of the high extraction variance, which appears to be the highest contributor to method variance. These variances are within the expected tolerances for low part-per-billion analysis [44].

Experiments with the charcoal-cooked sample showed AαC was not detected by the GC/MS method. Further investigations showed that liquid/liquid extraction did not recover the AαC. Therefore, the apolar extracts from the solid-phase-extraction method normally used for the HPLC analysis were derivatized

Table 2 Comparison of GC/MS and HPLC/UV Analysis Methods for Charcoal-Cooked Hamburgers

Compound	Method	Trials, n	Amount mean, ng/g	Std. dev.
MeIQx	HPLC/UV	10	2.8	1.1
	GC/MS	18	5.0	2.6
DiMeIQx	HPLC/UV	10	0.3	0.3
	GC/MS	18	1.2	1.1
PhIP	HPLC/UV	10	30	4.3
	GC/MS	18	25	4.5
AαC	HPLC/UV	10	18	5.4

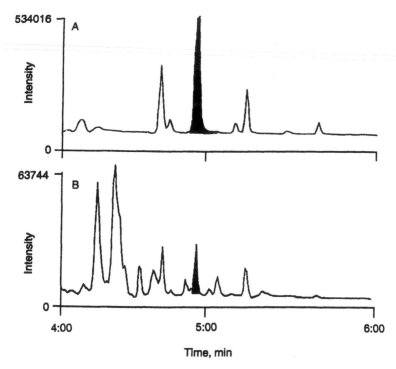

FIGURE 5   GC chromatograms (m/z 408) of derivatized AαC standard (upper) and charcoal-cooked meat (lower) extracted by the solid-phase method of Gross [16].

and analyzed by GC/MS as shown in the bottom of Figure 5. AαC was detected, showing that this compound can be derivatized and quantified at the expected mass of 408 with a retention time of 4.8 minutes.

Comparison of the precision between the HPLC/UV and GC/MS methods shows that each is suitable for general food sample analysis. We prefer the HPLC/UV method for routine analysis because of the fully automated analysis possible using an autosampler and confirmation of peaks using absorbance spectra. The GC/MS method is useful for peak confirmation because of the sensitivity and specificity of the mass spectrometry detection. Both methods show a high variance probably due to the nature of the samples and the low analyte levels. A combination of the solid-phase extraction procedure with the derivatization and GC/MS analysis enables AαC to be analyzed along with the other commonly detected heterocyclic amines. A combined procedure may improve sample recovery and detection and should be explored for those samples needing an improved method for heterocyclic amine determination at low levels or in complex samples.

B. Amounts of PhIP and MeIQx in Foods and Related Samples

Table 3 shows amounts of PhIP and MeIQx in some foods and related products taken from the published literature. Only MeIQx and PhIP are listed, although most studies found additional compounds, but almost always in lower amounts

Table 3 Literature Values of PhIP and MeIQx in Foods, Food Flavors, and Cooking Fumes from Studies Examining Multiple Samples per Category

Sample category	PhIP, ng/g	MeIQx, ng/g	Extraction-chromatography-detection method	Ref.
Bacon	nd–53	nd–18	SPE-LC/MS	33
	1.6–2.7	0.9–1.2	L/L-GC/MS	42
Beef	nd–32	nd–7.3	SPE-HPLC-UV	6
	nd–16.8	nd–4.6	SPE-HPLC-UV	57
Chicken	nd–480	nd–9.0	SPE-HPLC-UV	58
Fish	nd–73	nd–5.0	SPE-HPLC-UV	23
Various meats	nd–16	0.3–2.2	L/L-GC/MS	41
	0.56–69.2	0.64–6.44	BC-HPLC-UV	58
Fast-food meats	nd–0.6	nd–0.3	SPE-HPLC-UV	34
Swedish meats	nd–12.7	nd–23.7	SPE-HPLC-UV	47
	nd–4.5	nd–7.3	SPE-HPLC-UV	60
Restaurant cooked meats	nd–13	0.42–0.89	SPE-HPLC-UV	61
Meats, commercially processed	nd–5.5	nd–0.42	L/L-GC/MS	62
Beer and wine	80–65.7 ng/l	not done	combination	63
Food flavors	nd	nd–21.2	SPE-HPLC-UV	64
	nd	nd	SPE-HPLC-UV	23
	nd–0.3	0.1–0.6	L/L-GC/MS	42
	nd	nd–44	SPE-HPLC-UV	17
Cooking fumes	not done	0.014–0.26	L/L-GC/MS	42
	0.007–1.8 ng/g of meat cooked	nd–1.1 ng/g of meat cooked	SPE-HPLC-UV	65

nd, not detected.
L/L, liquid/liquid extraction.
BC, blue cotton extraction.

than these two. Interested readers should check the original source for cooking information (heterocyclic amine formation is greatly dependent on cooking conditions) and specifics of the extraction and chromatographic analysis. For most of the reports the amount of heterocyclic amines are in the low nanogram per gram range, with few having tens or hundreds of nanograms per gram. A large database containing heterocyclic amine amounts in food along with cooking information is compiled from the open literature and our laboratory, and is available from our internet home page: http://www-bio.llnl.gov/bbrp/mole_tox/toxicology.html.

V. FUTURE NEEDS AND DIRECTIONS

The analysis of foods for heterocyclic amines is important because there is widespread human exposure to these compounds, there is good suggestive epidemiology for cause and effect, and these chemicals are potent mutagens and animal carcinogens. Exposures vary among individuals, since dietary preferences and food preparation variations can greatly influence individual exposures. This area of research provides a unique opportunity in cancer etiology, the chance to evaluate a class of carcinogens in human populations. The variability in heterocyclic amine formation also provides the opportunity for intervention, to reduce exposure if it is warranted from risk evaluation.

Table 3 shows results for many types of foods, yet there are large ranges of values given for each category, so it is difficult to predict the content of foods without doing the analysis. Analysis is needed to develop schemes for reducing heterocyclic amine formation, and some promising work has been done in adding components to meat [45], microwave precursor removal [46], and simply reducing the temperature and time of cooking [6,47]. We recently found that the application of a marinade to chicken breast meat before grilling can greatly decrease PhIP, although MeIQx is increased at the longest cooking time [48]. At least one cooking grill designed to reduce the formation of heterocyclic amines has been patented [49]. And studies on mutagen formation in different meat types [50] and in model systems can lead to ideas to reduce their formation in foods [51,52]. Thus many experiments remain to be done and many samples need to be analyzed.

Related studies can also use the techniques for heterocyclic amine analysis developed for foods. The heterocyclic amines and their metabolites are being studied in model systems, in animals, and in humans to understand the biological consequences of exposure to these chemicals in the human population.

Current methods could still be improved. Sample extraction for some samples is a problem and is undergoing further research. Supercritical fluid extraction was investigated to extract heterocyclic amines from solid supports. This work showed recoveries greater than 80% for all heterocyclic amines tested except PhIP, for which less than 23% was recovered by the supercritical CO_2 with 10%

methanol [53]. More work could be done in this area. For all of the analysis methods discussed, improved detector specificity would improve quantitation. Sensitive mass-selective detectors or less expensive, and therefore more widely available, mass spectrometers would improve the detection after the chromatography step for many users.

Which compounds are important is another unresolved problem. All of the heterocyclic amines tested are rodent carcinogens, and potencies are within about one order of magnitude, so it could be argued that the total mass of heterocyclic amines is the most important factor. They differ in tumor-site specificity in rats, with most heterocyclic amines causing liver tumors (a rare tumor site in humans), but PhIP causes many colon and mammary tumors (sites relevant in human dietary exposure studies). Therefore, it could be argued that PhIP is the most important to measure and that others could be possibly ignored or given a lower priority.

If it is necessary to determine only a surrogate marker compound, extractions could be optimized for just a single analyte. That may not be possible though, because for MeIQx and PhIP, the relative amounts change greatly from sample to sample. Many meats show greater MeIQx than PhIP at lower temperatures, but much more PhIP at higher temperatures and longer cooking times [6,47]. We do not know of any meat samples that contain other heterocyclic amines that do not also include MeIQx or PhIP.

An additional group of less well studied compounds has been found in various meats and meat extracts by purification guided by the Ames/*Salmonella* mutagenicity assay [27]. Among these are amino-dimethyl-, amino-furo-, and amino-trimethyl-imidazo pyridines [54,55]. They have been reported in only a few instances in foods or food extracts, because analytical standards have not been available for method development. We believe some of these may be important because of their mass amounts present in foods and their structural similarity to the rat breast and colon carcinogen PhIP, and therefore their analysis in foods needs to be undertaken.

Related to the heterocyclic amines in cooked meats is the finding of mutagenic activity in non-meat foods. The biological response and the gross chemical separations suggest that the activity is caused by heterocyclic amines. These have been reported in a few baked grain foods [31], roasted grain-based beverages [30], and cooking oils [56]. Consumption of these foods also results in human exposure to bacterial mutagens that should be investigated for their chemical relationship to the heterocyclic amines found in meat.

The variety and complexity of foods and the small amounts of the heterocyclic amines present make it unlikely that the analysis will ever be as simple or as inexpensive as one would like. Still, great progress has been made in heterocyclic amine analysis, and a variety of extraction and chromatographic methods have been successfully used. The experience of the chromatographer and the instrumentation on hand will play a large role in the methods chosen.

VI. SUMMARY

A series of heterocyclic amines that are potent mutagens in laboratory test systems and carcinogenic in animals have been discovered in heated foods. These are formed most notably in meats derived from muscle. Determining the heterocyclic amine content in foods and food products is required for toxicological research, industry quality control, and possible future regulatory control. The low levels present require that chromatographic efficiency and both detector sensitivity and selectivity be optimized. Analysis methods require extensive sample preparation using a liquid/liquid or a solid-phase extraction scheme, and several methods have been developed. Chromatographic analysis by HPLC or GC has been applied to heterocyclic amines in foods. Peak detection has been successful using UV, fluorescence, electrochemical, and mass spectrometric detection and bacterial mutagenicity.

Results show that extraction recoveries range from 20 to 80% and that detection below 1 nanogram per gram of sample is possible by several methods. Amounts of heterocyclic amines in foods range from undetectable levels to hundreds of nanograms per gram in well-done grilled chicken samples. The biggest improvements in speed and accuracy will probably come from improved extraction methods as part-per-billion analysis of complex food samples will always be a challenge.

ACKNOWLEDGMENTS

The authors thank Donald Eades and Kathleen Dewhirst for the food extraction method comparison results and Cynthia Salmon for technical support. This work was performed under the auspices of the U.S. Department of Energy by Lawrence Livermore National Laboratory under contract no. W-7405-Eng-48 and supported by the NCI grant CA55861.

REFERENCES

1. R. Doll and R. Peto, *The Causes of Cancer, Oxford Medical Publications*, Oxford University Press, Oxford, 1981, p. 1226.
2. H. Ohgaki, S. Takayama, and T. Sugimura, *Mutat. Res. 259*: 399 (1991).
3. T. Sugimura, S. Sato, and K. Wakabayashi, in *Chemical Induction of Cancer* (Y.-T. Woo, D. Y. Lai, J. C. Arcos, and M. F. Arcos, eds.), Academic Press, New York, 1988, p. 681.
4. T. Sugimura, S. Takayama, H. Ohgaki, K. Wakabayashi, and M. Nagao, in *The Maillard Reaction in Food Processing, Human Nutrition and Physiology* (P. A. Finot, H. U. Aeschbacher, R. F. Hurrell, and R. Liardon, eds.), Birkhäuser, Basel, 1990, p. 232.
5. M. Jägerstad, K. Skog, S. Grivas, and K. Olsson, *Mutat. Res. 259*: 219 (1991).
6. M. G. Knize, F. A. Dolbeare, K. L. Carroll, D. H. Moore II, and J. S. Felton, *Fd. Chem. Toxic. 32*: 595 (1994).

7. K. Skog, *Fd. Chem. Toxic. 31*: 655 (1993).

8. D. W. Layton, K. T. Bogen, M. G. Knize, F. T. Hatch, V. M. Johnson, and J. S. Felton, *Carcinogenesis 16*: 39 (1995).

9. B. Stavric, *Fd. Chem. Toxic. 32*: 977 (1994).

10. S. Robbana-Baranat, M. Rabache, E. Railland, and J. Fradlin, *Environ. Health Perspectives 104*: 280 (1996).

11. IARC, *IARC Monographs on the Evaluation of Carcinogenic Risk of Chemicals to Man 56*, Lyon, 1993, p. 165.

12. E. Övervik, M. Kleman, I. Berg, and J.-Å. Gustafsson, *Carcinogenesis 10*: 2293 (1989).

13. J. S. Felton, S. K. Healy, D. H. Stuermer, C. Berry, H. Timourian, F. T. Hatch, M. Morris, and L. F. Bjeldanes, *Mutat. Res. 88*: 33 (1981).

14. C. A. Krone and W. T. Iowoka, *Cancer Lett. 14*: 93 (1981).

15. S. Murray, N. J. Gooderham, A. R. Boobis, and D. S. Davies, *Carcinogenesis 9*: 321 (1988).

16. G. A. Gross, G. Philippossian, and H. U. Aeschbacher, *Carcinogenesis 10*: 1175 (1989).

17. G. A. Gross, *Carcinogenesis 11*: 1597 (1990).

18. H. Hayatsu, Y. Matsui, Y. Ohara, T. Oka, and T. Hayatsu, *Gann. 74*: 472 (1983).

19. H. Hayatsu, S. Arimoto, and K. Wakabayashi, in *Mutagens in Food, Detection and Prevention* (H. Hayatsu, ed.), CRC Press, Boca Raton, 1991, p. 101.

20. M. G. Knize, J. S. Felton, and G. A. Gross, *J. Chromat. 624*: 253 (1992).

21. M. T. Galceran, P. Pais, and L. Puignou, *J. Chromatog. A 719*: 203 (1996).

22. G. A. Gross and A. Grüter, *J. Chromatography 592*: 271 (1992).

23. R. Schwarzenbach and D. Gubler, *J. Chromatography 624*: 491 (1992) 23.

24. G. A. Gross, A. Grüter, and S. Heyland, *Fd. Chem. Toxicol. 30*: 491 (1992).

25. L. Rivera, M. J. C. Churto, P. Pais, M. T. Galceran, and L. Puignou, *J. Chromatog. A 731*: 85 (1996).

26. R. J. Turesky, C. M. Forster, H. U. Aeschbacher, H. P. Würzner, P. L. Skipper, L. J. Trudel, and S. R. Tannenbaum, *Carcinogenesis 10*: 151 (1989).

27. M. G. Knize, N. H. Shen, S. K. Healy, F. T. Hatch, and J. S. Felton, *J. Industrial Microbiology*, Suppl. 2: 171 (1987).

28. M. G. Knize, B. D. Andresen, S. K. Healy, N. H. Shen, P. R. Lewis, L. F. Bjeldanes, F. T. Hatch, and J. S. Felton, *Food Chem. Toxicol. 23*: 1035 (1985).

29. R. T. Taylor, E. Fultz, and M. G. Knize, *Environ. Health Perspect. 67*: 41 (1986).

30. M. A. E. Johansson, M. G. Knize, M. Jägerstad, and J. S. Felton, *Environ. Molec. Mutag. 25*: 154 (1995).

31. M. G. Knize, P. L. Cunningham, A. L. Jones, E. A. Griffin, and J. S. Felton, *Fd. Chem. Toxic. 32*: 15 (1993).

32. M. T. Galceran, P. Pais, and L. Puignou, *J. Chromatog. A 655*: 101 (1993).

33. G. A. Gross, R. J. Turesky, L. B. Fay, W. G. Stillwell, P. L. Skipper, and S. R. Tannenbaum, *Carcinogenesis 14*: 2313 (1993).

34. M. G. Knize, R. Sinha, N. Rothman, E. D. Brown, C. P. Salmon, O. A. Levander, P. L. Cunningham, and J. S. Felton, *Fd. Chem. Toxic. 33*: 545 (1995).

35. M. Takahashi, K. Wakabayashi, M. Nagao, M. Yamamoto, T. Masui, T. Goto, N. Kinae, I. Tomita, and T. Sugimura, *Carcinogenesis 6*: 1195 (1985).

36. M. M. C. Van Dyck, B. Rollman, and C. De Meester, *J. Chromatog. A 697*: 377 (1995).

37. R. J. Turesky, C. M. Forster, H. U. Aeschbacher, H. P. Würzner, P. L. Skipper, L. J. Trudel, and S. R. Tannenbaum, *Carcinogenesis 10*: 151 (1989).
38. Z. Yamaizumi, H. Kasai, S. Nishimura, C. G. Edmonds, and J. A. McCloskey, *Mutation Res. 173*: 1 (1986).
39. M. T. Galceran, E. Moyano, L. Puignou, and P. Pais, *J. Chromatog. A 730*: 185 (1995).
40. S. Murray, N. J. Gooderham, A. R. Boobis, and D. S. Davies, *Carcinogenesis 10*: 763 (1989).
41. M. D. Friesen, K. Kaderlik, D. Lin, L. Garren, H. Barsch, N. P. Lang, and F. F. Kadluber, *Chem. Res. Toxicol. 7*: 733 (1994).
42. S. Vainiotalo, K. Matveinen, and A. Reunamen, *Fresenius J. Anal. Chem. 345*: 462 (1993).
43. S. Murray, A. M. Lynch, M. G. Knize, and N. J. Gooderham, *J. Chromatogr. (Biomedical Applications) 616*: 211 (1993).
44. W. Horowitz, *Anal. Chem. 54*: 67A (1982).
45. K. Skog, M. Jägerstad, and A. Laser-Reuterswärd, *Fd. Chem. Toxic. 30*: 681 (1992).
46. J. S. Felton, E. Fultz, F. A. Dolbeare, and M. G. Knize, *Fd. Chem. Toxic. 32*: 897 (1994).
47. K. Skog, G. Steineck, K. Augustsson, and M. Jägerstad, *Carcinogenesis 16*: 861 (1995).
48. C. P. Salmon, M. G. Knize, and J. S. Felton, *Fd. Chem. Toxic. 35*: 433 (1997).
49. R. M. Basel, U.S. patent 5,439,691 (1995).
50. R. Vikse and P. E. Joner, *Acta Vet. Scand. 34*: 1 (1993).
51. M. A. E. Johnsson, L. B. Fay, G. A. Gross, K. Olsson, and M. Jägerstad, *Carcinogenesis 16*: 2553 (1995).
52. S.-J. Tsai, S. N. Jenq, and H. Lee, *Mutagenesis 11*: 235 (1996).
53. H. P. Thiebaud, M. G. Knize, P. A. Kuzmicky, J. S. Felton, and D. P. Hsieh, *J. Agric. Fd. Chem. 42*: 1502 (1994).
54. G. Becher, M. G. Knize, I. F. Nes, and J. S. Felton, *Carcinogenesis 9*: 247 (1988).
55. M. G. Knize, M. Roper, N. H. Shen, and J. S. Felton, *Carcinogenesis 11*: 2259 (1990).
56. P. G. Sheilds, G. X. Xu, W. J. Blot, J. F. Fraumeni, Jr., G. E. Trivers, E. D. Pellizzari, Y. H. Qu, Y. T. Gao, and C. C. Harris, *J. Natl. Cancer Inst. 87* 836 (1995).
57. R. Sinha, N. Rothman, C. P. Salmon, M. G. Knize, E. D. Brown, C. Swanson, D. Rhodes, S. Rossi, J. S. Felton, and O. A. Levander, *Cancer Research*, submitted.
58. R. Sinha, N. Rothman, E. Brown, O. Levander, C. P. Salmon, M. G. Knize, and J. S. Felton, *Cancer Research 55*: 4516 (1995).
59. K. Wakabayashi, H. Ushiyama, M. Takahashi, H. Nukaya, S. B. Kim, M. Hirose, M. Ochiai, T. Sugimura, and M. Nagao, *Environ. Health Perspect. 99*: 129 (1993).
60. M. A. E. Johansson and M. Jägerstad, *Carcinogenesis 15*: 1511 (1994).
61. M. G. Knize, C. P. Salmon, S. S. Mehta, and J. S. Felton, *Mutat. Res. 376*: 129 (1997).
62. L. M. Tikkanen, T. M. Sauri, and K. J. Latva-Kala, *Fd. Chem. Toxic. 31*: 717 (1993).
63. S. Manabe, H. Suzuki, O. Wada, and A. Ueki, *Carcinogenesis 14*: 899 (1993).
64. L. S. Jackson, W. A. Hargraves, W. H. Stroup, and G. W. Diachenko, *Mutat. Res. 320*: 113 (1994).
65. H. P. Thiébaud, M. G. Knize, P. A. Kuzmicky, D. P. Hsieh, and J. S. Felton, *Food Chem. Toxic. 33*: 821 (1995).

5

Chromatographic Analysis of Mycotoxins

JOE W. DORNER
National Peanut Research Laboratory, Agricultural Research Service,
United States Department of Agriculture, Dawson, Georgia

I. INTRODUCTION

Mycotoxins are secondary metabolites of fungi that are toxic to higher animals. They are typically low molecular weight compounds with diverse structures, and many possess functionalities making them good candidates for chromatographic analysis. Knowledge of mycotoxins has existed for many years. It was recognized in the nineteenth century that the disease ergotism was caused by infection of grain with *Claviceps purpurea* [1]. Early in the 1900s, human stachybotryotoxicosis and alimentary toxic aleukia were associated with contamination of grains with various species of *Fusarium*. However, the birth of modern mycotoxicology is associated with the discovery of the aflatoxins as the causative agents of turkey X disease in the early 1960s [2]. It is not coincidental that it was during this same period that thin layer chromatography (TLC) was becoming well developed and standardized.

The science of mycotoxicology grew rapidly after the discovery of the aflatoxins, and chromatography played an important role in this growth. During the 1960s and early 1970s TLC was the method of choice for mycotoxin analysis. Methods were developed for a variety of mycotoxins as they were discovered, and many were studied in interlaboratory collaborative studies and adopted as "official" methods by organizations such as the Association of Official Analytical Chemists. By the mid 1970s, high-performance liquid chromatography (HPLC) was becoming the method of choice for mycotoxin analysis because of its resolving power and greater sensitivity for many mycotoxins, particularly the aflatoxins. In the last 15 years, HPLC has gained much wider usage, but TLC remains a very popular method, primarily because it is much less expensive than HPLC. Gas chromatography (GC) is the other widely used chromatographic technique for the analysis of mycotoxins, particularly certain chemical classes of mycotoxins that are not easily detected in TLC or HPLC analytical systems. The trichothecenes are in a chemical class that fits this description, and GC is the method of choice for the analysis of the majority of these compounds. The GC can be coupled to a mass spectrometer (GC-MS) to produce an analytical system that provides for specific detection of the analyte as well as confirmation of its identity.

This chapter will provide a general discussion of these chromatographic techniques followed by applications to specific mycotoxin groups. It is not possible in this space to cover all the mycotoxins nor all the methods that have been developed for those mycotoxins that will be covered; but an effort is made to focus on mycotoxins that are important contaminants of food or feed and those methods that have been vigorously studied, widely used, or represent the latest in technological advances in the field.

In actuality, the chromatographic method represents the final, determinative step in the analytical process. Prior to that, a commodity that is to be analyzed must be properly sampled, the sample must be prepared for analysis, and it must be extracted and, usually, partially purified before the chromatography is carried out. While the primary focus of this chapter is on chromatographic methods and their applications to specific mycotoxin groups, the importance of these earlier steps in the analytical process cannot be overemphasized. Therefore a discussion of sampling, sample preparation, extraction, and cleanup is included prior to the main focus of the chapter.

II. SAMPLING AND SAMPLE PREPARATION

Mycotoxin contamination in an agricultural commodity or food is rarely homogeneous, except in the case of a liquid, such as milk, or a highly processed and blended food, such as peanut butter. This lack of homogeneity presents difficulties in determining the true concentration of a mycotoxin in the lot to be analyzed. In fact, even detecting the presence of a mycotoxin can be difficult if an appropriate sample is not taken. Several factors must be considered in determining what constitutes an appropriate sample. The characteristics of the analyte, the probable source or time of contamination, and the level of contamination that is a concern are among the more important factors. All of these considerations are related to the degree of homogeneity of the mycotoxin in the material to be analyzed. The less homogeneity there is, the more stringent the sampling needs to be.

The problem of sampling for mycotoxin contamination has been most extensively studied with regard to aflatoxin contamination of peanuts, particularly preharvest contamination. It has been shown that when multiple samples from a lot are analyzed, the distribution of results is not normal but is a negative binomial distribution. This means that most of the values found are less than the mean of all the values. In a study of the variability of aflatoxin test results, Whitaker et al. [3] found that when 12 samples from 1 particular lot of peanuts were analyzed, the average of the 12 samples was 22.8 ppb. However, seven of the samples tested 0 ppb and the range of the 12 samples was 0–195 ppb. This illustrates that concentrating only on the analytical aspect of mycotoxin analysis can easily lead to erroneous results if proper sampling is not done.

The primary methods of sampling are continuous stream sampling and probe sampling [4]. When material is moving in a continuous stream, small samples can be collected from the moving stream periodically and combined to produce one representative sample. This can be done manually or automatically, as with a crosscut sampler. Probe sampling is used when the material is contained, such as in a box, bag, boxcar, storage bin, or truck. The key to effective probe sampling is that the probe must reach through all areas of the bulk storage. It is also important to probe several areas of the container to ensure that the sample is representative of the entire lot.

Once the sample has been drawn, it is still too large (in most cases) to be analyzed. A smaller subsample must be prepared from which the mycotoxin can be extracted. The same heterogeneity that exists in a lot of peanuts, for example, also exists in the sample that has been drawn. Therefore, before obtaining a subsample that can actually be extracted the sample should be ground as finely as possible and blended as much as practical in order to homogenize the mycotoxin in the sample. A vertical cutter mixer (VCM) was shown to be quite effective in homogenizing aflatoxin in samples of peanuts [5]. However, it was necessary to use a VCM of a size that was appropriate for the size of the sample to be ground. In many cases, the size of the sample that is taken is dependent on the equipment that is available to produce the subsample. Some mills have been designed to grind and subsample simultaneously, producing a subsample that is representative of the entire sample. One such mill, the Dickens subsampling mill [6], has been used for many years in the U.S. peanut industry to prepare samples of peanuts for analysis. The Romer mill is another subsampling mill that is used extensively in the corn industry to accomplish the same purpose [7].

The heterogeneity of mycotoxin contamination in commodities and the effects of different sampling and subsampling regimes have been shown by the variability seen when multiple samples from the same lot are analyzed. The total error involved in the analytical process is the sum of the sampling error, the subsampling error, and the analytical error [8]. Whitaker et al. [9] demonstrated that in analyzing peanuts for aflatoxin, sampling accounted for 92.7% of the total error; subsampling accounted for 7.2% when samples were ground in a VCM for 7 min; and analytical error accounted for 0.1% when HPLC was used as the analytical technique. Slightly different results would be obtained under different sampling, subsampling, and analytical scenarios, but other studies have also shown that sampling is by far the greatest source of error in the analysis of commodities for mycotoxins. Therefore, when planning a mycotoxin analytical procedure, the analyst must recognize the importance of good sampling and sample preparation procedures in achieving reliable analytical results, regardless of the technique used in the determinative step.

III. EXTRACTION AND CLEANUP

Mycotoxins are generally extracted from commodities and foods by blending or shaking the sample material with an appropriate solvent, the selection of which is based on the properties of the mycotoxin and the sample. Typical solvents used in the extraction of mycotoxins include chloroform, ethyl acetate, methanol, acetone, acetonitrile, water, and combinations of an organic solvent with water. Nonpolar solvents, such as hexane, may be included in the extraction solvent mixture for samples with a high oil content to partition the oils away during the extraction procedure. The pH of the extraction solvent is often adjusted to optimize the extraction of acidic or basic mycotoxins. The ratio of solvent to sample is usually in the range of 2–5 mL/g. Extractions using a blender are usually rapid (2–3 min) and result in additional particle size reduction, which may enhance the extraction. Mechanical shakers are also available that allow the simultaneous extraction of several samples, but the extraction time is usually on the order of 30 min. Extraction of liquid samples, such as milk, is usually accomplished by partitioning the toxin into an organic solvent in a separatory funnel, or the mycotoxin may be extracted by placing the sample directly on solid phase extraction (SPE) or immunoaffinity columns, which bind the toxin for subsequent elution with an appropriate solvent.

The extraction procedure usually produces a solution containing the myco- toxin of interest as well as many compounds that may interfere with subsequent chromatographic analysis. Removal of the interfering substances is a key element in being able to obtain accurate quantitative results. Cleanup methods usually employed include centrifugation, precipitation, liquid-liquid partition, column chromatography, SPE, and immunoaffinity column chromatography. Older methods relied heavily on liquid-liquid partition and column chromatography, which used large amounts of solvent. More modern methods often use SPE columns, which are commercially available as cartridges packed with high-efficiency sorbents. The specificities offered by the various packing materials coupled with the speed and low solvent usage of SPE cartridges make them an integral part of the chromatography laboratory today. The development of immunological assays for the determination of mycotoxins has brought about the development of immunoaffinity columns for the cleanup of specific mycotoxins. Antibodies specific to a particular mycotoxin are immobilized on a solid support (e.g., agarose) and packed into a cartridge. When the extract is passed through the cartridge, the antibodies bind only the specific toxin, which can then be eluted after impurities are washed through. This highly specific cleanup technique provides highly purified extracts, although usually at higher cost.

The final step in the extraction and cleanup process involves concentration of the toxin in a solvent that is appropriate for the chromatographic technique to be used for quantitation. This can be done with a rotary evaporator or by blowing a stream of nitrogen over the sample as it is being gently heated. Often the solvent is removed completely, and the sample is then redissolved in a small amount of the appropriate solvent. This step can result in some loss of toxin because it may be adsorbed onto the surface of a glass container, or excessive heating or exposure to light may cause degradation of the toxin. Therefore, care must be exercised during concentration to avoid losses that would not be detectable.

IV. CHROMATOGRAPHIC TECHNIQUES

A. Minicolumn Chromatography

Minicolumn chromatography has been a widely used chromatographic technique for the screening of mycotoxins since it was initially introduced by Holaday [10] for aflatoxin analysis of peanuts. In the procedure, an aliquot of a partially purified extract is applied to a small chromatographic column (4 to 8 mm i.d. × 100 to 150 mm length) that is packed with one or more layers of different adsorbents. The original Holaday minicolumn contained 50 mm of silica gel surrounded by 5 mm of glass fiber. One end of the minicolumn was placed in the filtered extract, and the filtrate was allowed to migrate through the column to the top of the silica gel layer. Aflatoxin was detected as a fluorescent band in the silica gel by exposing the column to UV light. Refinements of the method have included the incorporation of different bands of adsorbents into a single minicolumn, application of extract to the top of the column with elution aided by application of vacuum, and the use of different solvents for elution [11–13]. Two minicolumn methods, which have been adopted as official methods by AOAC International, are the Romer minicolumn method for screening for the presence of aflatoxins in a variety of foods and commodities [14] and the Holaday–Velasco method for determining aflatoxins in corn and peanuts [15]. This methodology has also been applied to ochratoxin A [16], zearalenone [16], and deoxynivalenol [17]. Dorner and Cole [18] used the minicolumn as a prescreen before HPLC analysis of peanuts for aflatoxin when contamination levels were expected to cover a wide range. The minicolumn prescreening enabled the proper dilution of extracts prior to HPLC analysis, thus saving time of rerunning many samples that would not have been properly diluted before the HPLC autorun.

B. Thin-Layer Chromatography

Thin layer chromatography (TLC) has been the most commonly used method for the analysis of mycotoxins since the discovery of the aflatoxins. It is still a very popular analytical technique today because it is relatively simple and inexpensive, allows the simultaneous analysis of many extracts, and offers great selectivity as well as the opportunity for multi toxin analyses.

1. Plates

The majority of TLC analyses are conducted on rigid glass plates to which a particular adsorbent is applied in a thin layer. Analyses can also be carried out on flexible plates that have a backing of aluminum or plastic. Typical adsorbents for normal phase TLC include silica gel, aluminum oxide, and cellulose. However, silica gel plates are by far the most popular for the analysis of mycotoxins. Plates can be prepared economically in the laboratory by making a slurry of the adsorbent in water, using a slurry-spreading apparatus to apply it to the plates, and then drying the plates in an oven [19]. However, commercially available precoated plates offer excellent uniformity and reproducibility and are usually preferred. Other substances, such as oxalic acid, can be incorporated into the silica gel layer to aid in the separation of acidic mycotoxins (e.g., citrinin, cyclopiazonic acid) [20].

Reversed phase TLC plates are also available commercially. They have a layer of silica gel to which has been bound any of various nonpolar functional groups, including C_2, C_8, C_{18}, and phenyl. In this type of TLC the mobile phase is more polar than the adsorbent, which is the reverse of normal phase, in which the silica gel adsorbent is used with more nonpolar solvents in the mobile phase.

The development of high-performance TLC (HPTLC) in recent years is an advancement that has kept TLC competitive with other, more sophisticated techniques. Plates are smaller than conventional TLC, and they are coated with small particle size sorbents that are extremely efficient. To perform HPTLC at peak efficiency, instrumentation for spotting small volumes of test samples consistently is advisable. Use of a densitometer is also necessary for accurate quantitation. Coker et al. [21] evaluated HPTLC instrumentation for the quantitation of aflatoxins. They recommended use of a fully automated TLC sampler, an unsaturated conventional TLC glass chamber, and a monochromatic fluorodensitometer.

2. Mobile Phases and Plate Development

After sample extracts are applied to a TLC plate, the plate is placed in a mobile phase or solvent system for chromatographic development. Mobile phases vary considerably depending on the mycotoxins to be separated, the type of plate used, and compounds that may interfere with detection or quantitation of the mycotoxin of interest. In normal phase TLC with silica gel plates, the basic component of the mobile phase is usually a relatively nonpolar solvent, such as chloroform. One or more polar solvents is usually added in order to produce the desired separation of metabolites. Common choices for this solvent include acetone, alcohols, water, and organic acids. The amount of polar solvent added usually totals 10% or less of the total mobile phase, but the amount is usually critical to producing the desired chromatography. In reversed phase TLC, the mobile phase is usually composed of water in combination with methanol or acetonitrile. Although reversed-phase TLC has been applied to the analysis of mycotoxins, the vast majority of TLC analyses of mycotoxins is accomplished with normal-phase silica gel plates.

Development of TLC plates is usually carried out with a single mobile phase. However, the use of two mobile phases and the development of plates in different directions gives TLC more flexibility and greater resolving power. With two-solvent TLC, the plate can be developed first with a nonpolar solvent to move nonpolar interferences away from the mycotoxin(s) of interest. The plate is then dried and developed with the second solvent to separate the mycotoxins. Bidirectional TLC accomplishes a similar purpose. Samples are spotted in the middle of the plate, which is then developed with a nonpolar solvent. The top of the plate is then cut off, the plate is turned upside down, and developed with the second, more polar solvent. A fourth type of plate development is two-dimensional TLC, which provides the greatest resolving power in TLC. The sample is spotted in one corner of the plate with standards spotted in the two adjacent corners. The plate is developed in one direction with one solvent and then rotated 90° and developed with a second solvent. This type of plate development has been particularly useful for the determination of aflatoxins in various foods [22–25].

3. Detection and Quantitation

The detection of mycotoxins in TLC is based primarily on the R_f of the spot in conjunction with its color or fluorescence. Without a known standard to compare with an unknown sample, the R_f value associated with specific solvent systems is a major guide in identification. Many mycotoxins, including the aflatoxins, exhibit a natural fluorescence when exposed to ultraviolet light. The intense fluorescence of some of these mycotoxins enables the detection of nanogram, and in some cases, picogram quantities in a single spot [26,27]. Quantitation can be achieved by comparing the intensity of the fluorescence of the sample spot with that of a series of known standard amounts. This can be done visually or instrumentally with a densitometer. Although more errors are associated with visual than with instrumental quantitation, individuals can be trained to be quite proficient doing visual quantitation.

Chemical derivatization of spots on developed TLC plates is another commonly used method for detection and quantitation of mycotoxins. Plates may be exposed to chemical vapors, which react with particular mycotoxin functional groups causing development of fluorescence or a characteristic color. Aluminum chloride has been used to increase the fluorescence intensity of deoxynivalenol [28] and sterigmatocystin [29]. Treatment of plates with ammonia or other bases has been used to enhance fluorescence of ochratoxins [30] and patulin [31]. Indole-containing mycotoxins, such as cyclopiazonic acid, can be sprayed with p-dimethylaminobenzaldehyde and exposed to HCl vapor to produce a lavender to purple spot. The same compound produces a reddish-orange color when sprayed with an ethanolic solution of ferric chloride [32]. Production of a characteristic chemical derivative is also an important method of confirming the identity of particular mycotoxins [19].

C. High-Performance Liquid Chromatography

High-performance liquid chromatography (HPLC) is quite similar to TLC in principle, but more expensive instrumentation is used to achieve greater resolution, sensitivity, accuracy, and precision. Whitaker et al. [33] studied the variability associated with TLC, HPLC, and enzyme-linked immunosorbent assays (ELISA) in the determination of aflatoxins in various agricultural commodities. Estimates of method precision were pooled from a number of collaborative studies, and the predicted coefficients of variation (CV) for the three types of methods were determined. The CV for HPLC were considerably lower than for TLC or ELISA, indicating that it is the most precise method available for aflatoxin quantitation.

Use of HPLC in the laboratory has increased greatly since its inception in the early 1970s. The majority of papers reporting new methods for mycotoxin analysis utilize HPLC. The primary advantages of HPLC in mycotoxin analysis include objectivity, sensitivity, versatility, and automation. The primary disadvantages are the cost of the instrumentation and the fact that rigorous cleanup is usually required in order to achieve accurate, reproducible results. However, the advantages usually outweigh the disadvantages, particularly when one considers that to achieve similar results with TLC, expensive instrumentation such as spotters and densitometers are also required.

1. Instrumentation

The versatility and sensitivity of HPLC derive from the fact that it is a system composed of various component instruments that can be interchanged to optimize the system for the types of compounds being analyzed. The first component to be considered in an HPLC system is the solvent pump. Several types of pumps are available, and these have been reviewed by Shepherd [34]. The basic requirement of an HPLC pump is that it deliver solvent at a constant rate reliably. The cost of pumps is quite variable, but selection is usually governed by the particular application.

The injector introduces the prepared extract into the HPLC system, and it is one of the more critical components in achieving efficient, reproducible separations. Manual injection is usually accomplished with an injection valve, which is highly reproducible when used properly. The extract is loaded into a fixed-volume loop with a syringe, and the injector diverts the solvent flow through the loop carrying the extract to the column. The best technique is to completely fill the loop with extract. Because the loop contains mobile phase, it is best to use a volume of extract that equals about five times the loop volume to ensure that all mobile phase is removed from the loop and replaced entirely by sample extract. In applications where excess extract is not available, partial filling of the loop can be used, but it is best to use a volume of extract that is not more than 50% of the total loop volume. Autosamplers equipped with automatic injection systems allow unattended injection of numerous samples and provide a high degree of flexibility with many programmable features. Although not as precise as the valve with a fixed volume loop, good autosamplers can provide quite acceptable reproducibility and are extremely advantageous in laboratories requiring a large number of analyses daily. In such a situation, however, it is important to maintain checks on reproducibility by periodically injecting a known standard to ensure that the injection system is functioning properly.

When the analyte elutes from the column (discussed below), it passes through a detector, the next component in the HPLC system. Several types of detectors are available for use in HPLC, and this provides for great selectivity and sensitivity. The most common type of detector for HPLC of mycotoxins is the UV detector, because most mycotoxins, with the notable exception of the trichothecenes, absorb in the UV. The fluorescence detector is commonly used to detect either mycotoxins that fluoresce or those that can be derivatized to form fluorescent derivatives. Fluorescence detectors offer high selectivity and are much more sensitive than UV detectors. They are the detectors of choice for the analysis of the aflatoxins and the fumonisins. The UV photodiode array detector has become very popular in recent years because numerous wavelengths can be monitored simultaneously and UV spectra of eluting compounds can be viewed. This detector offers several other advantages, particularly the ability to compare the UV spectrum of a peak of interest with a library of UV spectra for confirmation or aiding in identification [35]. Electrochemical detectors have received some limited use in the detection of mycotoxins, including aflatoxins [36], roquefortine [37], and xanthomegnin [38]. Finally, mass spectrometers can now be coupled with HPLC to provide immediate confirmation of the identity of eluting compounds.

The final basic component of HPLC instrumentation is the recorder/integrator, which receives the signal from the detector and produces the chromatograph. This task can also be handled with a computer equipped with appropriate software. When using a photodiode array detector, the computer is necessary to take full advantage of the capabilities of the detector.

2. Columns and Mobile Phases

Successful HPLC depends on selecting the right combination of column and mobile phase. The column contains the stationary phase on which the chromatographic separation takes place. In normal phase HPLC, the stationary phase is typically silica, and the mobile phase consists of a combination of less polar organic solvents. In reversed phase HPLC the stationary phase is silica to which has been bound any number of relatively nonpolar functional groups, and the mobile phase is a more polar combination of water with an organic solvent modifier. In the early days of HPLC, normal phase HPLC was dominant, but the vast majority of HPLC methods published today utilize the reversed phase method. Reversed phase packings are quite numerous and include C_{18}, C_8, phenyl, amino, cyano, and other bonded phases. Common organic solvent modifiers for reversed phase HPLC include methanol, acetonitrile, and tetrahydrofuran. These can be used in binary, ternary, or quaternary combinations with water, and slight changes in the combinations can result in significant changes in selectivity. Isocratic elution is usually desirable wherever practical, such as in the analysis of one or a few closely related mycotoxins. However, when it is necessary to analyze compounds with a wide variety of retention times, gradient elution may be required.

HPLC offers great flexibility in the manipulation of chromatographic conditions (primarily columns and mobile phases) to achieve difficult separations. Shepherd [34] presented a detailed discussion that provided many good examples of how variations in the stationary and mobile phases affect chromatography.

3. Derivatization

In many cases detection of a mycotoxin can be enhanced through derivatization, either before injection (precolumn derivatization) or in line after the toxin elutes from the column (postcolumn derivatization). Derivatization is usually performed to make a fluorescent derivative (as with the fumonisins) or to increase the fluorescence of a weakly fluorescent compound (as with aflatoxins B_1 and G_1 in aqueous mobile phases). This technique is discussed in greater detail in the afla-toxin application section.

D. Gas Chromatography

Whereas TLC and HPLC are the most commonly used chromatographic methods for the majority of mycotoxins, gas chromatography (GC) is the method of choice for mycotoxins that exhibit little or no UV absorption and are not fluorescent. Although GC methods have been published for numerous mycotoxins, the great-est practical use of GC is in the analysis of the trichothecenes, for which TLC and HPLC methods are quite limited. In GC analysis, the analytes are volatilized and passed by a carrier gas through a heated column that contains a stationary phase of high molecular weight polymer. Since most mycotoxins are not volatile at tem-peratures suitable for GC (30° to 350°C), volatile derivatives must be formed. Primary detection methods include flame ionization, electron capture, and coup-ling the GC to a mass spectrometer (MS) in a technique known as GC-MS. Beaver [39,40] has presented detailed reviews covering all practical aspects of GC.

1. Columns

Columns used in GC are of two types, packed and capillary. Packed columns are usually borosilicate glass tubes that are packed with an inert support that is coated with the polymer to be used as the stationary phase. The columns may range in length from 3 to 6 m with internal diameters of 2 to 6 mm. In capillary columns, the stationary phase is deposited as a thin film on the inner surface of the column tubing, which is usually made of fused silica. Fused silica capillaries are quite flexible and rugged; therefore, lengths of 25 to 50 m are common with internal diameters of 0.2 to 0.5 mm. Capillary columns offer great efficiency but less loading capacity than packed columns. The only AOAC International official method using GC [41] specifies a packed column, but the majority of methods published in recent years for the determination of trichothecenes use capillary columns. The choice of column really depends on the particular application, since each type offers certain advantages.

2. Derivatization

Because most mycotoxins, particularly the trichothecenes, contain numerous polar functional groups, they are not readily volatile, but they are good candidates for derivatization to compounds that are easily volatilized. The ethers, esters, amides, etc. that result from derivatization not only increase volatility but also may lead to increased thermal stability [40].

Many reagents have been used for various derivatization reactions prior to GC. The reagent and derivatization technique chosen depend on several factors, such as the volatility and stability of the product, the sensitivity required, and the separation of the derivatives formed.

Trimethylsilyl (TMS) ethers are probably the most common derivatives overall, and they have been the most frequently used for GC of trichothecenes [42]. These derivatives are generally quite stable and can be detected by any of the detection techniques. The various reagents used in making TMS derivatives require different reaction conditions and can be chosen on the basis of reactions with similar compounds in the literature or by trial and error.

Heptafluorobutyrate (HFB) derivatives are very sensitive to measurement with the electron capture detector (ECD) and are also detected by the MS. These derivatives are commonly formed by reactions with heptafluorobutyryl anhydride. When multiple HFB groups can be added to a molecule, the sensitivity of the ECD or MS can be in the low picogram range [42]. The high molecular weight of HFB derivatives also offers the opportunity for increased specificity for GC-MS. Pentafluoropropionyl (PFP) derivatives formed with pentafluoropropionic anhydride or pentafluoropropionylimidazole are similar to HFB derivatives and offer similar advantages.

Trifluoroacetyl (TFA) esters are usually formed through reaction with trifluoroacetic anhydride. These derivatives are of relatively low molecular weight and provide good volatility. They have been used with all the major GC detectors. Scott [42] summarized several potential problems associated with production of TFA derivatives. They are often relatively unstable, poor yielding, sensitive to presence of moisture or excess reagent, and have been shown to exhibit thermal instability.

3. Detection

Several types of detectors are available for GC, but for mycotoxin analysis, the flame ionization detector (FID), the ECD, and MS are most often utilized. The FID could be considered the universal GC detector. Carbon-containing compounds are burned in a hydrogen–air flame as they elute from the column, and the resulting ions are collected on a charged electrode that produces a current proportional to the number of ions present. The detector is rugged and sensitive but not at all selective, because nearly all organic compounds produce a response. Therefore, rigorous cleanup of extracts is usually required to obtain reliable results. Nevertheless, it is probably the most commonly used detector overall for GC of mycotoxins.

The ECD is much more selective than the FID, and it may or may not be more sensitive. It contains a β-radiation source that ionizes carrier gas molecules, producing a steady current. When another ionizable compound enters the detector, its electrons are captured and a reduction in the current is realized. Compounds, such as hydrocarbons, that have low electron affinities produce very little signal in the ECD. However, halogenated compounds have high electron affinities and can produce strong signals. Thus quite low concentrations of HFB and PFP derivatives can often be detected by the ECD.

Perhaps the ultimate detector for use with GC is the MS because it provides sensitivity, selectivity, and confirmation of the identity of an eluting compound.

The full mass spectrum of compounds eluting from a GC column can be used to aid in the structure determination of unknowns or provide unequivocal identification of particular peaks. Selectivity and sensitivity of GC-MS can be greatly increased by using select ion monitoring (SIM). In this mode the MS monitors a single ion mass; therefore, one can be chosen that is unique to the analyte of interest. However, one loses the ability to confirm the identity of a GC peak when using SIM because the full spectrum is not available. The primary disadvantage of this technique is the cost of the instrumentation. However, it is a powerful analytical tool, particularly for the analysis of trichothecenes and samples that may contain multiple closely related trichothecenes. The review by Vesonder and Rohwedder [43] fully describes this specialized technique.

V. APPLICATIONS

A. Aflatoxins

A collection of chromatographic methods used in the analysis of aflatoxins is given in Table 1. Minicolumn methods [10–13,16,44–46] are used primarily for qualitative screening of commodities such as corn and peanuts, but they can also be used by an experienced analyst for semiquantitation. Some of these methods have been subjected to interlaboratory collaborative studies, and two have been adopted by AOAC International as official methods [14,15].

The first TLC method for the aflatoxins was published in 1966 [47], and TLC has continued as a widely used technique for aflatoxin quantitation. Most methods use silica gel as the stationary phase, but there is much variety in choice of mobile phases and plate development techniques, including 1-dimensional [48–50], 2-dimensional [24,25], bidirectional [51–53], and multidevelopment [54]. The need for such a variety is brought about by the variety of interfering compounds found in the extracts of various commodities. Detection of aflatoxins in TLC is accomplished by visualization or densitometry of the fluorescent spots under longwave UV light (365 nm). Confirmation of the identity of the aflatoxins is often accomplished by adding trifluoroacetic acid (TFA) to the spots prior to plate development and observing the change in R_f [55].

Early HPLC methods for aflatoxin used normal phase columns with mobile phases consisting of water-saturated chloroform modified with other solvents [56,57]. To obtain the desired sensitivity for B_1 and B_2, the fluorescence detector flow cell was packed with silica gel, but this technique was not without drawbacks as the silica gel absorbed materials in the matrix and produced elevated noise in the detector. Reversed phase HPLC provides good separation of the toxins, and excellent sensitivity can be achieved without a packed flowcell if the B_1 and G_1 are derivatized prior to detection. The derivatization can be achieved by reaction with TFA prior to injection [58–64], or postcolumn derivatization can be achieved with iodine [18,65–70], bromine [71–73], or exposure to UV light [74,75].

Most methods for postcolumn iodination require a separate pump for

Table 1 Selected Chromatographic Methods for the Determination of Aflatoxins

Toxins detected	Analytical matrix	Extraction solvent	Cleanup	Chromatographic conditions	Detection limit (μg/kg)	Ref.
Total aflatoxins	Corn, peanuts	Methanol/water (80:20)	Precipitation and liquid-liquid partition	*Minicolumn* Packing: Layers of Florisil, silica gel, alumina Elution solvent: 9:1 chloroform/acetone Detection: Blue fluorescent band at top of Florisil layer when viewed under longwave UV light	10	45,46
Total aflatoxins	Various	Acetone/water (85:15)	Precipitation and liquid-liquid partition	*Minicolumn* Packing: Layers of Florisil, silica gel, alumina Elution solvent: 9:1 chloroform/acetone Detection: Blue fluorescent band at top of Florisil layer when viewed under longwave UV light	5–15	13,44
B_1, B_2, G_1, G_2	Peanuts, peanut products	Chloroform/water (10:1)	Silica gel column chromatography	*TLC* Plate: Silica gel GHR Solvent: 90:10 chloroform/acetone Detection: Visualization under longwave UV light		168
B_1, B_2, G_1, G_2	Peanuts, peanut products	Methanol/water (55:45)	Liquid-liquid partition	*TLC* Plate: Silica gel 60 Solvent: 90:10 chloroform/acetone Detection: Visualization under longwave UV light		48

B_1, B_2, G_1, G_2	Corn, peanuts, peanut butter	Methanol/water (85:15)	Liquid-liquid partition and silica gel column chromatography	*TLC* Plate: Silica gel 60 Solvent: 90:10 chloroform/acetone Detection: Visualization or densitometry under longwave UV light	3–50	50
M_1	Cheese	Chloroform/saturated NaCl solution (100:1)	Silica gel column chromatography	*TLC* Plate: Silica gel GHR Solvent: 87:10:3 chloroform/acetone/isopropanol Detection: Visualization or densitometry under longwave UV light		169
M_1	Milk, milk products	Methanol/acetone (50:20); Chloroform	Silica gel column chromatography	*TLC* Plate: Silica gel 60 Solvent: 85:10:5 chloroform/acetone/isopropanol Detection: Visualization under longwave UV light	0.02–0.1	170
B_1, B_2, G_1, G_2	Vegetable oils	Hexane	Silica gel column chromatography	*Two-Dimensional TLC* Plate: Silica gel 60 Solvent: (a) 9:1 chloroform/acetone; (b) 96:3:1 ethyl ether/methanol/water Detection: Visualization under longwave UV light		25
M_1	Cheese	Acetone/water (3:1)	C_{18} and silica cartridges	*Two-Dimensional TLC* Plate: Silica gel 60 Solvent: (a) 95:4:1 ethyl ether/methanol/water (b) 70:30 chloroform/acetone Detection: Fluorodensitometry (ex: 366 nm; em: >430 nm)	0.01	171

Table 1 Continued

Toxins detected	Analytical matrix	Extraction solvent	Cleanup	Chromatographic conditions	Detection limit (µg/kg)	Ref.
B_1, B_2, G_1, G_2	Peanut butter	Acetone/water	Phenyl SPE cartridge	*Bidirectional HPTLC* Plate: Aluminum-backed silica gel 60 Solvent: (a) diethyl ether; (b) 6:3:1 chloroform/xylene/acetone Detection: Fluorodensitometry		51
B_1, B_2, G_1, G_2	Figs, Brazil nuts, pistachio nuts	Methanol/water (71:29)	Liquid-liquid partition	*Bidirectional TLC* Plate: Silica gel sheet Solvent: (a) diethyl ether; cut sheet above aflatoxins; (b) 80:12:0.2 chloroform/acetone/water in the opposite direction Detection: Fluorodensitometry (ex: 365 nm; em: >400 nm)	0.2–100	52
B_1, M_1	Beef liver	Methanol/water (80:20)	C_{18} reversed phase SPE cartridge	*HPLC* Column: Spherisorb silica (5 µm, 25 cm × 4.6 mm) Mobile phase: 95:1:4 water-saturated chloroform/2-propanol/ tetrahydrofuran Detection: Fluorescence (silica gel packed flow cell)	0.1	172

Toxin	Matrix	Extraction	Cleanup	Method	Detection limit	Ref.
B_1, B_2, G_1, G_2	Peanut products	Methanol/0.1N HCl (4:1)	Liquid-liquid partition and silica gel column chromatography	*HPLC* Column: Silica gel (10 μm, 30 cm × 4 mm) Mobile phase: 25:7.5:1.0:1.5 water-saturated dichloromethane/cyclohexane/acetonitrile/ethanol Detection: UV (360–365 nm) or fluorescence (ex: 360–365 nm; em: 400–410 nm)	0.3–1.0	56
B_1, B_2, G_1, G_2	Peanuts, peanut butter, corn, tree nuts	Chloroform/water (10:1)	Silica gel column chromatography	*HPLC* Column: Reversed phase C_{18} (10 μm, 25 cm × 3.2 mm) Mobile phase: 75:15:10 water/acetonitrile/methanol Detection: Fluorescence (ex: 365 nm; em: 450 nm) Derivatization: Precolumn with TFA		59
B_1, B_2, G_1, G_2	Various	Acetonitrile/water (90:10)	Multifunctional column	*HPLC* Column: Reversed phase C_8 (5 μm, 10 cm × 4.6 mm) Mobile phase: 80:20 water/acetonitrile Detection: Fluorescence Derivatization: Precolumn with TFA	<0.5	63
M_1, M_2	Milk	Elute from C_{18} SPE cartridge with ether	Silica gel column chromatography	*HPLC* Column: Reversed phase C_{18} (5 μm. 25 cm × 4 mm) Mobile phase: 80:12:8 water/isopropanol/acetontrile Detection: Fluorescence (ex: 360 nm; em: 440 nm) Derivatization: Precolumn with TFA	0.07	173

Table 1 Continued

Toxins detected	Analytical matrix	Extraction solvent	Cleanup	Chromatographic conditions	Detection limit (μg/kg)	Ref.
B_1, B_2, G_1, G_2	Corn, peanuts, peanut butter	Methanol/water (70:30)	Immunoaffinity column	*HPLC* Column: Reversed phase C_{18} (5 μm, 25 cm × 4.6 mm) Mobile phase: 60:20:20 water/acetonitrile/methanol Detection: Fluorescence (ex: 360 nm; em: >420 nm) Derivatization: Postcolumn with iodine	10.0	68
B_1, B_2, G_1, G_2	Peanut butter	Methanol/water (55:45)	Liquid-liquid partition	*HPLC* Column: Reversed phase C_{18} (5 μm, 20 cm × 4.6 mm) Mobile phase: 60:24:16 water/methanol/acetonitrile Detection: Fluorescence (ex: 365 nm; em: 440 nm) Derivatization: Postcolumn split-flow iodine from a solid phase iodine column	1.0	67
B_1, B_2, G_1, G_2	Peanuts	Methanol/water (80:20)	Liquid-liquid partition	*HPLC* Column: Reversed phase C_{18} (4 μm, 15 cm × 3.9 mm) Mobile phase: 60:40 water/methanol plus 100 μL HNO_3 and 60 mg KBr/L Detection: Fluorescence (ex: 365 nm; em: 440 nm) Derivatization: Postcolumn with electrochemically generated bromine	0.1	71,174

Toxin	Matrix	Extraction	Cleanup	Conditions	LOD	Ref.
B_1, B_2, G_1, G_2	Spices	Methanol/water (80:20)	Immunoaffinity column	*HPLC* Column: Reversed phase C_{18} (5 μm, 25 cm × 4.6 mm) Mobile phase: 60:22.2:17.8 water/acetonitrile/methanol Detection: Fluorescence (ex: 362 nm; em: 418 nm) Derivatization: Postcolumn with pyridinium bromide perbromide	1.0	175
B_1, B_2, G_1, G_2	Corn	Chloroform/water (10:1)	Silica gel column chromatography	*HPLC* Column: μBondapak C_{18} (30 cm × 3.9 mm) Mobile phase: 1:1 methanol/0.005 g/mL solution of β-cyclodextrin in water Detection: Fluorescence (ex: 365 nm; em: 418 nm) Fluorescence enhancement: postcolumn addition of 0.015 g/ml β-cyclodextrin solution	1.0	78
B_1, B_2, G_1, G_2	Peanuts	Methanol/water (56:44)	Liquid-liquid partition	*HPLC* Column: Reversed phase C_{18} (4 μm, 15 cm × 3.9 mm) Mobile phase: 70:35:1.2 water/methanol/butanol Detection: Fluorescence (ex: 365 nm; em: 440 nm) Derivatization: Postcolumn photochemical	0.1	74,75

aqueous iodine, which is introduced to the column effluent in a low dead-volume tee. The reaction takes place in a coil of teflon tubing held at ca. 70°C. Jansen et al. [67] achieved postcolumn iodination using one pump by splitting the mobile phase flow between the analytical column and a column packed with solid iodine. The iodine-containing solution was recombined with the column effluent before passing through a knitted open tubular reactor at 60°C. The solid iodine column could be used for long periods because of the low solubility of iodine in the mobile phase (60:24:16 water/methanol/acetonitrile). Postcolumn bromination accomplishes about the same enhancement of fluorescence as postcolumn iodination, but it offers several advantages. A separate pump is not required because bromine is produced from potassium bromide present in the mobile phase. The flow from the column passes into an electrochemical cell (KOBRA cell) that releases free bromine. The reaction takes place at room temperature in a shorter coil than with the iodine reaction, and following the detector the flow passes back into the counterelectrode compartment of the KOBRA cell, where the bromine is converted back into the bromide form. Detailed information concerning these derivatization techniques is provided in a review by Kok [76], and a comparison of precolumn TFA derivatization with postcolumn iodination was recently published [77].

Postcolumn photochemical derivatization can be accomplished by passing the flow from the column through a knitted reactor coil (0.25 mm i.d. × 25 m) housed in a photochemical reactor containing a mercury lamp ("PHRED") [74,75]. The result is an enhancement of the fluorescence of B_1 and G_1 on the same order as that achieved with iodine or bromine.

Excitation of B_1 and G_1 has also been accomplished with postcolumn addition of an aqueous solution of cyclodextrins [78]. This produces a complex in which the fluorescence of B_1 and G_1 is enhanced without any change in the fluorescence of B_2 and G_2. This reaction was recently studied by Cepeda et al. [79], and they found that 2,6-β-*o*-dimethyl-cyclodextrin produced a fluorescence response 2–3 times larger than that produced by β-cyclodextrin and 8 times larger than that of α-cyclodextrin.

Methods for GC analysis of aflatoxins have been very limited, primarily because the need does not exist with their successful analysis by the other methods. Trucksess et al. [49] used a methyl silicone–coated fused silica column in a GC-MS method to confirm the identity of aflatoxin B_1. All four aflatoxins were analyzed by Goto [80] on a 5% phenylmethylsilicone column with flame ionization detection.

B. Fumonisins

The fumonisins are a relatively new group of aliphatic mycotoxins that are of great interest because they cause leukoencephalomalacia in horses, pulmonary edema in swine, renal lesions and hepatic cancer in rats, and are associated with a high incidence of esophageal cancer in southern Africa [81–90]. They are produced by

Fusarium moniliforme and *F. proliferatum* and are natural contaminants of corn worldwide. At least seven analogues are currently known (fumonisins B_1, B_2, B_3, B_4, A_1, A_2, and C_1), but only FB_1, FB_2, and FB_3 are recognized as natural contaminants of corn, while the others have been found only in culture [91].

The fumonisins do not absorb UV or visible light and they do not fluoresce; therefore, derivatization is normally required for detection. Chromatographic methods for fumonisins include TLC and GC, but the most extensive work has been done with HPLC. A collection of representative methods for fumonisin analysis appears in Table 2. A TLC method for screening samples for the presence of FB_1 and FB_2 uses reversed phase C_{18} plates developed in 3:2 methanol-1% aqueous KCl. Plates are sprayed with 0.1 M sodium borate buffer, fluorescamine, and 40:60 0.01 M boric acid/acetonitrile for visualization under UV light [92]. Plattner et al. [93] reported a GC-MS method for hydrolyzed fumonisin that had good sensitivity but less precision than HPLC methods. However, use of electron capture negative chemical ionization with deuterium-labeled fumonisin B_1 added as an internal standard provided both high sensitivity and precision [94].

Numerous HPLC methods have been reported for fumonisin analysis, and this has become the technique of choice, although several different cleanup and derivatization methods are used. Cleanup of extracts prior to derivatization has been accomplished with strong anion-exchange columns or SPE cartridges [94–101], C_{18} cartridges [84,102], and immunoaffinity columns [103–105]. Gelderblom et al. [81] formed a maleyl derivative for UV detection of fumonisins after separation by reversed phase HPLC. However, most methods rely on fluorescent derivatives for more sensitive detection. Perhaps the most widely used derivative is that formed by reaction with *o*-phthaldialdehyde (OPA) [95,96,99,102]. This was used in an AOAC IUPAC collaborative study in which fumonisins were derivatized prior to HPLC on a C_{18} reversed phase column with a mobile phase of 77:23 methanol/0.1 M sodium dihydrogen phosphate (pH 3.3) and fluorescence detection [106]. This study resulted in the adoption of the method as an AOAC International official method [107]. A disadvantage of this technique is that the OPA derivative is not stable and requires that samples be injected into the HPLC within minutes of derivatization. Therefore the method is not suitable for unattended HPLC with an autosampler. An improvement in derivative stability was observed after treatment with 4-fluoro-7-nitrobenzofurazan (NBD-F) [108], but peak heights started to deteriorate after 20–30 min. Further improvement in the stability of derivatives was achieved by reaction with 4-(*N,N*-dimethylamino-sulfonyl)-7-fluoro-2,1,3-benzoxadiazole (DBD-F) [101]. This reaction has the disadvantage that it requires heating at 60°C for 60 min, but the derivative was stable for 7 days. Derivatization with naphthalene-2,3-dicarboxaldehyde (NDA) has been used successfully [97,98,100,103] by reacting purified extracts with NDA in the presence of sodium borate buffer (pH 9.5) and sodium cyanide for 15 min at 60°C [97]. The fluorescent derivatives of FB_1 and FB_2 were stable for at least 24 h after preparation [97]. Miyahara et al. [109] recently reported a post-

Table 2 Selected Chromatographic Methods for the Determination of Fumonisins

Analytical matrix	Extraction solvent	Cleanup	Chromatographic conditions	Detection limit (μg/kg)	Ref.
Corn	Acetonitrile/water (50:50)	C₁₈ SPE cartridge	*TLC* Plate: Reversed phase C₁₈ (Whatman KC18) Solvent: 60:40 methanol/1% aqeous KCl Detection: Visualization under longwave UV light Derivatization: Spray with 0.1 M sodium borate buffer, fluorescamine, and 40:60 0.01 M boric acid/acetonitrile	100	92
Corn	Methanol/water (3:1)	Liquid-liquid partition	*1. TLC* Plate: Silica gel (Whatman K6F) Solvent: 60:30:10 chloroform/methanol/acetic acid Detection: Blue-violet spot at Rf 0.24 Derivatization: *p*-anisaldehyde spray and heat *2. GC-MS of hydrolyzed fumonisin* Column: DB1 fused-silica capillary (0.25 μm, 15 cm) Temp. profile: 120°C for 2 min; 20°C/min to 200°C; 10°C/min to 270°C Derivatization: TMS	4	93
Corn	Methanol/water (3:1)	Strong anion-exchange cartridge	*HPLC* Column: Reversed phase C₈ (5 μm; 12.5 cm × 4 mm) Mobile phase: 68:32 methanol/0.1 M NaH₂PO₄ Detection: Fluorescence (ex: 335 nm; em: 440 nm) Derivatization: Precolumn with OPA/mercaptoethanol	50	96

Sample	Extraction	Cleanup	Method	Recovery (%)	Ref.
Corn, corn-based foods	Methanol/water (3:1)	Strong anion-exchange column chromatography	*Gradient HPLC* Column: Reversed phase C_{18} (5 μm, 25 cm × 4.6 mm) Mobile phase: (a) 50:50 methanol/0.05 M NaH_2PO_4 (pH 4.4) (b) 80:20 acetonitrile/water Gradient: 100% (a) for 5 min; 50% (a) 50% (b) for 15 min Detection: Fluorescence (ex: 460 nm; em: 500 nm) Derivatization: Precolumn with NBD-F	100	98, 108
Corn	Methanol/water (3:1)	Strong anion-exchange column chromatography	*Gradient HPLC* Column: TSKgel ODS-80Ts (15 cm × 4.6 mm) Mobile phase: (a) 50:50 0.05 M NaH_2PO_4/methanol (b) 75:25 acetonitrile/water Detection: Fluorescence (ex: 450 nm; em: 590 nm) Derivatization: Precolumn with DBD-F	10	101
Corn	Acetonitrile/water (50:50)	Strong anion-exchange SPE column	*Gradient HPLC* Column: Reversed phase C_{18} (5 μm, 25 cm × 4.6 mm) Mobile phase: (a) 99:1 acetonitrile/acetic acid (b) 99:1 water/acetic acid Gradient: 60% (a) + 40% (b) for 8 min; 80% (a) + 20% (b) for 16 min Detection: Fluorescence (ex: 420 nm; em: 500 nm) Derivatization: Precolumn with NDA		97
Corn	Methanol/water (80:20)	Immunoaffinity column chromatography	*HPLC* Column: Reversed phase C_{18} (5 μm, 25 cm × 4.6 mm) Mobile phase: 55:45:1 acetonitrile/water/acetic acid Detection: Fluorescence (ex: 420 nm; em: 470 nm) Derivatization: Precolumn with NDA-NaCN	4–10	103

Table 2 Continued

Analytical matrix	Extraction solvent	Cleanup	Chromatographic conditions	Detection limit (µg/kg)	Ref.
Corn	Acetonitrile/water (50:50)	C18 and strong anion-exchange SPE cartridges in tandem	*HPLC* Column: Reversed phase C18 (5 µm, 30 cm × 4.7 mm) Mobile phase: 50:50 acetonitrile/phosphate buffer (pH 2.5) Detection: Fluorescence (ex: 336 nm; em: 460 nm) Derivatization: Postcolumn with OPA/N-acetyl-L-cysteine	20 ng	109
Aqueous solution			*Gradient HPLC* Column: Base-deactivated C8 (5 µm, 25 cm × 4.6 mm) Mobile phase: (a) 5:95:0.025 acetonitrile/water/TFA (b) 90:10:0.025 acetonitrile/water/TFA Detection: Evaporative light scattering		110
Corn	Acetonitrile/water (50:50)	Strong anion-exchange cartridge	*GC-MS of hydrolyzed fumonisin* Column: DB5 fused-silica capillary (0.25 µm, 30 m) Temp. profile: 80°C for 1 min; 20°C/min to 200°C; 5°C/min to 270°C Detection: Electron capture negative chemical ionization Derivatization: TFA	10	94

column derivatization method with OPA following ion pair chromatography. The reaction was carried out at 40°C in a 5-m stainless steel reaction coil (0.25 mm i.d.). HPLC detection of fumonisins without derivatization has been achieved with evaporative light-scattering detection (110) after separation on a base-deactivated C_8 column with a gradient of trifluoroacetic acid buffer (pH 2.7) and acetonitrile.

C. Trichothecenes

The trichothecenes are a very large group of related sesquiterpenoids comprising approximately 150 compounds [42] that are produced by various fungal genera, notably species of *Fusarium, Stachybotrys, Myrothecium,* and *Trichothecium* [111]. The majority of these compounds have been detected in fungal cultures, but a few have been found naturally occurring in foodstuffs [42]. The most commonly found and analyzed for is deoxynivalenol (DON; vomitoxin), which is a common contaminant of cereal grains. Other important members of the group that naturally occur in grains include T-2 toxin, HT-2 toxin, nivalenol (NIV), and diacetoxyscirpenol (DAS).

Detection of trichothecenes is most often carried out with GC, although several TLC, HPLC, and minicolumn methods have been reported. Because of its widespread natural occurrence in grains (particularly wheat) [112], many methods have been reported specifically for DON, and two have been collaboratively studied and adopted by AOAC International [28,113]. One is a TLC method using silica gel plates with a mobile phase of 8:1:1 chloroform/acetone/isopropanol. Developed plates are sprayed with $AlCl_3$ (or plates can be predipped in $AlCl_3$) and DON appears as a blue fluorescent spot under longwave UV light [28]. The other is a GC method in which DON is derivatized with heptafluorobutyric anhydride and detected by electron capture [113]. An example of using HPLC for determination of DON in milk is the method of Vudathala et al. [114], which used a C_{18} column eluted with 4% acetonitrile in water and UV detection at 220 nm. The detection limit was 5 ng/ml of milk.

Numerous methods have been published for the determination of multiple trichothecenes. Furlong and Soares [111] separated seven trichothecenes (T-2 toxin, HT-2 toxin, T-2 tetraol, T-2 triol, DON, NIV, and DAS) on a capillary column and used flame ionization detection of HFB derivatives with limits of detection ranging from 0.1 to 0.5 μg/g. Kang et al. [115] used GC-MS in the selected ion mode to detect TMS derivatives of four naturally occurring trichothecenes in corn. Selected examples of chromatographic conditions used in trichothecene analysis are presented in Table 3. Scott [42] has provided a thorough review of the analysis of trichothecenes by GC.

D. Zearalenone

Zearalenone is an estrogenic mycotoxin produced by species of *Fusarium* and often occurs with trichothecenes in cereal crops [116,117]. Chromatographic

Table 3 Selected Chromatographic Methods for the Determination of Trichothecenes

Toxins detected	Analytical matrix	Extraction solvent	Cleanup	Chromatographic conditions	Detection limit (μg/kg)	Ref.
DON	Corn, wheat	Acetonitrile/water (85:25)	Preparative minicolumn chromatography	*Minicolumn* (SAM-DON, Rialdon Diagnostics, College Station, TX) Elution solvent: 95:5 Toluene/acetone Detection: Blue florescent band	≥500	176
DON	Wheat	Acetonitrile/water (84:16)	Column chromatography	*TLC* Plate: Silica gel 60 Solvent: 80:10:10 chloroform/acetone/ isopropanol Detection: Visualization or densitometry under longwave UV light Derivatization: Spray with aluminum chloride solution; heat 7 min at 120°C	300	28
DON, NIV, DAS, T-2	Cereals, rice foods	Chloroform/ethanol (80:20)	Charcoal-alumina-Celite column chromatography	*2-Dimensional TLC* Plate: Silica gel 60 Solvent: (a) 98:2 chloroform/methanol (b) 70:30 toluene/ethanol Detection: Visualization under longwave UV light Derivatization: Spray with ethanolic aluminum chloride	37–100	177
DON, NIV	Cereals	Acetonitrile/water (85:15)	Cation exchange chromatography; minicolumn chromatography	*HPLC* Column: Reversed phase C_{18} (5 μm, 25 cm × 4 mm) Mobile phase: 86:14 water/methanol Detection: UV (222 nm)	15–50	178

DON, NIV, fusarenon-X	Corn, wheat, barley	Acetonitrile/water (3:1)	Liquid-liquid partition, Florisil column chromatography, and CN SPE cartridge	*HPLC* Column: Reversed phase C_{18} (10 µm, 25 cm × 4 mm) Mobile phase: 85:15 water/acetonitrile Detection: Fluorescence (ex: 370 nm; em: 460 nm) Derivatization: Postcolumn with methyl acetoacetate and ammonium acetate	20–50	179
DON	Milk		Precipitation; centrifugation; liquid-liquid extraction; column chromatography	*HPLC* Column: Reversed phase C_{18} (5 µm, 25 cm × 4.6 mm) Mobile phase: 96:4 water/acetonitrile Detection: UV (220 nm)	5	114
DON, NIV	Cereals	Methanol/water (70:30)	Liquid-liquid partition and silica gel column chromatography	*GC* Column: 3% OV-3 on 80-100 mesh Chromosorb W (183 cm × 2 mm glass) or DB-5 capillary (0.25 µm film, 30 m × 0.32 mm) Detection: Electron capture Derivatization: TMS		180
DON	Wheat	Chloroform/ethanol/ water	Silica gel H column chromatography	*GC* Column: 3% OV-101 on 80-100 mesh Chromosorb W (183 cm × 2 mm glass) Carrier gas: 5:95 Methane-argon Detection: Electron capture Derivatization: HFB	10–29	113

Table 3 Continued

Toxins detected	Analytical matrix	Extraction solvent	Cleanup	Chromatographic conditions	Detection limit (μg/kg)	Ref.
DAS, NIV, DON, T-2 triol, T-2 tetraol	Wheat	Methanol/4% aqueous KCl (90:10)	Precipitation, liquid-liquid partition, column chromatography	GC Column: BD-255 capillary (0.25 μm film, 15 m × 0.33 mm) Carrier gas: Hydrogen Detection: Flame ionization Derivatization: HFB Temp. profile: 100°C for 1 min; 30°C/min to 155°C; 5°C/min to 220°C	100–500	111
DAS, T-2, NMA	Grains	Methanol/water (50:50)	Partition chromatography and silica SPE cartridge	GC Column: SE-30 capillary (0.25 μm film, 25 m × 0.32 mm) Carrier gas: Helium Detection: Electron capture Derivatization: HFB Temp. profile: 180–220°C at 4°C/min; 15°C/min to 250°C; 250°C for 5 min	200	181

Multi	Cereals	Ethyl acetate/acetonitrile/water	Liquid-liquid partition and Florisil SPE cartridge	*GC* Column: OV-1701 capillary (0.21 μm film, 25 m × 0.32 mm) Carrier gas: Helium Detection: Electron capture Derivatization: TMS Temp. profile: 60°C for 2 min; 40°C/min to 150°C; 5°C/min to 250°C; hold for 10 min	182
Multi	Corn	Acetonitrile/water (3:1)	Liquid-liquid partition and Florisil column chromatography	*GC-MS* Column: DB-5 fused silica capillary (0.25 μm film, 30 m × 0.25 mm) Carrier gas: Helium Detection: Mass spectrometry with selected ion monitoring Derivatization: TMS Temp. profile: 120°C for 5 min; 5°C/min to 270°C	115
Multi	Grains	Methanol	Liquid-liquid partition and silica SPE column chromatography	*GC-MS* Column: DB-5 fused silica capillary (0.25 μm film, 30 m × 0.25 mm) Carrier gas: Helium Detection: Negative ion chemical ionization mass spectrometry Derivatization: HFB Temp. profile: 150°C for 1 min; 10°C/min to 250°C; 25°C/min to 300°C; hold for 10 min	183

Abbreviations: DON, deoxynivalenol; NIV, nivalenol; DAS, diacetoxyscirpenol; NMA, neosolaniol monoacetate; TMS, trimethylsilylation; HFB, heptafluorobutyrylation.

methods for analysis of zearalenone include TLC, HPLC, and GC, but HPLC seems to be the method of choice if available. Table 4 includes examples of each type. TLC is usually carried out on silica gel plates using a variety of mobile phases. Zearalenone can be visualized as a greenish blue spot under shortwave UV light, or plates can be sprayed with various reagents to enhance fluorescence or produce colored reaction products. Aluminum chloride spray reagent enhanced the fluorescence of zearalenone [118], while reaction with Fast Violet B salt solution produced pink spots in visible light [119]. HPLC of zearalenone is most often accomplished in the reversed phase mode with a C_{18} column and mobile phases primarily composed of combinations of water, methanol, and acetonitrile [114,120–124]. One normal phase method uses a silica gel column and a mobile phase of water-saturated chloroform with cyclohexane, acetonitrile, and ethanol [125]. Zearalenone is usually detected by fluorescence with excitation at either 236 or 285 nm and emission at 418 or 440 nm. Another method uses postcolumn derivatization with aluminum chloride to enhance the response of zearalenone [123]. Although HPLC is the preferred method for zearalenone analysis, many GC methods have also been published and are quite often used when analyzing for zearalenone in combination with the trichothecenes. Therefore, several of the methods reported for the trichothecenes can also be used for zearalenone analysis. Most methods use TMS for derivatization usually coupled with flame ionization detection (42).

E. Ochratoxins

Ochratoxins (A, B, and C) are mycotoxins produced by several species of *Aspergillus* (particularly *A. ochraceus*) and *Penicillium verrucosum* [126,127]. Ochratoxin A is an important nephrotoxin that is found naturally as a contaminant of corn [128] and small grains, particularly wheat and barley [112]. One minicolumn screening method has been reported [16], but most analysis of ochratoxins is done exclusively with TLC or HPLC (Table 5). Methods using each technique have been collaboratively studied and adopted by AOAC International [129–132]. TLC of ochratoxin A is usually carried out on silica gel plates with an acidified mobile phase or on oxalic acid-treated silica gel plates with a neutral solvent mixture [129,131,133,134]. Detection of a green or blue-green spot under long-wave UV light can be accomplished visually or with a densitometer. Frohlich et al. [135] used reversed phase TLC as a cleanup for determination of ochratoxin A by either HPLC or direct spectrofluorometric measurement. Most HPLC methods use the reversed phase mode with C_{18}, C_8, or C_{22} columns and mobile phases consisting of acidified water in combination with acetonitrile or methanol. Detection is usually by fluorescence with an excitation wavelength around 330 or 365 nm and emission wavelengths in the 460–470 range. A recently published method [136] that determined both ochratoxin A and citrinin used a C_{18} column with a mobile phase of 70:30 methanol/water containing tetrabutylammonium hydroxide. Post-column reaction with a butanolic solution of terbium, trioctylphosphine oxide, and triethylamine was performed prior to time-resolved luminescence (TRL) detection. Compared with UV and fluorescence detection, the TRL detection method resulted in a significant increase in selectivity for the two mycotoxins [136]. Another postcolumn derivatization method mixed the column effluent with a 25% sodium hydroxide solution [137]. Reaction of ochratoxin A with ammonia resulted in a sixfold increase in the fluorescence signal using excitation and emission wavelengths of 390 and 440 nm, respectively.

F. Patulin

Patulin is a β-unsaturated lactone produced by species of *Aspergillus* and *Penicillium*, particularly *P. expansum* [138]. Production of patulin in rotting apples can lead to quite high levels in apple juice or cider, but natural intoxications involving patulin have not been substantiated [139]. Several TLC and GC methods have been used for patulin analysis, and these have been thoroughly reviewed [42,117]. However, HPLC now appears to be the most popular analytical tool for patulin analysis. Brause et al. [140] recently reported a collaborative study of a reversed phase HPLC method that has been adopted as an official AOAC International method. The method specifies either of two C_{18} columns (15 or 25 cm × 3.9 mm, 5 μm) and a mobile phase of 100:0.8 water/tetrahydrofuran (THF) or water with various amounts of acetonitrile or THF with pH adjusted to 4 with acetic acid. UV detection was carried out at 276 nm (deuterium or xenon lamps) or 254 nm (mercury lamps).

TLC methods generally rely on silica gel plates developed with a variety of solvent systems [117]. Scott and Kennedy [141] developed plates in 5:4:1 toluene/ethyl acetate/90% formic acid and used 3-methyl-2-benzothiazolinone hydrazone hydrochloride spray followed by heat to produce a yellow-brown spot under longwave UV light. This method was subjected to collaborative study [142] and adopted by AOAC as an official method. Other representative chromatographic procedures for patulin analysis appear in Table 6.

G. Sterigmatocystins

Sterigmatocystins are a group of related fungal metabolites of which sterigmatocystin is the most economically important [143]. Sterigmatocystin is produced by several species of *Aspergillus* and *Emericella* as well as other fungal genera [127]. TLC has been carried out primarily on silica gel plates with a variety of mobile phases [117]. The AOAC International official TLC method prescribes a solvent system of 90:5:5 benzene/methanol/acetic acid with detection of sterigmatocystin as a bright yellow fluorescent spot under shortwave UV light [144]. It can also be detected as a brick-red spot under longwave (365 nm) UV light, and

Table 4 Selected Chromatographic Methods for the Determination of Zearalenone

Analytical matrix	Extraction solvent	Cleanup	Chromatographic conditions	Detection limit (µg/kg)	Ref.
Corn; wheat; sorghum	Methanol/water (80:20)	Liquid-liquid partition	*Minicolumn* Packing: Layers of Florisil and alumina Elution solvent: 90:10 hexane/acetone Detection: Blue fluorescent band at the Florisil/alumina interface when viewed under longwave UV light	35	16
Corn	Chloroform/water (10:1)	Silica gel column chromatography and liquid-liquid partition	*TLC* Plate: Silica gel GHR Solvent: 95:5 chloroform/alcohol Derivatization: Spray plate with 20% aluminum chloride solution and heat 5 min at 130°C Detection: Visualization as greenish-blue fluorescent spot under shortwave UV light before spraying and blue fluorescent spot under longwave UV after spraying	<300	184
Corn	Acetonitrile/4% KCl (90:10)	Precipitation and liquid-liquid partition	*TLC* Plate: Silica gel 60 aluminum sheet Solvent: 90:10 chloroform/acetone Detection: Visualization of fluorescent spot under 254 nm UV light	50	185
Cereals	Methanol/water (75:25)	Precipitation and liquid-liquid partition	*HPTLC* Plate: Whatman LHP-K Solvent: 95:5 chloroform/ethanol Derivatization: Spray plate with Fast Violet B, borate buffer, and sulfuric acid solution; heat Detection: Visualization as violet spots	80	186

Matrix	Extraction	Cleanup	Method		Ref.
Cereals	Acetonitrile/water (3:1)	Florisil column chromatography	*HPLC* Column: Silica gel (Nucleosil 50-10, 30 cm × 4 mm) Mobile phase: 50:15:2:1 water-saturated chloroform/cyclohexane/acetonitrile/ethanol Detection: Fluorescence (ex: 276 nm; em: 460 nm)	2.0	125
Corn	Chloroform	Liquid-liquid partition	*HPLC* Column: Reversed phase C_8 or C_{18} Mobile phase: 1.0:1.6:2.0 methanol/acetonitrile/water Detection: Fluorescence (ex: 236; em: 418 nm)	>50	122
Corn	Acetonitrile/water (3:1)	Liquid-liquid partition and Florisil column chromatography	*HPLC* Column: Reversed phase C_{18} (5 µm, 15 cm × 4.6 mm) Mobile phase: 70:30 methanol/water Detection: Fluorescence (ex: 236 nm; em: 418 nm)	2.0	115
Barley	Methanol/water (60:40)	Liquid-liquid partition and piperidinohydroxypropyl-Sephadex LH-20 column chromatography	*HPLC* Column: Reversed phase C_{18} (5 µm, 25 cm × 4.6 mm) Mobile phase: 11:10 acetonitrile/0.3% sodium acetate (pH 3.2) Detection: Fluorescence (ex: 274 nm; em: 446 nm)	1.0	124
Cereals	Acetonitrile/4% KCl (90:10)	Liquid-liquid partition	*HPLC* Column: Reversed phase C_{18} (25 cm × 4.6 mm) Mobile phase: 80:20 methanol/water Detection: Fluorescence (ex: 285 nm; em: 440 nm) Derivatization: Postcolumn with 0.25M aluminum chloride in 3:1 methanol/water		123
Cereals	Ethyl acetate	Liquid-liquid partition	*GC-MS* Column: Fused-silica capillary (0.28 µm film, 25 m × 0.25 mm) Carrier gas: Helium Temp. profile: 180°C for 2 min; 4.5°C/min to 270°C; 15°C/min to 285°C Derivatization: TMS Detection: Electron impact ion-trap mass spectrometry	1.0	187

Table 5 Selected Chromatographic Methods for the Determination of Ochratoxins

Analytical matrix	Extraction solvent	Cleanup	Chromatographic conditions	Detection limit (μg/kg)	Ref.
Wheat; barley; corn	Methanol/water (80:20)	Liquid-liquid partition	*Minicolumn* Packing: Florisil Elution solvent: methanol Detection: Add 0.25 N H_2SO_4 to minicolumn; view as a blue fluorescent band under longwave UV light	20	16
Barley	Chloroform/0.1 M phosphoric acid (10:1)	Aqueous sodium bicarbonate-diatomaceous earth column	*TLC* Plate: Silica gel GHR Solvent: 18:1:1 benzene/methanol/acetic acid Detection: Visualization or densitometry under longwave (ochratoxin A) and shortwave (ochratoxin B) UV light	12	129,130
Green coffee	Chloroform	Basic diatomaceous earth column chromatography	*TLC* Plate: Silica gel GHR Solvent: 5:4:1 toluene/ethanol/formic acid Detection: Visualization of greenish-blue fluorescent spot under longwave UV light Confirmation: Spray with alcoholic $NaHCO_3$ or alcoholic $AlCl_3$; spot becomes blue and more intensively fluorescent		131
Corn, barley	Chloroform/0.1 M phosphoric acid (10:1)	Liquid-liquid partition and C_{18} column chromatography	*HPLC* Column: Reversed phase C_{18} (5 μm, 25 cm × 4.6 mm)		132

Barley, wheat	Chloroform/0.1 M phosphoric acid (12:1)	Reversed phase TLC Plate: Whatman KC-18 (200 μm, 5 × 10 cm) Solvent: (a) hexane (b) 70:30 methanol/water Detection: Visualization under longwave UV light; elute blue fluorescent spot at Rf 0.76 for quantitative analysis	Mobile phase: 99:99:2 water/acetonitrile/acetic acid Detection: Fluorescence (ex: 333 nm; em: 460 nm) 1. *Spectrofluorometry* (ex: 340 nm; em: 429 nm) 2. HPLC Column: Reverse phase C_{18} (5 μm, 25 cm × 4.6 mm) Mobile phase: 70:30 methanol/water Detection: Fluorescence (ex: 333 nm; em: 418 nm)	100–200	135
Cheese	Methanol		*HPLC* Column: Reversed phase C_{18} (5 μM, 25 cm × 4.6 mm) Mobile phase: 70:30 methanol/water with tetrabutylammonium hydroxide (pH 5.5 with HCl) Detection: Time-resolved luminescence (ex: 331 nm; em: 545 nm) Derivatization: Postcolumn chelation with terbium		136
Various	Immunoaffinity column	Chloroform	*HPLC* Column: Spherisorb ODS-1 (5 μm, 25 cm × 4.6 mm) Mobile phase: 18:7 methanol/9% glacial acetic acid in water Detection: Fluorescence (ex: 390 nm; em: 440 nm) Derivatization: Postcolumn with ammonia		137

Table 6 Selected Chromatographic Methods for the Determination of Patulin

Analytical matrix	Extraction solvent	Cleanup	Chromatographic conditions	Detection limit (µg/kg)	Ref.
Apple juice	Ethyl acetate	Silica gel column chromatography	*TLC* Plate: Adsorbosil-5 Solvent: 5:4:1 toluene/ethyl acetate/90% formic acid Derivatization: Spray plate with 0.5% 3-methyl-2-benzothiazolinone hydrazone hydrochloride solution; heat 15 min at 130°C Detection: Yellow brown fluorescent spot under longwave UV light or yellow spot in visible light		142
Apple juice	Ethyl acetate	Liquid-liquid partition	*HPLC* Column: Reversed phase C$_{18}$ (5 µm, 25 cm × 3.9 mm) Mobile phase: 0.8% tetrahydrofuran in water Detection: UV (276 nm)		140
Apple juice	Ethyl acetate (with diphasic dialysis)	Silica gel SPE cartridge	*HPLC* Column: Reversed phase C$_{18}$ (15 cm × 3.9 mm) Mobile phase: 99:1 water/tetrahydrofuran Detection: UV (275 nm)	1.0	188
Apple juice	Ethyl acetate	Liquid-liquid partition	*HPLC* Column: Reversed phase C$_{18}$ (5 µm, 25 cm × 3 mm) Mobile phase: Distilled water Detection: UV (276 nm)	5.0	189

visualization can be increased by spraying with aluminum chloride followed by heat to produce a yellow fluorescent spot [29,145]. Several HPLC methods for sterigmatocystin have been published, but because it is only weakly fluorescent, UV has been the primary choice for detection [146–148]. However, interferences with UV detection in barley extracts prompted Abramson and Thorsteinson [149] to use fluorescence detection (excitation: 256 nm; emission filter: 418 nm) of a precolumn acetate derivative. Extracts were heated with pyridine and acetic anhydride for 3 h at 100°C to produce a stable derivative that was separated on a C_{18} column with a water/methanol gradient. Neely and Emerson [150] used postcolumn derivatization with aluminum chloride and fluorescence detection to enhance selectivity and sensitivity compared with UV detection. They used a C_{18} column and a methanol/water mobile phase, and the postcolumn method was easily automated for routine analysis. Although not often used for sterigmatocystin analysis, GC methods have been reported [42]. Salhab et al. [151] used GC-MS with SIM to detect down to 5 ppb of sterigmatocystin in grains. Selected chromatographic conditions for sterigmatocystin analysis are summarized in Table 7.

H. *Alternaria* Toxins

Many species of *Alternaria* cause decay of fruits and vegetables, and some produce a variety of metabolites that are toxic to animals. They include alternariol (AOH), alternariol methyl ether (AME), altenuene (ALT), altertoxins (ATX) I and II, tenuazonic acid (TA), and AAL toxins [112,152]. Grabarkiewicz-Szczesna et al. [153] used silica gel and solvent systems consisting of toluene/ethyl acetate/ formic acid (6:3:1) or chloroform/ethanol/ethyl acetate (90:5:5) to separate all but the AAL toxins. TA was detected by the quenching of fluorescence under short-wave (254 nm) UV light. After spraying with 20% ethanolic aluminum chloride, ATX showed yellow-orange fluorescence under longwave UV (365 nm), while AOH, AME, and ALT were violet-blue. HPLC methods for this group have been reviewed, and it was noted that TA is more difficult to analyze for than other members of the group because, as an acid and strong metal chelator, it tends to give very broad, tailing peaks [127]. However, Ozcelik et al. [154] achieved good results for TA, AOH, AME, and ALT using normal phase HPLC with a silica gel column and a mobile phase of 95:5 chloroform/methanol and UV detection at 280 nm. Stack et al. [155] used a C_{18} column eluted with methanol/water (85:15) containing 300 mg zinc sulfate per liter to separate TA, AOH, and AME. Although all toxins could be detected by UV at 280 nm, a fluorescence detector was connected in series to achieve enhanced detection of AOH and AME.

The AAL toxins are produced by *Alternaria alternata* f. sp. *lycopersici* and are important in the development of *Alternaria* stem canker disease of tomato [156]. They are of particular interest at present because of their structural sim-

Table 7 Selected Chromatographic Methods for the Determination of Sterigmatocystin

Analytical matrix	Extraction solvent	Cleanup	Chromatographic conditions	Detection limit (μg/kg)	Ref.
Barley wheat	Acetonitrile/4% KCl (9:1)	Liquid-liquid partition and silica gel column chromatography	*TLC* Plate: Adsorbosil-1 Solvent: 90:5:5 benzene/methanol/acetic acid Detection: Visualization as bright yellow fluorescent spot under shortwave UV light Derivatization: Spray plate with 20% aluminum chloride solution and heat 10 min at 80°C		144
Cheese	Acetonitrile/4% KCl (85:15)	Liquid-liquid partition and cupric carbonate-diatomaceous earth (1:2) column chromatography	*TLC* Plate: Precoated Sil G-25 HR Solvent: 85:10:5 benzene/methanol/acetic acid Detection: Visualization or densitometry under longwave UV light Derivatization: Spray plate with 15% aluminum chloride solution and heat 5 min at 110°C; spray with 18:82 silicone-anhydrous ether and heat 3 min at 110°C	2.0–5.0	29

Barley	Acetonitrile/4% KCl (90:10)	Liquid-liquid partition and silica gel column chromatography	*Gradient HPLC* Column: Reversed phase C_{18} (10 μm, 25 cm × 4 mm) Mobile phase: (a) 50:50 methanol/water (b) methanol Detection: Fluorescence (ex: 256 nm; em: 418 nm) Derivatization: Precolumn with pyridine/acetic anhydride		149
Fermentation broths	Methanol	Filtration	*HPLC* Column: Reversed phase C_{18} (5 μm, 25 cm × 4.6 mm) Mobile phase: 88:12 methanol/water Detection: Fluorescence (ex: 254 nm; em: 455 nm) Derivatization: Postcolumn with aluminum chloride	90	150
Grains	Acetonitrile/4% KCl (90:10)	Liquid-liquid partition and column chromatography	*GC-MS* Column: 3% OV-17 on Chromosorb G packed glass (25 cm × 2 mm) Detection: Mass spectrometry with SIM at m/e 295, 306, 324 Derivatization: None	5	151

Table 8 Selected Chromatographic Methods for the Determination of *Alternaria* Toxins

Toxins detected	Analytical matrix	Extraction solvent	Cleanup	Chromatographic conditions	Detection limit (μg/kg)	Ref.
AOH, AME, ALT, ATX-I, TA	Olives	Methanol/water/hexane/HCl (45:40:40:1)	Centrifugation and liquid-liquid partition	*1. Two-Dimensional TLC* Plate: Silica gel with fluorescence indicator Solvent: (a) 88:12 chloroform/acetone (b) 60:30:10 toluene/ethyl acetate/formic acid Detection: Visualization under longwave and shortwave UV light *2. HPLC* Column: Reversed phase C_{18} (7 μm, 25 cm × 4 mm) Mobile phase: 80:20 methanol/water plus 300 mg $ZnSO_4 \cdot 7H_2O$/L or 60:40 methanol/water Detection: UV at 280 nm or 257 nm	30–200	190
AOH, AME, ALT, TA	Tomatoes and apples	Chloroform/ethanol (80:20; pH 2)	Centrifugation and filtration of organic phase	*HPLC* Column: Silica (7.5 μm, 30 cm × 5 mm) Mobile phase: 95:5 chloroform/methanol Detection: UV (280 nm)		154
AOH, AME, ALT	Fruits and vegetables	Chloroform	Dry extract over sodium sulfate and concentrate	*HPLC* Column: Reversed phase C_{18} (10 μm, 30 cm × 3.9 mm) Mobile phase: 65:35 acetone/water Detection: UV (324 nm)		191

Toxin	Matrix	Extraction	Cleanup	Method and conditions	Detection limit	Ref.
AME, TA	Tomatoes and tomato products	Chloroform	Silica gel column chromatography	HPLC Column: Reversed phase C_{18} Mobile phase: 85:15 methanol/water with 300 mg $ZnSO_4 \cdot 7H_2O$/L Detection: UV (280 nm) for TA; Fluorescence (ex: 278 nm; em: 389 nm) for AME	3–25	155
ATX-I, ATX-II	Corn, rice, tomatoes	Methanol	Precipitation and liquid-liquid partition	HPLC Column: Reversed phase C_{18} (5 μm, 12.5 cm × 4 mm) Mobile phase: 60:40 methanol/water plus 0.1 M sodium nitrate and 1 mM HNO_3 Detection: Electrochemical with dual in-series electrodes	<100	192
AOH, AME, ALT, ATX-I, ATX-II	Corn, rice, tomatoes	Methanol	Precipitation and liquid-liquid partition	Gradient HPLC Column: Reversed phase C_{18} (5 μm, 12.5 cm × 4.6 mm) Mobile phase: (a) Methanol; (b) H_3PO_4-Acidified water (pH 3) Gradient: 50 to 70% (a) in 10 min; 70 to 85% (a) in 0.1 min; isocratic for 8 min Detection: Photodiode array	<100	193
AAL Toxins	Corn cultures	Water	C_{18} SPE cartridge	HPLC Column: Reversed phase C_{18} (5 μm; 25 cm × 4.6 mm) Mobile phase: 70:30 methanol/0.1 M NaH_2PO_4 (pH 6.1) Detection: Fluorescence (ex: 335 nm; em: 440 nm) Derivatization: Precolumn with OPA/mercaptoethanol	2 ng/inj.	157

AOH = Alternariol; AME = Alternariol methyl ether; ALT = Altenuene; ATX-1 = Altertoxin I; TA = Tenuazonic acid; ALT-II = Altertoxin II.

Table 9 Chromatographic Methods for the Determination of Cyclopiazonic Acid

Analytical matrix	Extraction solvent	Cleanup	Chromatographic conditions	Detection limit (μg/kg)	Ref.
Corn; peanuts	Acidic chloroform/ methanol (80:20)	Liquid-liquid partition	*TLC* Plate: Silica gel 60 F-254 HPTLC Solvent: 50:15:10 ethyl acetate/2-propanol/ammonium hydroxide Derivatization: Spray with 1% *p*-dimethylaminobenzaldehyde in 75:25 ethanol-HCl Detection: Reflection densitometry at 540 nm	125	159
Milk; eggs	Acidic chloroform/ methanol (80:20)	Liquid-liquid partition	*TLC* Plate: Silica gel 60 F-254 HPTLC Solvent: 85:15:10 ethyl acetate/methanol/ammonium hydroxide Derivatization: Spray with 1% *p*-dimethylaminobenzaldehyde in ethanol followed by 50% ethanolic sulfuric acid Detection: Bluish-purple spot in visible light at R_f 0.36	5	160
Corn; peanut meal; rice	Chloroform/85% phosphoric acid (100:1)	Silica SPE cartridge	*HPLC* Column: Silica (10 cm × 4 mm) Mobile phase: 55:20:5 ethyl acetate/2-propanol/25% aqueous ammonia Detection: UV (284 nm)	0.2 ng	162
Corn; peanuts	Methanol/2% NaHCO$_3$ (70:30)	Liquid-liquid partition and silica SPE cartridge	*Gradient HPLC* Column: Reversed phase C$_{18}$ (5 μm, 15 cm × 3.9 mm) Mobile phase: (a) 85:15 methanol/water (b) 85:15 Methanol/water with 4 mM ZnSO$_4$·7H$_2$O Gradient: 100% (a) to 100% (b) in 10 min; 100% (b) for 10 min Detection: UV (279 nm)	50–100	163

Table 10 Selected Chromatographic Methods for the Determination of Citrinin

Analytical matrix	Extraction solvent	Cleanup	Chromatographic conditions	Detection limit (µg/kg)	Ref.
Corn; barley	Acetonitrile/10% glycolic acid solution (90:10)	Liquid-liquid partition	*TLC* Plate: Silica gel G-25HR impregnated with glycolic acid. Solvent: (a) 70:90:40:2 ethyl ether/hexane/ethyl acetate/90% formic acid (b) 70:50:50:20 toluene/ethyl acetate/chloroform/90% formic acid (c) 50:100:50 ethyl ether/hexane/ethyl acetate; Develop a different plate with each solvent. Detection: Visualization of fluorescent spot under 254 nm and 366 nm UV light before and after spraying with aluminum chloride or treating with NH_3 vapor	15–20	165
Cereals	Chloroform/0.1 M phosphoric acid (75:10)	Partition column chromatography	*HPLC* Column: Silica buffered with 0.2 M citric acid (5 µm, 30 cm × 4.6 mm) Mobile phase: 60:40 hexane/chloroform Detection: Fluorescence (ex: 360 nm; em: 500 nm)	0.1	166
Fungal cultures; cheese			*HPLC* Column: Reversed phase C_{18} (5 µm, 25 cm × 4.6 mm) Mobile phase: methanol with 5.7×10^{-4} M tetrabutylammonium hydroxide (adjust pH to 5.5 with 1 M HCl) Detection: Fluorescence (ex: 331 nm; em: 500 nm) Fluorescence enhancement: Postcolumn acidification with 1 M HCl	0.45 ng	167

ilarity to the fumonisins and the fact that they have been shown to be toxic to certain cultured mammalian cell lines [157]. Methods used for fumonisin analysis can be tailored for AAL toxin analysis. An example is the method of Shephard et al. [157] in which precolumn derivatization with OPA was coupled with reversed phase HPLC and fluorescence detection to detect the toxins in corn cultures.

I. Cyclopiazonic Acid

Cyclopiazonic acid (CPA) is an indole tetramic acid produced by several species of *Aspergillus* and *Penicillium* [158]. Chromatographic analysis has been carried out primarily by TLC and HPLC (Table 9). TLC on silica gel plates results in a spot that streaks badly; but this can be overcome by using plates that have been treated with oxalic acid [134], or ammonium hydroxide can be added to the solvent system to avoid using oxalic acid-treated plates [159,160]. Visualization of CPA is accomplished by spraying with Erlich's reagent (p-dimethylaminobenz-aldehyde in ethanol and HCl) to produce a blue spot. Reversed phase HPLC of CPA has been reported using C_8 or C_{18} columns and a mobile phase consisting of acetonitrile, isopropyl alcohol, and water containing ammonium acetate, 4-do-decyldiethylenetriamine, and zinc acetate [161]. The limit of detection with UV at 284 nm was approximately 4 ng of pure CPA. Goto et al. [162] detected 0.2 ng of pure CPA with a normal phase system consisting of a silica gel column and a mobile phase of 55:20:5 ethyl acetate/2-propanol/25% aqueous ammonia, also with UV detection at 284 nm. Urano et al. [163] achieved good separation of CPA in peanut and corn extracts on a C_{18} column with a linear gradient of 0–4 mM zinc sulfate in methanol/water (60:40), but the limit of quantitation was 20 ng.

J. Citrinin

Citrinin is an antibiotic and nephrotoxin produced by many species of *Penicillium* and *Aspergillus* [164]. TLC of citrinin is usually carried out on oxalic acid-impregnated silica gel plates with a variety of solvent systems. It can be visualized as a yellow fluorescent spot or sprayed with reagents to produce various colors [117]. Gimeno [165] found that citrinin produced a more compact spot on glycolic acid–impregnated plates than on oxalic acid plates, thus increasing sensitivity to 15–20 μg/kg. Greater sensitivity has been achieved in recently published HPLC methods. Zimmerli et al. [166] used an acid-buffered silica gel column with fluorescence detection to detect as little as 0.1 ng of citrinin per injection (Table 10). Franco et al. [167] achieved enhanced detection of citrinin in reversed phase HPLC with postcolumn addition of 1 M hydrochloric acid to the column effluent. As mentioned in the discussion of ochratoxin A methods, the method of time-resolved luminescence detection of the terbium chelate was also applicable to citrinin [136].

REFERENCES

1. R. P. Sharma and D. K. Salunkhe, Introduction to mycotoxins, *Mycotoxins and Phytoalexins* (R. P. Sharma and D. K. Salunkhe, eds.), CRC Press, Boca Raton, FL, 1991, p. 3.
2. R. J. Cole, H. G. Cutler, and J. W. Dorner, Biological screening methods for mycotoxins and toxigenic fungi, *Modern Methods in the Analysis and Structural Elucidation of Mycotoxins* (R. J. Cole, ed.), Academic Press, Orlando, FL, 1986, p. 1.
3. T. B. Whitaker, J. W. Dickens, and R. J. Monroe, Variability of aflatoxin test results, *J. Amer. Oil Chem. Soc. 51*: 214 (1974).
4. J. W. Dickens and T. B. Whitaker, Sampling and sample preparation methods for mycotoxin analysis, *Modern Methods in the Analysis and Structural Elucidation of Mycotoxins* (R. J. Cole, ed.), Academic Press, Orlando, FL, 1986, p. 29.
5. J. W. Dorner and R. J. Cole, Variability among peanut subsamples prepared for aflatoxin analysis with four mills, *J. AOAC Int. 76*: 983 (1993).
6. J. W. Dickens and J. B. Satterwhite, Subsampling mill for peanut kernels, *Food Technol. 23*: 90 (1969).
7. *Aflatoxin Handbook*, USDA, Federal Grain Inspection Service, Washington, D.C., 1990, p. 4-1.
8. T. B. Whitaker and D. L. Park, Problems associated with accurately measuring aflatoxin in food and feeds: errors associated with sampling, sample preparation, and analysis, *The Toxicology of Aflatoxins* (D. L. Eaton and J. D. Groopman, eds.), Academic Press, San Diego, CA, 1994, p. 433.
9. T. B. Whitaker, F. E. Dowell, W. M. Hagler, Jr., F. G. Giesbrecht, and J. Wu, Variability associated with sampling, sample preparation, and chemical testing for aflatoxin in farmers' stock peanuts, *J. AOAC Int. 77*: 107 (1994).
10. C. E. Holaday, Rapid method for detecting aflatoxins in peanuts, *J. Amer. Oil Chem. Soc. 45*: 680 (1968).
11. J. Velasco, Detection of aflatoxin using small columns of florisil, *J. Amer. Oil Chem. Soc. 49*: 141 (1972).
12. C. E. Holaday and J. A. Lansden, Rapid screening method for aflatoxin in a number of products, *J. Agric. Food Chem. 23*: 1134 (1975).
13. T. R. Romer, Screening method for the detection of aflatoxins in mixed feeds and other agricultural commodities with subsequent confirmation and quantitative measurement of aflatoxins in positive samples, *J. Assoc. Off. Anal. Chem. 58*: 500 (1975).
14. *Official Methods of Analysis*, 16th ed., AOAC International, Arlington, VA, Sec. 975.36 (1995).
15. *Official Methods of Analysis*, 16th ed., AOAC International, Arlington, VA, Sec. 979.18 (1995).
16. C. E. Holaday, Minicolumn chromatography: state of the art, *J. Amer. Oil Chem. Soc. 58*: 931A (1981).
17. Y. Ramakrishna, R. B. Sashidhar, and R. V. Bhat, Minicolumn technique for the detection of deoxynivalenol in agricultural commodities, *Bull. Environ. Contam. Toxicol. 42*: 167 (1989).

18. J. W. Dorner and R. J. Cole, Rapid determination of aflatoxins in raw peanuts by liquid chromatography with postcolumn iodination and modified minicolumn cleanup, *J. Assoc. Off. Anal. Chem. 71:* 43 (1988).

19. S. Nesheim and M. W. Trucksess, Thin-layer chromatography/high-performance thin-layer chromatography as a tool for mycotoxin determination, *Modern Methods in the Analysis and Structural Elucidation of Mycotoxins* (R. J. Cole, ed.), Academic Press, Orlando, FL, 1986, p. 239.

20. P. S. Steyn, P. G. Thiel, and D. W. Trinder, Detection and quantification of mycotoxins by chemical analysis, *Mycotoxins and Animal Foods* (J. E. Smith and R. S. Henderson, eds.), CRC Press, Boca Raton, FL, 1991, p. 165.

21. R. D. Coker, K. Jewers, K. I. Tomlins, and G. Blunden, Evaluation of instrumentation used for high performance thin layer chromatography of aflatoxins, *Chromatographia 25:* 875 (1988).

22. M. W. Trucksess, L. Stoloff, W. A. Pons, Jr., A. F. Cucullu, L. S. Lee, and A. O. Franz, Jr., Thin layer chromatographic determination of aflatoxin B$_1$ in eggs, *J. Assoc. Off. Anal. Chem. 60:* 795 (1977).

23. R. D. Stubblefield and O. L. Shotwell, Determination of aflatoxins in animal tissues, *J. Assoc. Off. Anal. Chem. 64:* 964 (1981).

24. T. Takahashi, Aflatoxin contamination in nutmeg: analysis of interfering TLC spots, *J. Food Sci. 58:* 197 (1993).

25. N. Miller, H. E. Pretorius, and D. W. Trinder, Determination of aflatoxins in vegetable oils, *J. Assoc. Off. Anal. Chem. 68:* 136 (1985).

26. R. D. Coker, A. E. John, and J. A. Gibbs, Techniques of thin layer chromatography, *Chromatography of Mycotoxins* (V. Betina, ed.), Elsevier, Amsterdam, 1993, p. 12.

27. F. S. Chu, Detection and determination of mycotoxins, *Mycotoxins and Phytoalexins* (R. P. Sharma and D. K. Salunkhe, eds.), CRC Press, Boca Raton, FL, 1991, p. 33.

28. R. M. Eppley, M. W. Trucksess, S. Nesheim, C. W. Thorpe, and A. E. Pohland, Thin layer chromatographic method for determination of deoxynivalenol in wheat: collaborative study, *J. Assoc. Off. Anal. Chem. 69:* 37 (1986).

29. O. J. Francis, Jr., G. E. Ware, A. S. Carman, and S. S. Kuan, Thin layer chromatographic determination of sterigmatocystin in cheese, *J. Assoc. Off. Anal. Chem. 68:* 643 (1985).

30. F. S. Chu, Studies on ochratoxins, *CRC Crit. Rev. Toxicol. 2:* 499 (1977).

31. T. F. Salem and B. G. Swanson, Fluorodensitometric assay of patulin in apple products, *J. Food Sci. 41:* 1237 (1976).

32. R. T. Gallagher, J. L. Richard, H. M. Stahr, and R. J. Cole, Cyclopiazonic acid production by aflatoxigenic and non-aflatoxigenic strains of *Aspergillus flavus*, *Mycopathologia 66:* 31 (1978).

33. T. Whitaker, W. Horwitz, R. Albert, and S. Nesheim, Variability associated with analytical methods used to measure aflatoxin in agricultural commodities. *J. AOAC Int. 79:* 476 (1996).

34. M. J. Shepherd, High-performance liquid chromatography and its application to the analysis of mycotoxins, *Modern Methods in the Analysis and Structural Elucidation of Mycotoxins* (R. J. Cole, ed.), Academic Press, Orlando, FL, 1986, p. 293.

35. P. Kuronen, Techniques of liquid column chromatography, *Chromatography of Mycotoxins* (V. Betina, ed.), Elsevier, Amsterdam, 1993, p. 36.

36. B. T. Duhart, S. Shaw, M. Wooley, T. Allen, and G. Grimes, Determination of aflatoxins B_1, B_2, G_1, and G_2 by high-performance liquid chromatography with electrochemical detection, *Anal. Chim. Acta 208*: 343 (1988).

37. G. M. Ware, C. W. Thorpe, and A. E. Pohland, Determination of roquefortine in blue cheese and blue cheese dressing by high pressure liquid chromatography with ultraviolet and electrochemical detectors, *J. Assoc. Off. Anal. Chem. 63*: 637 (1980).

38. A. S. Carman, S. S. Kuan, O. J. Francis, G. M. Ware, and A. E. Luedtke, Determination of xanthomegnin in grains and animal feeds by liquid chromatography with electrochemical detection, *J. Assoc. Off. Anal. Chem. 67*: 1095 (1984).

39. R. W. Beaver, Gas chromatography in mycotoxin analysis, *Modern Methods in the Analysis and Structural Elucidation of Mycotoxins* (R. J. Cole, ed.), Academic Press, Orlando, FL, 1986, p. 265.

40. R. W. Beaver, Techniques of gas chromatography, *Chromatography of Mycotoxins* (V. Betina, ed.), Elsevier, Amsterdam, 1993, p. 78.

41. *Official Methods of Analysis*, 16th ed., AOAC International, Arlington, VA, Sec. 986.18 (1995).

42. P. M. Scott, Gas chromatography of mycotoxins, *Chromatography of Mycotoxins: Techniques and Applications* (V. Betina, ed.), Elsevier, Amsterdam, 1993, p. 373.

43. R. F. Vesonder and W. K. Rohwedder, Gas chromatographic–mass spectrometric analysis of mycotoxins, *Modern Methods in the Analysis and Structural Elucidation of Mycotoxins* (R. J. Cole, ed.), Academic Press, Orlando, FL, 1986, p. 335.

44. T. R. Romer and A. D. Campbell, Collaborative study of a screening method for the detection of aflatoxins in mixed feeds, other agricultural products, and foods, *J. Assoc. Off. Anal. Chem. 59*: 110 (1976).

45. G. M. Shannon and O. L. Shotwell, Minicolumn detection methods for aflatoxin in yellow corn: collaborative study, *J. Assoc. Off. Anal. Chem. 62*: 1070 (1979).

46. O. L. Shotwell and C. E. Holaday, Minicolumn detection methods for aflatoxin in raw peanuts: collaborative study, *J. Assoc. Off. Anal. Chem. 64*: 674 (1981).

47. R. M. Eppley, A versatile procedure for assay and preparatory separation of aflatoxin from peanut products, *J. Assoc. Off. Anal. Chem. 49*: 1218 (1966).

48. A. E. Waltking, Collaborative study of three methods for determination of aflatoxin in peanuts and peanut products, *J. Assoc. Off. Anal. Chem. 53*: 104 (1970).

49. M. W. Trucksess, W. C. Brumley, and S. Nesheim, Rapid quantitation and confirmation of aflatoxins in corn and peanut butter, using a disposable silica gel column, thin layer chromatography, and gas chromatography/mass spectrometry, *J. Assoc. Off. Anal. Chem. 67*: 973 (1984).

50. D. L. Park, M. W. Trucksess, S. Nesheim, M. Stack, and R. F. Newell, Solvent-efficient thin-layer chromatographic method for the determination of aflatoxins B_1, B_2, G_1, and G_2 in corn and peanut products: collaborative study, *J. AOAC Int. 77*: 637 (1994).

51. M. P. K. Dell, S. J. Haswell, O. G. Roch, R. D. Coker, V. F. P. Medlock, and K. Tomlins, Analytical methodology for the determination of aflatoxins in peanut butter: comparison of high-performance thin-layer chromatographic, enzyme-linked immunosorbent assay and high-performance liquid chromatographic methods, *Analyst 115*: 1435 (1990).

52. W. E. Steiner, K. Brunschweiler, E. Leimbacher, and R. Schneider, Aflatoxins and fluorescence in Brazil nuts and pistachio nuts, *J. Agric. Food Chem. 40*: 2453 (1992).

53. P. Majerus and Z. Zakaria, A rapid, sensitive and economic method for the detection, quantification and confirmation of aflatoxins, *Z. Lebensm. Unters. Forsch. 195*: 316 (1992).

54. M. S. Madhyastha and R. V. Bhat, Aflatoxin-like fluorescent substance in spices, *J. Food Safety 7*: 101 (1985).

55. M. W. Trucksess and G. W. Wood, Recent methods of analysis for aflatoxins in foods and feeds, *The Toxicology of Aflatoxins: Human Health, Veterinary and Agricultural Significance* (D. L. Eaton and J. D. Groopman, ed.), Academic Press, Inc., San Diego, CA, 1994, p. 409.

56. W. A. Pons, Jr. and A. O. Franz, Jr., High pressure liquid chromatographic determination of aflatoxins in peanut products, *J. Assoc. Off. Anal. Chem. 61*: 793 (1978).

57. O. J. Francis, Jr., L. J. Lipinski, J. A. Gaul, and A. D. Campbell, High pressure liquid chromatographic determination of aflatoxins in peanut butter using a silica gel-packed flowcell for fluorescence detection, *J. Assoc. Off. Anal. Chem. 65*: 672 (1982).

58. D. M. Takahashi, Reversed-phase high-performance liquid chromatographic analytical system for aflatoxins in wines with fluorescence detection, *J. Chromatogr. 131*: 147 (1977).

59. R. M. Beebe, Reverse phase high pressure liquid chromatographic determination of aflatoxins in foods, *J. Assoc. Off. Anal. Chem. 61*: 1347 (1978).

60. E. J. Tarter, J. P. Hanchay, and P. M. Scott, Improved liquid chromatographic method for determination of aflatoxins in peanut butter and other commodities, *J. Assoc. Off. Anal. Chem. 67*, 597 (1984).

61. J. E. Hutchins, Y. J. Lee, K. Tyczkowska, and W. M. Hagler, Jr., Evaluation of silica cartridge purification and hemiacetal formation for liquid chromatographic determination of aflatoxins in corn, *Arch. Environ. Contam. Toxicol. 18*: 319 (1989).

62. D. L. Park, S. Nesheim, M. W. Trucksess, M. E. Stack, and R. F. Newell, Liquid chromatographic method for determination of aflatoxins B_1, B_2, G_1, and G_2 in corn and peanut products: collaborative study, *J. Assoc. Off. Anal. Chem. 73*: 260 (1990).

63. T. J. Wilson and T. R. Romer, Use of the mycosep multifunctional cleanup column for liquid chromatographic determination of aflatoxins in agricultural products, *J. Assoc. Off. Anal. Chem. 74*: 951 (1991).

64. O. G. Roch, G. Blunden, R. D. Coker, S. Nawaz, The development and validation of a solid phase extraction/HPLC method for the determination of aflatoxins in groundnut meal, *Chromatographia 33*: 208 (1992).

65. P. G. Thiel, S. Stockenström, and P. S. Gathercole, Aflatoxin analysis by reverse phase HPLC using post-column derivatization for enhancement of fluorescence, *J. Liq. Chromatogr. 9*: 103 (1986).

66. M. J. Shepherd and J. Gilbert, An investigation of HPLC postcolumn iodination conditions for the enhancement of aflatoxin B_1 fluorescence, *Food Addit. Contam. 1*: 325 (1984).

67. H. Jansen, R. Jansen, U. A. Th. Brinkman, and R. W. Frei, Fluorescence enhancement for aflatoxins in HPLC by post-column split-flow iodine addition from a solid-phase iodine reservoir, *Chromatographia 24*: 555 (1987).

68. M. W. Trucksess, M. E. Stack, S. Nesheim, S. W. Page, R. H. Albert, T. J. Hansen, and K. F. Donahue, Immunoaffinity column coupled with solution fluorometry or liquid chromatography postcolumn derivatization for determination of aflatoxins in corn, peanuts, and peanut butter: collaborative study, *J. Assoc. Off. Anal. Chem. 74*: 81 (1991).

69. A. L. Patey, M. Sharman, and J. Gilbert, Liquid chromatographic determination of aflatoxin levels in peanut butters using an immunoaffinity column cleanup method: internal collaborative trial, *J. Assoc. Off. Anal. Chem. 74*: 76 (1991).

70. M. Sharman and J. Gilbert, Automated aflatoxin analysis of foods and animal feeds using immunoaffinity column clean-up and high-performance liquid chromatographic determination, *J. Chromatogr. 543*: 220 (1991).

71. W. Th. Kok, Th. C. H. Van Neer, W. A. Traag, and L. G. M. Th. Tuinstra, Determination of aflatoxins in cattle feed by liquid chromatography and post-column derivatization with electrochemically generated bromine, *J. Chromatogr. 367*: 231 (1986).

72. K. Reif and W. Metzger, Determination of aflatoxins in medicinal herbs and plant extracts, *J. Chromatogr. A 692*: 131 (1995).

73. T. Urano, M. W. Trucksess, and S. W. Page, Automated affinity liquid chromatography system for on-line isolation, separation, and quantitation of aflatoxins in methanol-water extracts of corn or peanuts, *J. Agric. Food Chem. 41*: 1982 (1993).

74. H. Joshua, Determination of aflatoxins by reversed-phase high-performance liquid chromatography with post-column in-line photochemical derivatization and fluorescence detection, *J. Chromatogr. A 654*: 247 (1993).

75. J. W. Dorner and R. J. Cole, A method for determining kernel moisture content and aflatoxin concentrations in peanuts, *J. Amer. Oil Chem. Soc. 74*: 285 (1997).

76. W. Th. Kok, Derivatization reactions for the determination of aflatoxins by liquid chromatography with fluorescence detection, *J. Chromatogr. B 659*: 127 (1994).

77. O. G. Roch, G. Blunden, J. H. Duncan, R. D. Coker, and C. Gay, Determination of aflatoxins in groundnut meal by high-performance liquid chromatography: a comparison of two methods of derivatisation of aflatoxin B_1, *Br. J. Biomed Sci. 52*: 312 (1995).

78. O. J. Francis, Jr., G. P. Kirschenheuter, G. M. Ware, A. S. Carman, and S. S. Kuan, β-Cyclodextrin post-column fluorescence enhancement of aflatoxins for reverse-phase liquid chromatographic determination in corn, *J. Assoc. Off. Anal. Chem. 71*: 725 (1988).

79. A. Cepeda, C. M. Franco, C. A. Fente, B. I. Vazquez, J. L. Rodriguez, P. Prognon, and G. Mahuzier, Postcolumn excitation of aflatoxins using cyclodextrins in liquid chromatography for food analysis, *J. Chromatogr. A 721*: 69 (1996).

80. T. Goto, M. Matsui, and T. Kitsuwa, Determination of aflatoxins by capillary column gas chromatography, *J. Chromatogr. 447*: 410 (1988).

81. W. C. A. Gelderblom, K. Jaskiewicz, W. F. O. Marasas, P. G. Thiel, R. M. Horak, R. Vleggaar, and N. P. J. Kriek, Fumonisins—novel mycotoxins with cancer-promoting activity produced by *Fusarium moniliforme*, *Appl. Environ. Microbiol. 54*: 1806 (1988).

82. S. C. Bezuidenhout, W. C. A. Gelderblom, C. P. Gorst-Allman, R. M. Horak, W. F. O. Marasas, G. Spiteller, and R. Vleggaar, Structure elucidation of the fumonisins, mycotoxins from *Fusarium moniliforme*, *J. Chem. Soc., Chem. Commun.* 743 (1988).

83. W. F. O. Marasas, T. S. Kellerman, W. C. A. Gelderblom, J. A. Coetzer, P. G. Thiel, and J. J. van der Lugt, Leukoencephalomalacia in a horse induced by fumonisin B_1 isolated from *Fusarium moniliforme, Onderstepoort J. Vet. Res. 55*: 197 (1988).

84. P. F. Ross, P. E. Nelson, J. L. Richard, G. D. Osweiler, L. G. Rice, R. D. Plattner, and T. M. Wilson, Production of fumonisins by *Fusarium moniliforme* and *Fusarium proliferatum* isolates associated with equine leukoencephalomalacia and a pulmonary edema syndrome in swine, *Appl. Environ. Microbiol. 56*: 3225 (1990).

85. L. R. Harrison, B. M. Colvin, J. T. Green, L. E. Newman, and J. R. Cole, Pulmonary edema and hydrothorax in swine produced by fumonisin B_1, a toxic metabolite of *Fusarium moniliforme, J. Vet. Diagn. Invest 2*: 217 (1990).

86. M. E. Cawood, W. C. A. Gelderblom, R. Vleggaar, Y. Behrend, P. G. Thiel, and W. F. O. Marasas, Isolation of the fumonisin mycotoxins—a quantitative approach, *J. Agric. Food Chem. 39*: 1958 (1991).

87. M. E. Stack and R. M. Eppley, Liquid chromatographic determination of fumonisins B_1 and B_2 in corn and corn products, *J. AOAC Int. 75*: 834 (1992).

88. J. P. Rheeder, W. F. O. Marasas, P. G. Thiel, E. W. Sydenham, G. S. Shephard, and D. J. van Schalkwyk, *Fusarium moniliforme* and fumonisins in corn in relation to human esophageal cancer in Transkei, *Phytopathology 82*: 353 (1992).

89. B. E. Branham and R. D. Plattner, Isolation and characterization of a new fumonisin from liquid cultures of *Fusarium moniliforme, J. Nat. Prod. 56*: 1630 (1993).

90. W. P. Norred and K. A. Voss, Toxicity and role of fumonisins in animal diseases and human esophageal cancer, *J. Food Prot. 57*: 522 (1994).

91. E. W. Sydenham, G. S. Shephard, P. G. Thiel, C. Bird, and B. M. Miller, Determination of fumonisins in corn: evaluation of competitive immunoassay and HPLC techniques, *J. Agric. Food Chem. 44*: 159 (1996).

92. G. E. Rottinghaus, C. E. Coatney, H. C. Minor, A rapid, sensitive thin layer chromatography procedure for the detection of fumonisin B_1 and B_2, *J. Vet. Diagn. Invest. 4*: 326 (1992).

93. R. D. Plattner, W. P. Norred, C. W. Bacon, K. A. Voss, R. Peterson, D. D. Shackelford, and D. Weisleder, A method of detection of fumonisins in corn samples associated with field cases of equine leukoencephalomalacia, *Mycologia 82*: 698 (1990).

94. R. D. Plattner and B. E. Branham, Labeled fumonisins: production and use of fumonisin B_1 containing stable isotopes, *J. AOAC Int. 77*: 525 (1994).

95. G. S. Shephard, E. W. Sydenham, P. G. Thiel, and W. C. A. Gelderblom, Quantitative determination of fumonisins B_1 and B_2 by high-performance liquid chromatography with fluorescence detection, *J. Liq. Chromatogr. 13*: 2077 (1990).

96. E. W. Sydenham, G. S. Shephard, and P. G. Thiel, Liquid chromatographic determination of fumonisins B_1, B_2, and B_3 in foods and feeds, *J. AOAC Int. 75*: 313 (1992).

97. G. A. Bennett and J. L. Richard, Liquid chromatographic method for analysis of the naphthalene dicarboxaldehyde derivative of fumonisins, *J. AOAC Int. 77*: 501 (1994).

98. P. M. Scott and G. A. Lawrence, Stability and problems in recovery of fumonisins added to corn-based foods, *J. AOAC Int. 77*: 541 (1994).

99. M. Holcomb, H. C. Thompson, Jr., G. Lipe, and L. J. Hankins, HPLC with electro-

chemical and fluorescence detection of the OPA/2-methyl-2-propanethiol derivative of fumonisin B$_1$, *J. Liq. Chromatogr. 17*: 4121 (1994).

100. C. M. Maragos and J. L. Richard, Quantitation and stability of fumonisins B$_1$ and B$_2$ in milk, *J. AOAC Int. 77*: 1162 (1994).

101. H. Akiyama, M. Miyahara, M. Toyoda, and Y. Saito, Liquid chromatographic determination of fumonisins B$_1$ and B$_2$ in corn by precolumn derivatization with 4-(*N,N*-dimethylaminosulfonyl)-7-fluoro-2,1,3-benzoxadiazole (DBD-F), *J. Food Hyg. Soc. Japan 36*: 77 (1995).

102. L. G. Rice and P. F. Ross, Methods for detection and quantitation of fumonisins in corn, cereal products and animal excreta, *J. Food Protect. 57*: 536 (1994).

103. G. M. Ware, P. P. Umrigar, A. S. Carman, Jr., and S. S. Kuan, Evaluation of fumonitest immunoaffinity columns, *Anal. Lett. 27*: 693 (1994).

104. M. W. Trucksess, M. E. Stack, S. Allen, and N. Barrion, Immunoaffinity column coupled with liquid chromatography for determination of fumonisin B$_1$ in canned and frozen sweet corn, *J. AOAC Int. 78*: 705 (1995).

105. P. M. Scott and G. A. Lawrence, Analysis of beer for fumonisins, *J. Food Protect. 58*: 1379 (1995).

106. E. W. Sydenham, G. S. Shephard, P. G. Thiel, S. Stockenström, P.W. Snijman, and D. J. Van Schalkwyk, Liquid chromatographic determination of fumonisins B$_1$, B$_2$, and B$_3$ in corn: AOAC-IUPAC collaborative study, *J. AOAC Int. 79*: 688 (1996).

107. *Official Methods of Analysis*, 16th ed., AOAC International, Arlington, VA, Sec. 995.15 (1995).

108. P. M. Scott and G. A. Lawrence, Liquid chromatographic determination of fumonisins with 4-fluoro-7-nitrobenzofurazan, *J. AOAC Int. 75*: 829 (1992).

109. M. Miyahara, H. Akiyama, M. Toyoda, and Y. Saito, New procedure for fumonisins B$_1$ and B$_2$ in corn and corn products by ion pair chromatography with *o*-phthaldialdehyde postcolumn derivatization and fluorometric detection, *J. Agric. Food Chem. 44*: 842 (1996).

110. J. G. Wilkes, J. B. Sutherland, M. I. Churchwell, and A. J. Williams, Determination of fumonisins B$_1$, B$_2$, B$_3$, and B$_4$ by high-performance liquid chromatography with evaporative light-scattering detection, *J. Chromatogr. A 695*: 319 (1995).

111. E. B. Furlong and L. M. V. Soares, Gas chromatographic method for quantitation and confirmation of trichothecenes in wheat. *J. AOAC Int. 78*: 386 (1995).

112. T. Yoshizawa, Natural occurrence of mycotoxins in small grain cereals (wheat, barley, rye, oats, sorghum, millet, rice), *Mycotoxins and Animal Foods* (J. E. Smith and R. S. Henderson, eds.), CRC Press, Boca Raton, FL, 1991, p. 301.

113. G. M. Ware, O. J. Francis, A. S. Carman, and S. S. Kuan, Gas chromatographic determination of deoxynivalenol in wheat with electron capture detection: collaborative study, *J. Assoc. Off. Anal. Chem. 69*: 899 (1986).

114. D. K. Vudathala, D. B. Prelusky, and H. L. Trenholm, Analysis of trace levels of deoxynivalenol in cow's milk by high pressure liquid chromatography, *J. Liq. Chromatogr. 17*: 673 (1994).

115. H.-J. Kang, J.-C. Kim, J.-A. Seo, Y.-W. Lee, and D.-H. Son, Contamination of *Fusarium* mycotoxins in corn samples imported from China, *Agric. Chem. Biotechnol. 37*: 385 (1994).

116. C. J. Mirocha, S. V. Pathre, and C. M. Christensen, Zearalenone, *Mycotoxins in*

Human and Animal Health (J. V. Rodricks, C. W. Hesseltine, and M. A. Mehlman, eds.), Pathotox Publishers, Park Forest South, IL, 1977, p. 345.

117. V. Betina, Thin-layer chromatography of mycotoxins, *Chromatography of Mycotoxins* (V. Betina, ed.), Elsevier, Amsterdam, 1993, p. 141.

118. R. M. Eppley, L. Stoloff, M. W. Trucksess, and C. W. Chung, Survey of corn for Fusarium toxins, *J. Assoc. Off. Anal. Chem. 57*: 632 (1974).

119. P. M. Scott, T. Panalaks, S. Kanhere, and W. F. Miles, Determination of zearalenone in cornflakes and other corn-based foods by thin layer chromatography, high pressure liquid chromatography, and gas-liquid chromatography/high resolution mass spectrometry. *J. Assoc. Off. Anal. Chem. 61*: 593 (1978).

120. H. L. Chang and J. W. DeVries, Short liquid chromatographic method for determination of zearalenone and alpha-zearalenol, *J. Assoc. Off. Anal. Chem. 67*: 741 (1984).

121. H. L. Trenholm, R. M. Warner, and D. W. Fitzpatrick, Rapid, sensitive liquid chromatographic method for determination of zearalenone and α- and β-zearalenol in wheat, *J. Assoc. Off. Anal. Chem. 67*: 968 (1984).

122. G. A. Bennett, O. L. Shotwell, and W. F. Kwolek, Liquid chromatographic determination of α-zearalenol and zearalenone in corn: collaborative study, *J. Assoc. Off. Anal. Chem. 68*: 958 (1985).

123. M. T. Hetmanski and K. A. Scudamore, Detection of zearalenone in cereal extracts using high-performance liquid chromatography with post-column derivatization, *J. Chromatogr. 588*: 47 (1991).

124. T. Tanaka, R. Teshima, H. Ikebuchi, J. Sawada, T. Terao, and M. Ichinoe, Sensitive determination of zearalenone and α-zearalenol in barley and job's-tears by liquid chromatography with fluorescence detection, *J. AOAC Int. 76*: 1006 (1993).

125. T. Tanaka, A. Hasegawa, Y. Matsuki, U.-S. Lee, and Y. Ueno, Rapid and sensitive determination of zearalenone in cereals by high-performance liquid chromatography with fluorescence detection, *J. Chromatogr. 328*: 271 (1985).

126. R. W. Wannemacher, Jr., D. L. Bunner, and H. A. Neufeld, Toxicity of trichothecenes and other related mycotoxins in laboratory animals, *Mycotoxins and Animal Foods* (J. E. Smith and R. S. Henderson, eds.), CRC Press, Boca Raton, FL, 1991, p. 499.

127. J. C. Frisvad and U. Thrane, Liquid column chromatography of mycotoxins, *Chromatography of Mycotoxins: techniques and applications* (V. Betina, ed.), Elsevier, Amsterdam, 1993, p. 253.

128. O. L. Shotwell, Natural occurrence of mycotoxins in corn, *Mycotoxins and Animal Foods* (J. E. Smith and R. S. Henderson, eds.), CRC Press, Boca Raton, FL, 1991, p. 325.

129. S. Nesheim, N. F. Hardin, O. J. Francis, Jr., and W. S. Langham, Analysis of ochratoxins A and B in their esters in barley, using partition and thin layer chromatography. I. Development of method. *J. Assoc. Off. Anal. Chem. 56*: 817 (1973).

130. S. Nesheim, Analysis of ochratoxins A and B and their esters in barley, using partition and thin layer chromatography. II. collaborative study, *J. Assoc. Off. Anal. Chem. 56*: 822 (1973).

131. C. P. Levi, Collaborative study of a method for the determination of ochratoxin A in green coffee, *J. Assoc. Off. Anal. Chem. 58*: 258 (1975).

132. S. Nesheim, M. E. Stack, M. W. Trucksess, R. M. Eppley, and P. Krogh, Rapid solvent-efficient method for liquid chromatographic determination of ochratoxin A in corn, barley, and kidney: collaborative study, *J. AOAC Int. 75*: 481 (1992).

133. B. Le Tutour, A. Tantaoui-Elaraki, and A. Aboussalim, Simultaneous thin layer chromatographic determination of aflatoxin B_1 and ochratoxin A in black olives, *J. Assoc. Off. Anal. Chem. 67*: 611 (1984).

134. C. P. Gorst-Allman and P. S. Steyn, Screening methods for the detection of thirteen common mycotoxins, *J. Chromatogr. 175*: 325 (1979).

135. A. A. Frohlich, R. R. Marquardt, and A. Bernatsky, Quantitation of ochratoxin A: use of reverse phase thin-layer chromatography for sample cleanup followed by liquid chromatography or direct fluorescence measurement, *J. Assoc. Off. Anal. Chem. 71*: 949 (1988).

136. B. I. Vazquez, C. Fente, C. Franco, A. Cepeda, P. Prognon, and G. Mahuzier, Simultaneous high-performance liquid chromatographic determination of ochratoxin A and citrinin in cheese by time-resolved luminescence using terbium, *J. Chromatogr. A 727*: 185 (1996).

137. B. Zimmerli and R. Dick, Determination of ochratoxin A at the ppt level in human blood, serum, milk and some foodstuffs by high-performance liquid chromatography with enhanced fluorescence detection and immunoaffinity column cleanup: methodology and Swiss data, *J. Chromatogr. B 666*: 85 (1995).

138. J. I. Pitt and L. Leistner, Toxigenic *Penicillium* species, *Mycotoxins and Animal Foods* (J. E. Smith and R. S. Henderson, eds.), CRC Press, Boca Raton, FL, 1991, p. 81.

139. E. R. McKinley and W. W. Carlton, Patulin, *Mycotoxins and Phytoalexins* (R. P. Sharma and D. K. Salunkhe, eds.), CRC Press, Boca Raton, FL, 1991, p. 191.

140. A. R. Brause, M. W. Trucksess, F. S. Thomas, and S. W. Page, Determination of patulin in apple juice by liquid chromatography: collaborative study, *J. AOAC Int. 79*: 451 (1996).

141. P. M. Scott and B. P. C. Kennedy, Improved method for the thin layer chromatographic determination of patulin in apple juice, *J. Assoc. Off. Anal. Chem. 56*: 813 (1973).

142. P. M. Scott, Collaborative study of a chromatographic method for determination of patulin in apple juice, *J. Assoc. Off. Anal. Chem. 57*: 621 (1974).

143. R. J. Cole and R. H. Cox, *Handbook of Toxic Fungal Metabolites*, Academic Press, New York, 1981, 937 pages.

144. M. E. Stack and J. V. Rodricks, Method for analysis and chemical confirmation of sterigmatocystin, *J. Assoc. Off. Anal. Chem. 54*: 86 (1971).

145. A. Gimeno, Thin layer chromatographic determination of aflatoxins, ochratoxins, sterigmatocystin, zearalenone, citrinin, T-2 toxin, diacetoxyscirpenol, penicillic acid, patulin and penitrem A, *J. Assoc. Off. Anal. Chem. 62*: 579 (1979).

146. M. E. Stack, S. Nesheim, N. L. Brown, and A. E. Pohland, Determination of sterigmatocystin in corn and oats by gel permeation and high pressure liquid chromatography, *J. Assoc. Off. Anal. Chem. 59*: 966 (1976).

147. R. Schmidt, K. Neunhoeffer, and K. Dose, Quantitative determination of sterigmatocystin in mouldy food of plant origin, *Fresenius' J. Anal. Chem. 299*: 382 (1979).

148. K. Berry, M. F. Dutton, and M. S. Jeenah, Separation of aflatoxin biosynthetic intermediates by high performance liquid chromatography, *J. Chromatogr. 283*: 421 (1984).

149. D. Abramson and T. Thorsteinson, Determination of sterigmatocystin in barley by acetylation and liquid chromatography, *J. Assoc. Off. Anal. Chem. 72*: 342 (1989).

150. F. L. Neely and C. S. Emerson, Determination of sterigmatocystin in fermentation broths by reversed-phase high-performance liquid chromatography using post-column fluorescence enhancement, *J. Chromatogr. 523*: 305 (1990).

151. A. S. Salhab, G. F. Russell, J. R. Coughlin, and D. P. Hsieh, Gas-liquid chromatography and mass spectrometric ion selective detection of sterigmatocystin in grains, *J. Assoc. Off. Anal. Chem. 59*: 1037 (1976).

152. S. D. Clouse and D. G. Gilchrist, Interaction of the *asc* locus in F_8 paired lines of tomato with *Alternaria alternata* f. sp. *lycopersici* and AAL-toxin, *Phytopathology 77*: 80 (1987).

153. J. Grabarkiewicz-Szczesna, J. Chelkowski, and P. Zajkowski, Natural occurrence of *Alternaria* mycotoxins in the grain and chaff of cereals, *Mycotoxin Res. 5*: 77 (1989).

154. S. Ozcelik, N. Ozcelik, and L. R. Beuchat, Toxin production by *Alternaria alternata* in tomatoes and apples stored under various conditions and quantitation of the toxins by high-performance liquid chromatography, *Int. J. Food Microbiol. 11*: 187 (1990).

155. M. E. Stack, P. B. Mislivec, J. A. G. Roach, and A. E. Pohland, Liquid chromatographic determination of tenuazonic acid and alternariol methyl ether in tomatoes and tomato products, *J. Assoc. Off. Anal. Chem. 68*: 640 (1985).

156. S. D. Clouse, A. N. Martensen, and D. G. Gilchrist, Rapid purification of host-specific pathotoxins from *Alternaria alternata* f. sp. *lycopersici* by solid-phase adsorption on octadecylsilane, *J. Chromatogr. 350*: 255 (1985).

157. G. S. Shephard, P. G. Thiel, W. F. O. Marasas, E. W. Sydenham, and R. Vleggaar, Isolation and determination of AAL phytotoxins from corn cultures of the fungus *Alternaria alternata* f. sp. *lycopersici*, *J. Chromatogr. 641*: 95 (1993).

158. J. W. Dorner, R. J. Cole, and L. G. Lomax, The toxicity of cyclopiazonic acid, *Trichothecenes and Other Mycotoxins* (J. Lacey, ed.), John Wiley, New York, 1985, p. 529.

159. J. A. Lansden, Determination of cyclopiazonic acid in peanuts and corn by thin layer chromatography, *J. Assoc. Off. Anal. Chem. 69*: 964 (1986).

160. J. W. Dorner, R. J. Cole, D. J. Erlington, S. Suksupath, G. H. McDowell, and W. L. Bryden, Cyclopiazonic acid residues in milk and eggs, *J. Agric. Food Chem. 42*: 1516 (1994).

161. J. A. Lansden, Liquid chromatographic analysis system for cyclopiazonic acid in peanuts, *J. Assoc. Off. Anal. Chem. 67*: 728 (1984).

162. T. Goto, E. Shinski, K. Tanaka, and M. Manabe, Analysis of cyclopiazonic acid by normal phase high-performance liquid chromatography, *Agric. Biol. Chem. 51*: 2581 (1987).

163. T. Urano, M. W. Trucksess, J. Matusik, and J. W. Dorner, Liquid chromatographic determination of cyclopiazonic acid in corn and peanuts, *J. AOAC Int. 75*: 319 (1992).

164. R. V. Reddy and W. O. Berndt, Citrinin, *Mycotoxins and Phytoalexins* (R. P. Sharma and D. K. Salunkhe, eds.), CRC Press, Boca Raton, FL, 1991, p. 237.

165. A. Gimeno, Determination of citrinin in corn and barley on thin layer chromatographic plates impregnated with glycolic acid, *J. Assoc. Off. Anal. Chem. 67*: 194 (1984).

166. B. Zimmerli, R. Dick, and U. Baumann, High-performance liquid chromatographic determination of citrinin in cereals using an acid-buffered silica gel column, *J. Chromatogr. 462*: 406 (1989).

167. C. M. Franco, C. A. Fente, B. Vazquez, A. Cepeda, L. Lallaoui, P. Prognon, and G. Mahuzier, Simple and sensitive high-performance liquid chromatography–fluorescence method for the determination of citrinin: application to the analysis of fungal cultures and cheese extracts, *J. Chromatogr. A 723*: 69 (1996).

168. R. M. Eppley, L. Stoloff, and A. D. Campbell, Collaborative study of "a versatile procedure for assay of aflatoxins in peanut products," including preparatory separation and confirmation of identity, *J. Assoc. Off. Anal. Chem. 51*: 67 (1968).

169. R. D. Stubblefield, H. P. van Egmond, W. E. Paulsch, and P. L. Schuller, Determination and confirmation of identity of aflatoxin M_1 in dairy products: collaborative study, *J. Assoc. Off. Anal. Chem. 63*: 907 (1980).

170. L. Dominguez, J. L. Blanco, E. Gomez-Lucia, E. F. Rodriguez, and G. Suarez, Determination of aflatoxin M_1 in milk and milk products contaminated at low levels, *J. Assoc. Off. Anal. Chem. 70*: 470 (1987).

171. J. P. Bijl, C. H. van Peteghem, and D. A. Dekeyser, Fluorimetric determination of aflatoxin M_1 in cheese, *J. Assoc. Off. Anal. Chem. 70*: 472 (1987).

172. G.-S. Qian and G. C. Yang, Rapid extraction and detection of aflatoxins B_1 and M_1 in beef liver, *J. Agric. Food Chem. 32*: 1071 (1984).

173. R. D. Stubblefield and W. F. Kwolek, Rapid liquid chromatographic determination of aflatoxins M_1 and M_2 in artificially contaminated fluid milks: collaborative study, *J. Assoc. Off. Anal. Chem. 69*: 880 (1986).

174. J. W. Dorner, P. D. Blankenship, and R. J. Cole, Performance of two immunochemical assays in the analysis of peanuts for aflatoxin at 37 field laboratories, *J. AOAC Int. 76*: 637 (1993).

175. R. C. Garner, M. M. Whattam, P. J. L. Taylor, and M. W. Stow, Analysis of United Kingdom purchased spices for aflatoxins using an immunoaffinity column clean-up procedure followed by high-performance liquid chromatographic analysis and post-column derivatisation with pyridinium bromide perbromide, *J. Chromatogr. 648*: 485 (1993).

176. W. C. Gordon and L. J. Gordon, Rapid screening method for deoxynivalenol in agricultural commodities by fluorescent minicolumn, *J. Assoc. Off. Anal. Chem. 73*: 266 (1990).

177. L. Czerwiecki and H. Giryn, Occurrence of trichothecenes in selected polish cereal and rice food products, *Pol. J. Food Nutr. Sci. 3/44*: 111 (1994).

178. D. R. Lauren and R. Greenhalgh, Simultaneous analysis of nivalenol and deoxynivalenol in cereals by liquid chromatography, *J. Assoc. Off. Anal. Chem. 70*: 479 (1987).

179. A. Sano, S. Matsutani, M. Suzuki, and S. Takitani, High-performance liquid chromatographic method for determining trichothecene mycotoxins by post-column fluorescence derivatization, *J. Chromatogr. 410*: 427 (1987).

180. P. M. Scott, S. R. Kanhere, and E. J. Tarter, Determination of nivalenol and deoxynivalenol in cereals by electron-capture gas chromatography, *J. Assoc. Off. Anal. Chem. 69*: 889 (1986).

181. E. W. Sydenham and P. G. Thiel, The simultaneous determination of diacetoxyscirpenol and T-2 toxin in fungal cultures and grain samples by capillary gas chromatography, *Food Addit. Contam. 4*: 277 (1987).

182. T. E. Möller and H. F. Gustavsson, Determination of type a and b trichothecenes in

cereals by gas chromatography with electron capture detection, *J. AOAC Int. 75*: 1049 (1992).

183. T. Krishnamurthy, M. B. Wasserman, and E. W. Sarver, Mass spectral investigations on trichothecene mycotoxins, *Biomed. Environ. Mass Spectrom. 13*: 503 (1986).

184. O. L. Shotwell, M. L. Goulden, and G. A. Bennett, Determination of zearalenone in corn: collaborative study, *J. Assoc. Off. Anal. Chem. 59*: 666 (1976).

185. N. M. Quiroga, I. Sola, and E. Varsavsky, Selection of a simple and sensitive method for detecting zearalenone in corn, *J. AOAC Int. 77*: 939 (1994).

186. S. P. Swanson, R. A. Corley, D. G. White, and W. B. Buck, Rapid thin layer chromatographic method for determination of zearalenone and zearalenol in grains and animal feeds, *J. Assoc. Off. Anal. Chem. 67*: 580 (1984).

187. K. Schwadorf and H.-M. Müller, Determination of α- and β-zearalenol and zearalenone in cereals by gas chromatography with ion-trap detection, *J. Chromatogr. 595*: 259 (1992).

188. J. Prieta, M. A. Moreno, J. Bayo, S. Díaz, G. Suárez, L. Domínguez, R. Canela, and V. Sanchis, Determination of patulin by reversed-phase high-performance liquid chromatography with extraction by diphasic dialysis, *Analyst 118*: 171 (1993).

189. T. E. Möller and E. Josefsson, Rapid high pressure liquid chromatography of patulin in apple juice, *J. Assoc. Off. Anal. Chem. 63*: 1055 (1980).

190. A. Visconti, A. Logrieco, and A. Bottalico, Natural occurrence of *Alternaria* mycotoxins in olives—their production and possible transfer into the oil, *Food Addit. Contam. 3*: 323 (1986).

191. E. G. Hiesler, J. Siciliano, E. E. Stinson, S. F. Osman, and D. D. Bills, High-performance liquid chromatographic determination of major mycotoxins produced by *Alternaria* molds, *J. Chromatogr. 194*: 89 (1980).

192. A. Visconti, A. Sibilia, and F. Palmisano, Selective determination of altertoxins by high-performance liquid chromatography with electrochemical detection with dual "in-series" electrodes, *J. Chromatogr. 540*: 376 (1991).

193. F. Palmisano, P. G. Zambonin, A. Visconti, and A. Bottalico, Profiling of *Alternaria* mycotoxins in foodstuffs by high-performance liquid chromatography with diode-array ultraviolet detection, *J. Chromatogr. 465*: 305 (1989).

6

Nitrobenzene and Related Compounds

Norimitsu Saito
Iwate Prefectural Institute of Public Health, Morioka, Japan

Kenji Yamaguchi
Yokogawa Analytical Systems, Inc., Tokyo, Japan

I. INTRODUCTION

Benzene and polynuclear aromatic hydrocarbons (PAH) with one or more of the hydrogens substituted with nitro groups are collectively called nitroarenes or nitro-PAHs. There are over 300 nitro-PAHs commercially produced in the chemical industry [1], and among them the 1 and 2 benzene-ring nitro-PAHs are manufactured and used in largest quantities [2]. The applications of nitro-PAHs are diverse and include explosives, dyes, pigments, pharmaceuticals, flavors, pesticides, rubbers, antiaging agents, fuel additives, solvents, and intermediates for synthetic chemicals. Nitroarenes in the industry are handled by well-trained workers in restricted areas under a strict handling protocol. Therefore problems with nitro-PAHs occur due to occupational exposure and environmental pollution from manufacturing plants.

In 1978 Pitts et al. [3] showed that when PAHs were exposed to a trace amount of NO_2, direct-acting mutagenic nitroarenes were formed, and it was suggested that this reaction could occur in the atmosphere. In 1980 Rosenkranz et al. [4] detected nitropyrenes from carbon black and toners used in copying machines, and showed that these compounds gave a strong mutagenic activity in *Salmonella* strain. Further, nitroarenes have been found in emissions from diesel engines, which had been known as suspect agents of lung cancer. When diesel particulate emissions were found to be the major source of the airborne nitro-arenes, each of the compounds contained in the diesel emissions was identified as a constituent of atmospheric pollutants [5–8].

Earlier airborne particulates and diesel emissions studies on nitro-PAHs were done by *Salmonella* bioassay as a mutagenicity test (Ames test), because chemical analysis gave poor sensitivities at that time [8–11]. Because logarithmic calibration curves were used in mutagenicity tests for quantification, the accuracy was insufficient and the data obtained were influenced by coexisting compounds, resulting in large data errors. Therefore the development of a new analytical technique with high analytical accuracy has been anticipated.

PAHs, which are also atmospheric pollutants, are fluorescent materials, and because of this characteristic, high-performance liquid chromatography (HPLC) can be used with a fluorescence detector (Fl) for a high-sensitivity analysis [12–14]. Further, because the concentration of atmospheric PAHs is high, on the order of $\mu g/m^3$, gas chromatography/mass spectrometry (GC/MS) or GC/flame ioniza-tion detector (FID) can be used for reliable measurements, and has been estab-lished as the official analytical technique.

On the other hand, nitro-PAHs are not fluorescent, and the concentration in the atmosphere is low, on the order of pg/m^3. Further, diesel emissions contain a variety of nitro-PAHs with varied toxicities, and the number is believed to be about 120 [5]. Actual particulate samples contain a number of other contaminants in addition to nitro-PAHs [15]. This makes the chemical analysis of nitro-PAHs difficult, and unified official analytical methods have not been established.

At present the most reliable analytical instrument is the gas chromatography/ mass-mass spectrometric detector (GC/MS-MS) [16,17] or high-resolution GC/ MS (GC/HRMS) [18–20]. However, the cost of these instruments is so high that not every laboratory can afford to acquire them. However, these leading edge instruments are not necessary to analyze nitro-PAHs, and continued efforts are needed only to be able to analyze some of the nitro-PAHs. In fact, GC and HPLC have mainly been used for analysis of nitro-PAHs with varied combinations of detectors and sample preparations.

Pretreatments for analysis differ in analytical instruments, detectors used, and sample forms. Many of the 1- and 2-benzene-ring nitro-PAHs used in the chemical industry are volatile and may be analyzed upon concentrating them without any pretreatment. On the other hand, 3-5 benzene-ring nitro-PAHs from diesel emissions found as atmospheric pollutants are particulates and require an analytical process with extensive pretreatment. Although analysis of particulates takes time and is complicated, the following four basic procedures are common in analysis with GC and HPLC: (1) sample collection, (2) extraction, (3) cleanup, and (4) chromatographic determination.

In this chapter, analysis of volatile compounds is discussed first, followed by analysis of particulates in order of the analytical procedures described above. To date, extractions of particulate samples have been done with Soxhlet extraction or ultrasonic extraction. Extraction techniques are being developed rapidly. Pressurized solvent extraction (PSE) and supercritical fluid extraction (SFE) are also described, although there have been no reports on the applications of these techniques to the analysis of nitro-PAHs. These may be important in the future.

II. NITRO-PAHs—ATMOSPHERIC POLLUTANTS

In the reports of WHO, cigarette smoke, asbestos, coal tar, coal tar pitch, and soot are listed as lung cancer–causing agents. However, they do not specify which of the chemical compounds contained in coal tars and soot are dangerous. Epidemiological studies done on lung cancer have indicated that diesel exhaust and soot generated by the combustion of fossil fuels increases the risk of lung cancer.

In studies on workers in the United States, a high occurrence of lung cancer was found in railroad diesel engine drivers, and the cause was attributed to long-term exposure to diesel emissions [21–23]. In China, investigations on lifestyle were focused on fuels used in cooking and heating, and it was found that the risk of lung cancer increased with wood, smokeless coal, and smoky coal in order of increasing risk. Based on the measurement of benzo(a)pyrene (B(a)H), it was suggested that the cause was PAH exposure [24]. Animal studies have demonstrated firm evidence that organics found in diesel emissions contained carcinogenic agents [25].

In the epidemiological studies mentioned above, evidence may not be adequate, so the carcinogenic ranking of the International Agency for Research on Cancer (IARC), which emphasized epidemiological results, ranks asbestos as the most risky chemical (group 1: carcinogenic chemicals) and chemicals in diesel emissions second (2A: probable carcinogenic chemicals) [26]. The issue is which of the compounds in the diesel emissions and soot generated in the combustion of fossil fuels are causing the increased risk of lung cancer. Upon repeated investigations of mutagenicity tests, the most risky organics were found to be nitropyrenes, a family of nitro-PAHs [27]. Later, a variety of nitro-PAHs were identified as high-risk chemicals. Diesel emissions are currently thought to contain as many as 120 nitro-PAHs. Figure 1 shows some of the important nitro-PAHs reported to date.

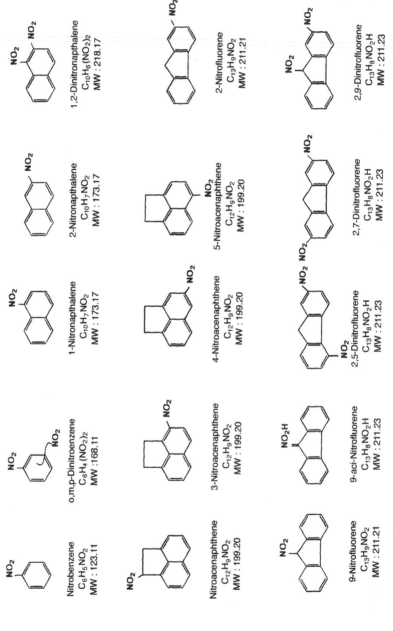

FIGURE 1 Names and chemical structures of nitro-PAHs.

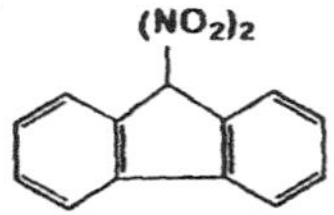

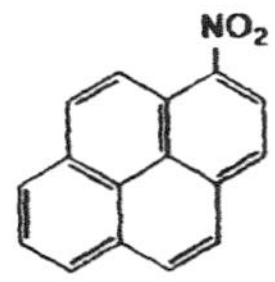

1-Nitropyrene
$C_{16}H_9NO_2$
MW : 247.24

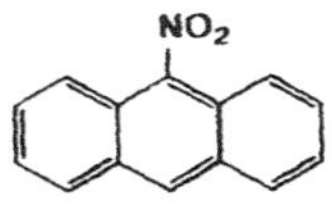

9-Nitroanthracene
$C_{14}H_9NO_2$
MW : 223.22

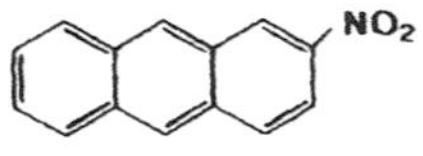

2-Nitroanthracene
$C_{14}H_9NO_2$
MW : 223.22

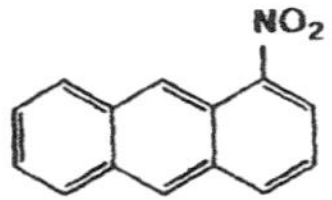

1-Nitroanthracene
$C_{14}H_9NO_2$
MW : 223.22

9,9-Dinitrofluorene
$C_{13}H_8(NO_2)_2$
MW : 256.21

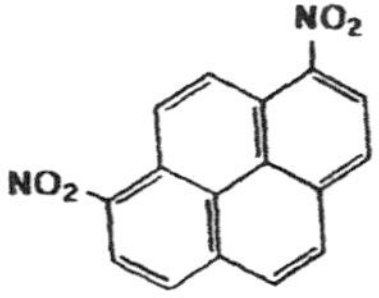

1,8-Dinitropyrene
$C_{16}H_8(NO_2)_2$
MW : 292.24

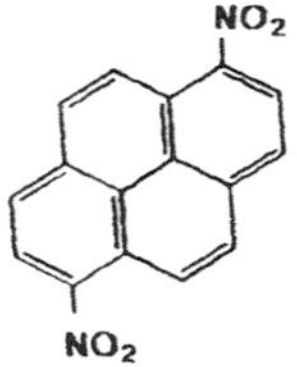

1,6-Dinitropyrene
$C_{16}H_8(NO_2)_2$
MW : 292.24

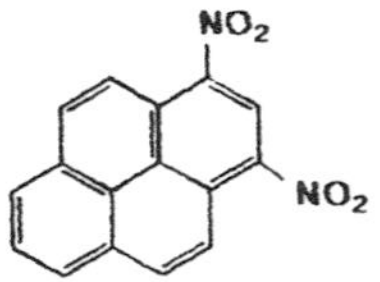

1,3-Dinitropyrene
$C_{16}H_8(NO_2)_2$
MW : 292.24

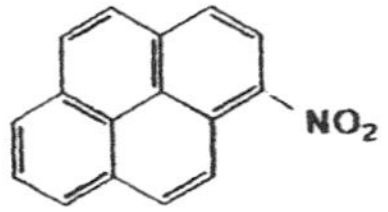

3-Nitropyrene
$C_{16}H_9NO_2$
MW : 247.24

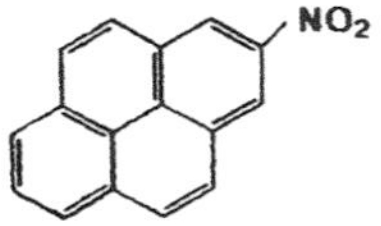

2-Nitropyrene
$C_{16}H_9NO_2$
MW : 247.24

FIGURE 1 Continued.

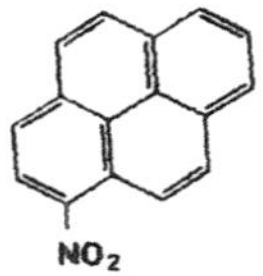

6-Nitropyrene
C_{16}H_9NO_2
MW : 247.24

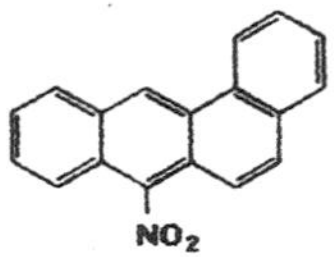

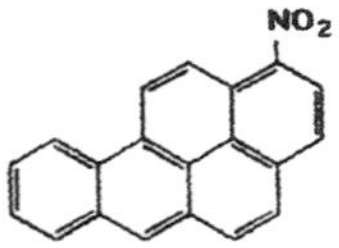

1-Nitrobenzo[a]pyrene
C_{20}H_{11}NO_2
MW : 297.30

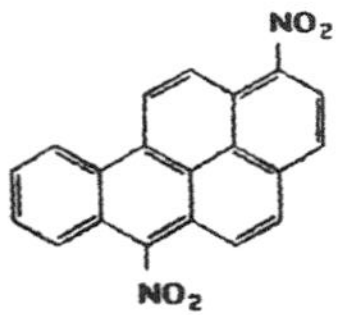

1,6-Dinitrobenzo[a]pyrene
C_{20}H_{10}(NO_2)_2
MW : 342.30

7-Nitrobenz[a]anthracene
C_{18}H_{11}NO_2
MW : 273.28

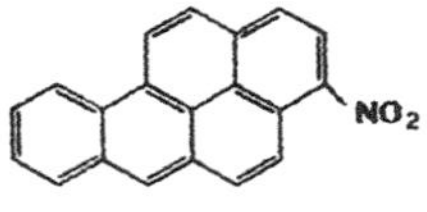

3-Nitrobenzo[a]pyrene
C_{20}H_{11}NO_2
MW : 297.30

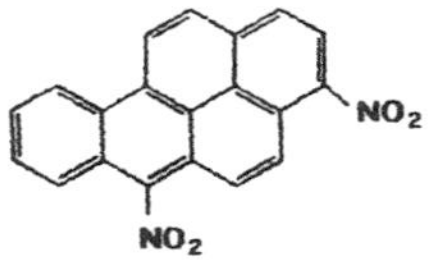

3,6-Dinitrobenzo[a]pyrene
C_{20}H_{10}(NO_2)_2
MW : 342.30

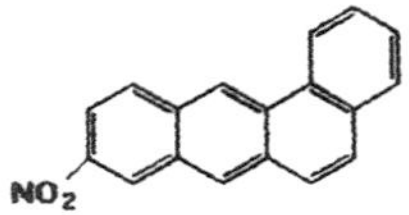

9-Nitrobenz[a]anthracene
C_{18}H_{11}NO_2
MW : 273.28

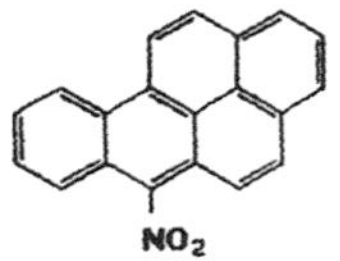

6-Nitrobenzo[a]pyrene
C_{20}H_{11}NO_2
MW : 297.30

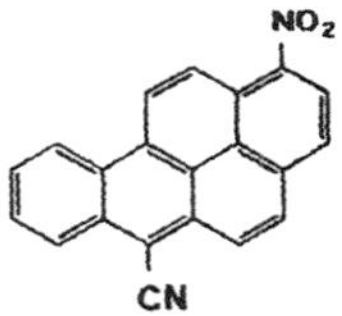

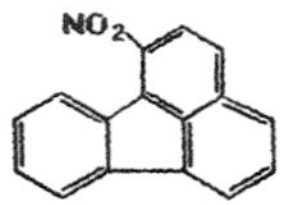

1-Nitrofluoranthene
C_{16}H_9NO_2
MW : 247.24

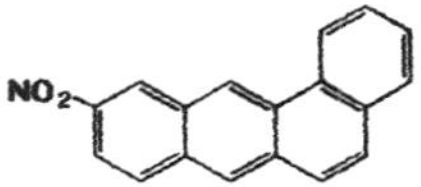

10-Nitrobenz[a]anthracene
C_{18}H_{11}NO_2
MW : 273.28

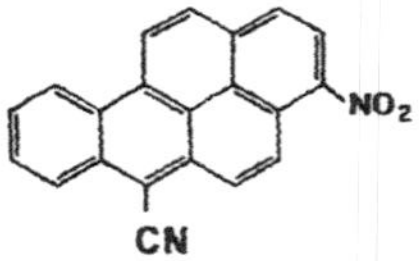

1-Nitro-6-cyanobenzo[a]pyrene
C_{20}H_{10}NO_2CN
MW : 322.31

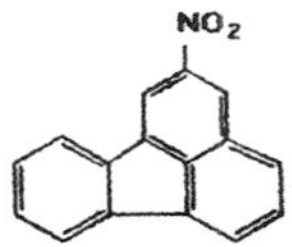

2-Nitrofluoranthene
C_{16}H_9NO_2
MW : 247.24

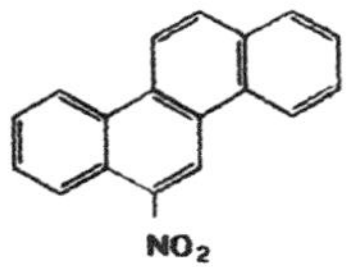

6-Nitrochrysene
C_{18}H_{11}NO_2
MW : 273.28

1-Nitro-6-cyanobenzo[a]pyrene
C_{20}H_{10}NO_2CN
MW : 322.31

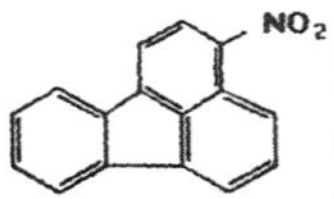

3-Nitrofluoranthene
C_{16}H_9NO_2
MW : 247.24

FIGURE 1 Continued.

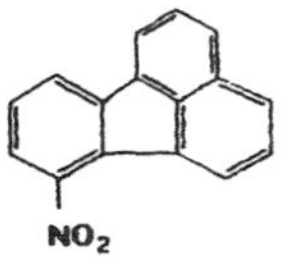

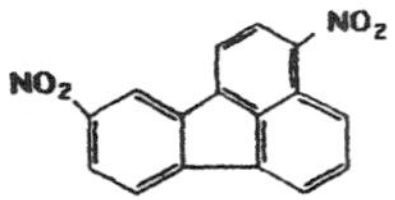

3,9-Dinitrofluoranthene
$C_{16}H_9(NO_2)_2$
MW : 292.24

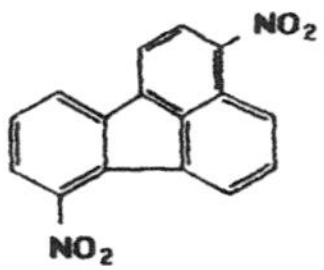

3,7-Dinitrofluoranthene
$C_{16}H_9(NO_2)_2$
MW : 292.24

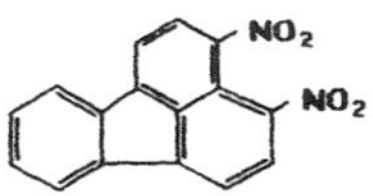

3,4-Nitrofluoranthene
$C_{16}H_9(NO_2)_2$
MW : 292.24

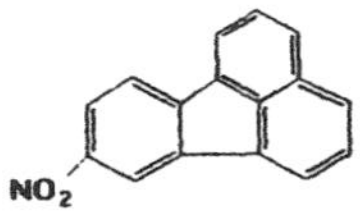

8-Nitrofluoranthene
$C_{16}H_9NO_2$
MW : 247.24

7-Nitrofluoranthene
$C_{16}H_9NO_2$
MW : 247.24

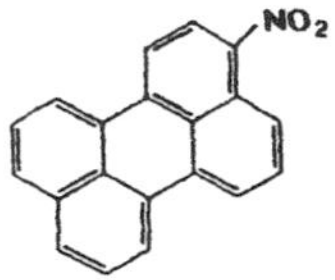

3-Nitroperylene
$C_{20}H_{11}NO_2$
MW : 297.30

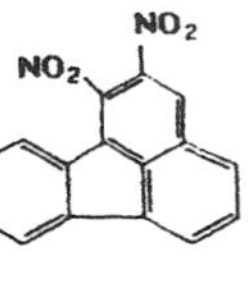

1,2-Dinitrofluoranthene
$C_{16}H_9(NO_2)_2$
MW : 292.24

FIGURE 1 Continued.

These nitro-PAHs are easily released to the atmosphere when fossil fuels are combusted. This is because PAHs generated by a series of radical reactions react with nitrogen and oxygen in air at high temperatures [16]. Nitro-PAHs exist as airborne particulates [28] and are also found in grilled foods [29] and used motor oils [30]. The origin of these compounds found to date is exhaust from automobiles, aircraft, and gas and kerosene heaters [11,27,31]. Among them, diesel engine-powered vehicles are the most important cause of atmospheric pollution [28,32].

In terms of 1-NP contents of emissions, the difference between gasoline- and diesel-powered automobiles is small [33]. But, the 1-nitropyrene (1-NP) per kilometer contents of particulate emissions from diesel engines are hundreds of times as large as gasoline-powered automobiles, because of a much larger amount of emissions in weight from diesel engines. The emissions from diesel engines are high not only in 1-NP but also in other nitro-PAHs.

Because diesel engines are the major sources of these atmospheric pollutants, they are widely distributed in the environment. This causes serious problems in urban areas, where large numbers of automobiles are concentrated in small regions [34] (see Table 1).

Table 1   The Effect of Engine Operating Conditions on the Concentration of Nitro-PAHs in Heavy-Duty Diesel Particulates

	PPM Concentration in Particulates		
Compound	A Idle	B High-Speed Zero Load	C High-Speed Full Load
2-nitrofluorene	84 (164)	62 (134)	1.9 (15)
3-nitro-9-fluorenone	18 (35)	7.9 (17)	8.0 (63)
2-nitro-9-fluorenone	10 (19)	4.8 (10)	3.7 (29)
9-nitroanthracene	94 (184)	16 (35)	5.1 (40)
9-nitro-1-methylanthracene	129 (252)	13 (28)	0.2 (1.6)
3-nitro-1,8-naphthalic acid anhydride	23 (46)	10 (22)	22 (174)
1-nitropyrene	14 (28)	3.0 (6.5)	0.13 (1.0)
2,7-dinitrofluorene	15 (30)	18 (39)	3.9 (31)
2,4-dinitro-9-fluorenone	5.5 (11)	8.0 (17)	2.1 (17)
2,4,7-trinitro-9-fluorenone	(<1.0) (<2.0)	0.4 (0.9)	NR
1,3-dinitropyrene	(<0.8) (<1.6)	0.6 (1.3)	0.4 (3.1)
1,6-dinitropyrene	(<0.8) (<1.6)	1.2 (2.6)	0.8 (6.3)
1,8-dinitropyrene	(<0.8) (<1.6)	1.2 (2.6)	0.8 (6.3)
6-nitrobenzo(a)pyrene	(<3.2) (<6.5)	1.6 (3.5)	0.3 (2.4)

Number in parentheses is ppm concentration in extract. Extraction of samples A, B, and C gave 51.0, 45.9, and 12.75% extractables, respectively. NR = not resolved.

III. TOXICITY OF NITRO-PAHs

1- and 2-benzene-ring nitro-PAHs used in the chemical industry become an issue when poisoning in the work environment occurs. These nitro-PAHs exhibit high percutaneous absorption, and workers are susceptible to inhalation exposure. This is the main reason for poisoning. The symptoms are characterized by met hemoglobin formation and anemia by acute exposure [35]. This is comparable to nitro-PAHs originating from diesel engine emissions which exhibit direct-acting mutagenic activity [36–38]. This mutagenic activity is so high that it is called supermutagenic.

If mutation is defined as an alteration of genes, 2-benzene ring systems such as nitronaphthalenes and nitroacenaphthenes are mutagens resulting from base displacement (transition and/or transversion), while 3-, 4-, and 5-benzene ring systems such as fluorene, fluoranthene, pyrene, and nitro derivatives of B(a)P are structural mutagens by frameshift (shifting of read frame by addition, insertion, removal of bases). The epidemiological conclusions mentioned above, in which diesel emissions increased the risk of lung cancer, have been supported by animal studies.

In particular, nitropyrenes and nitrofluoranthenes exhibited carcinogenicity in rodents of different species, and this was caused by genetic carcinogens due to the formation of DNA adducts [39–43]. If the vapors of these nitro-PAHs are ingested into the body, they are hydroxylated and excreted readily. However, it has been reported that if nitro-PAHs adsorbed in particulates are inhaled into the respiratory systems, a significantly increased risk of lung cancer results due to its long retention time [44,45]. In daily life, exposure to carcinogens occurs through airborne particulates; therefore nitro-PAHs in the atmosphere from diesel emissions have been considered a serious issue.

Rather than monitoring PAHs, a more realistic and accurate risk evaluation of atmospheric pollutants could be obtained by directly monitoring nitro-PAHs which have been proven to be carcinogens by animal studies. Accurate measurements of environmental nitro-PAHs and their data evaluations will be increasingly important in the future in order to assess the influence on the health of workers vulnerable to nitro-PAH exposure (diesel engine drivers, coal liquefaction workers, aluminum workers, etc.) and of citizens vulnerable to diesel emission exposure.

IV. ANALYSIS OF VOLATILE NITRO-PAHs

A. Air Samples

While 4-benzene-ring pyrenes are known to exit in the gas phase into the atmosphere, 1-NP in which one of hydrogens is substituted with NO_2 exists as particulates in the atmosphere [46]. This is because 1-NP has a large polar NO_2 group preventing the molecule from existing in the gas phase. The larger the number of

NO_2 groups in the molecule, the harder it is to be in the gas phase in the atmosphere. Only mononitro-PAHs with 3-benzene rings or less and dinitro-PAHs with 2-benzene rings can exist in the atmosphere in the gas phase.

The concentration of volatile nitro-PAHs found in the air or the working environment is generally low. Therefore, when analyzing these volatile compounds, adsorption tubes are generally required to concentrate samples. The packing material for these adsorption tubes is a polymer adsorbent such as Tenax, Chromosorb 102, or XAD 2 [47,48]. GC-MS analysis of the volatile nitro-PAH can be performed easily. To prepare an adsorption tube, for example, Tenax-TA, an adsorbent, is put into a glass-coated stainless tube, and both ends are filled with silica wool; then adaptors are attached at both ends. Before using the adsorption tube, conditioning must be done by passing nitrogen gas through at 40 ml/min while heating the adsorption tube at 280°C for 12 hours.

To collect air samples, a diaphragm pump is connected to one of the tube ends, and about 30 l of air is drawn through the tube. The adsorption tube containing the air sample is then stoppered and stored for analysis.

A 3-way cock is attached to the injection port of a GC/MS. A precolumn, used as a cold trap, is immersed in liquid oxygen or liquid nitrogen. A syringe needle is plugged into the adsorption tube, then attached at the GC/MS injection port. The adsorption tube is heated from room temperature to 280°C in 3 min so that the materials in the adsorption tube are desorbed and trapped in the precolumn. The precolumn, a cryofocusing capillary column, is required to narrow the band of the sample before introduction into the separation column. Liquid oxygen or liquid nitrogen is removed, and GC/MS analysis is initiated. Nonpolar capillary columns are used for the separation columns. A volatile nitro-PAH sample collected by Tenax-TA is stable at ambient temperature for 3 weeks.

B. Water Samples

1. Concentration of Nitro-PAHs in Water for GC/MS Analysis

Samples need to be concentrated before analyzing for a trace amount of volatile nitro-PAH. Sample concentrations are done by a glass column (15 cm × 1 cm φ) packed with 2–3 g of polymer adsorbent (Amberlite XAD, C18 reverse phase cartridge, etc). This adsorbent consists of spherical beads made of rigid and insoluble microporous resin. The surface of the adsorbent is hydrophobic; thus it requires conditioning. Conditioning of the adsorbent is performed as follows. The adsorbent is hydrated by immersing it in pure water overnight. It is then packed in a glass column and rinsed with pure water. A polar solvent such as an alcohol or acetone is run through the column to remove water in the resin beads. The amount of the solvent required is 2–3 times the amount of the adsorbent. It is allowed to stand overnight, and is then rinsed with pure water to remove impurities and bubbles contained in the resin surface. It is further treated with 3–5 times the amount of the resin of 4% NaOH solution for 1–2 h and then is rinsed with pure water until the eluent is no longer colored by phenolphthalein solution. The sample is passed through the column at 30 ml/min in order to adsorb nitro-PAHs. After 1–2 l of the sample water is run through the column, the resin is dried by blowing nitrogen through for 15 min. Nitro-PAHs are eluted off with dichloromethane. The eluent is then concentrated by a rotary evaporator down to 1 ml for GC/MS analysis.

Feltes et al. [49] used a 1:1:1 mixture of Amberlite XAD-2, -4, and -8 resins, and analyzed waters in ponds and rivers for 11 different nitro-PAHs such as nitrobenzene. They found that this method gave higher yields than the liquid-liquid extraction or cartridge column method.

2. Purge and Trap GC/MS

GC/MS equipped with a purge and trap (P&T) is used to analyze polluted environmental water for volatile nitro-PAHs which exist at relatively high concentrations. In the P&T-GC/MS, the material in a sample vial is purged and introduced into the injection port of the GC/MS for analysis (Fig. 2). The whole process is fully automated. Procedures are as follows. A sample is collected in a vial. A septum is attached, and after the vial cap is securely closed, the vial is loaded into the autosampler. All the vials and other apparatuses are heated at 130°C for 24 h prior to use so that they are free of contamination. A purified inorganic gas is used to purge the aqueous sample through a frit sparger. Volatile species are allowed to dissolve in the bubbles or strip from the aqueous sample. The volatile components are collected in a trap tube packed with an adsorbent. High purity He is generally used for purging the gas. Nitro-PAH is collected by a trap tube. The adsorbent used consists of low polarity porous polymer beads, silica gel, and activated carbon. These 3 adsorbents are packed in a tube in this order. The sample is trapped in the tube by introducing it into the tube end closest to the porous polymer beads. While gradually heating the tube, a desorbing gas is run through the tube from the other end to remove the trapped materials from the adsorbent. The volatile components are collected by a precolumn (cryofocusing capillary column).

As described in Sec. IV.A. above, when using a capillary column for analysis of volatile compounds, a cryofocusing technique that reduces the sample volume and its bandwidth is used in order to increase the peak resolution. While a carrier gas is allowed to flow into the cryofocusing column, rapid heating of the column will drive volatile compounds instantly into the GC-MS for analysis.

3. Head Space GC/MS

The head space technique utilizes the equilibrium between the liquid and gas phases of volatile spaces contained in a closed container under a certain condition;

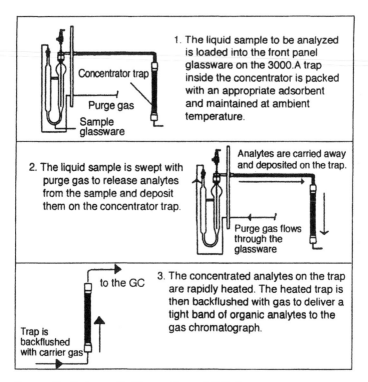

1. The liquid sample to be analyzed is loaded into the front panel glassware on the 3000.A trap inside the concentrator is packed with an appropriate adsorbent and maintained at ambient temperature.

2. The liquid sample is swept with purge gas to release analytes from the sample and deposit them on the concentrator trap.

Analytes are carried away and deposited on the trap.

Purge gas flows through the glassware

3. The concentrated analytes on the trap are rapidly heated. The heated trap is then backflushed with gas to deliver a tight band of organic analytes to the gas chromatograph.

Trap is backflushed with carrier gas

to the GC

Concentrator trap

Purge gas

Sample glassware

FIGURE 2   Schematic illustration of P & T.

it is used for analysis of volatile nitro-PAHs. The amount of material introduced into GC is 1/10–1/100 of that for the P&T technique, which uses almost all of the material in the sample for analysis; therefore, this is used only when the compounds of interest in the sample exist at high concentrations.

In actual analytical situations, the sample must be carefully transferred into a vial without producing bubbles. The amount of sample transferred into the vial must be 70–85% of the vial volume and must be held constant during the series of analyses. In order for the partition coefficient to be held equal for all the samples, 1 μl of methanol per 10 ml of sample is added to the sample. A piece of Teflon sheet (0.05 mm thick) is placed over the vial mouth, followed by a vial rubber stopper, and then an aluminum vial cap is placed over it. The rubber stopper is secured to the vial using a cap tightener. The vial is then loaded into the autosampler and the system is brought to an equilibrium at 25°C for 30–60 min. Using a gas-tight syringe or sampling tube, a volume is sampled out from the gas phase of the sample and is injected into GC/MS. The analysis by GC/MS is fully automated.

V. ANALYSIS OF PARTICULATE NITRO-PAHs

A. Sampling

1. Air Samples

When fossil fuels are burned, many kinds of PAHs and nitro-PAHs are generated. Among them, PAHs with 5 or more benzene rings, mononitro-PAHs with 4 or more benzene rings, and dinitro-PAHs with 3 or more benzene rings exist in particulate matter. Nitro-PAHs with 2 and 3 benzene rings are known to be detected both from particulates and the gas phase. Generally the concentrations of PAHs and nitro-PAHs in the environment are so low that they can hardly be detected directly, even with modern analytical instrumentation. Therefore, a large quantity of sample must be collected in order to quantify particulate nitro-PAHs in the atmosphere. A high-volume air sampler (HV air sampler) is often employed for this purpose. A low-volume air sampler (LV air sampler) collects about 40 m^3, while a HV air sampler collects about 2,000 m^3. One of the advantages to the use of the HV air sampler is that a large amount of particulates can be collected on a filter paper. When GC/MS or HPLC is used for analysis, the amount of sample collected by a middle volume air sampler or by a LV air sampler may not be sufficient. The filters that can be used with a HV air sampler are mainly glass fiber filters, quartz fiber filters, and Teflon-coated filters [20,50–53]. The choice of filter is important and requires careful consideration. One that best suits your experiments should be used [54] (Table 2).

    When using the HV air sampler, the following precautions should be exercised. Before and after the sample collection, the filter paper used with the HV air sampler should be weighed repeatedly until the weight reading becomes constant under identical drying conditions, e.g., in a desiccator or in a constant temperature oven. The filter should be weighed to 0.01 mg. Samples should be analyzed immediately following the sample collection. If this is not practical, cover the sample with aluminum foil to protect it from ambient light and store it in a freezer. This will prevent losses due to evaporation, secondary reactions, and photodegradation [55–59]. In short, samples collected by the HV air sampler must not be stored under circumstances that can alter PAHs or nitro-PAHs in any way [60].

2. Diesel Exhaust Particulate Samples

About 0.2% of diesel fuel combusted in a diesel engine is emitted as particulates. Particulates of 2 μm in diameter are called diesel particulates and are chains of particles with an average diameter of 0.1–0.2 μm. The nucleus of the diesel particulates is carbon and is covered by SO_4, metals (Fe, Ca, Zn, etc.), H_2O, soluble organic fraction (SOF), total organic extract (TOE), etc. (Fig. 3). SOF is a component that can be extracted with organic solvents and contains PAHs and nitro-PAHs. PAHs and nitro-PAHs in diesel particulate emissions are generally

Table 2 Selected Filter Types and Their Measured Efficiencies

Filter type	Efficiency (%)	Flow resistance (mm H$_2$O)	Humidity (mg)	Thermal stability (°C)	Filter strength (–)	Acidic gas absorption (–)	Application
Glass fiber filter	99.99<	30	0.06	500	Medium	Large	Moisture adsorption high; Used for analysis of particulate concentration and composition by HV-LV
Silica (quartz) fiber filter	99<	45	0.03	1000	Weak	Nothing	Moisture adsorption high; Filter strength low, handling needs care; Used for analysis of particulate concentration and composition by HV-LV
Teflon-coated glass fiber filter	99.9<	30	0.04	315	Medium	Nothing	Used for analysis of particulate concentration and composition by HV-LV
Supported PTFE type membrane filter	almost 100	120	0.03	150	Medium	Nothing	Used for analysis of particulate concentration in exhaust gas; Flow resistance high; Hard to obtain required flow
Polycarbonate type membrane filter	almost 100	120	##	140	Weak	Nothing	Flow resistance high; Filter positioning difficult; Difficulty in handling; Best for collecting samples for optical microscopes and SEM analysis; Poor yields in particulate collection
Polyflon filter	nearly 95	25	0.09	260	Strong	Small	Used for analysis of dust concentrations of a gas containing sulfuric acid mist when collecting particulates by HV-LV

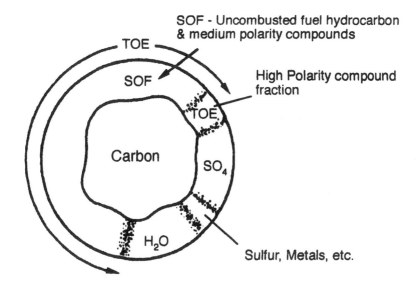

Solvent extracted compounds

Organic solvent soluble compounds (SOF) - Soxhlet extracts by dichloromethane

Total organic extract (TOE) - Soxhlet extracts by toluene/ethanol

Water soluble sulfates (SULFATES) - elution by water, ion chromatography analysis

FIGURE 3   Structure of diesel particulates.

collected on a filter by a dilution tunnel [15] (Fig. 4). Teflon-coated filters are most commonly used for sample collection [15,61]. Collected with this filter (47–102 mm φ) are species existing as particulates at ambient temperatures. Lower molecular weight PAHs and nitro-PAHs existing in the gas phase can be collected by an adsorbent (XAD-2, PUF, etc.) placed after the filter in series. Sample collection at high temperatures should be avoided. Schuetzle et al. [7] have shown that sample collections should be made at exhaust gas temperatures of 43°C or less and NO_2 concentrations of 3 ppm or less using a Teflon-coated filter, because the higher the temperature and NO_2 concentration, the higher the reactivity of PAHs and nitro-PAHs.

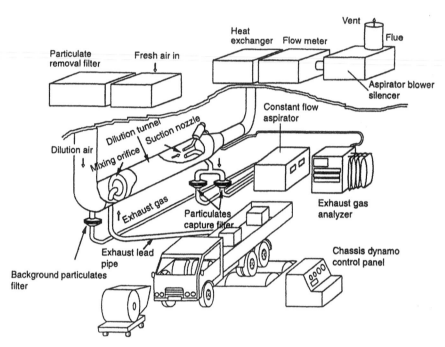

FIGURE 4 Schematic illustration of dilution tunneling diesel emission analyzer.

B. Extraction

1. Soxhlet Extraction

The most accepted and standard method for extracting nitro-PAHs from diesel emission particulates collected on filter paper is Soxhlet extraction [16,20,62]. A Soxhlet extractor is made of a glass chamber with a side tube acting as a siphon. Once the amount of solvent collected in the extracting chamber reaches a certain level, the solvent is transferred to the receiving flask by siphon action. Then fresh solvent gradually fills the extracting compartment to repeat the cycle. The extractor is available in various sizes. The size to be selected should be based on the amount of the sample and the size of the filter paper used for sample collection. Powder samples are placed in a cylindrical filter paper to be used with a Soxhlet extractor. Filter papers on which airborne particulates or diesel particulates have been collected should not be cut into thin strips and placed in a cylindrical filter paper. Doing so can cause some sample to be lost and generate errors. Instead, the filter paper should be carefully folded up with sample inside and tied with a metal wire to prevent unfolding during extraction. This method should not be applied to some filters of low strength, such as Tissuquartz.

Solvents used for Soxhlet extraction are dichloromethane, cyclohexane, acetonitrile, toluene, benzene-methanol mixture, etc. [50,63]. The choice of solvent is determined by the sample analyzed. One-half to one-third of the receiving flask should be filled with the solvent to start. Since nitro-PAHs are unstable at high temperatures, the extraction must be performed at low temperatures, possibly at 50–70°C or lower. The number of extraction cycles and times can be varied; however, 2–4 cycles/h for 6–24 h are commonly used. Nitro-PAHs are also unstable to light irradiation and should be protected from photodegradation by ambient light by covering the extractor with aluminum foils or by running the extraction in the dark, if it requires a long period of time.

2. Ultrasonic Extraction

Like Soxhlet extraction, ultrasonic extraction is a commonly used technique [64]. A powder sample can be placed directly in an Erlenmeyer flask with a stopper, while a filter paper on which sample has been collected is cut into small strips before being placed into the flask. Then a solvent is added to the flask. The solvent is stirred with a glass rod until the sample is completely wet and all bubbles are removed. The flask is then sonicated on a ultrasonicator [65].

Solvents used in ultrasonic extraction are benzene-methanol, benzene-ethanol, ethyl acetate, dichloromethane, acetone, ethanol, etc. The flask is sonicated first at a high frequency for about 30 s or so in order to remove the particulates from the filter paper. Sonication should result in a dark suspension. Then, after switching to a low frequency, sonication is continued for 15–30 min. In ultrasonic extraction, high and low frequencies are switched alternatively to promote an efficient extraction. The extract is then filtered through a membrane filter before analysis.

Lee and Schuetzle [111] have reported a modified ultrasonic extraction in which a Millipore glass fiber funnel and a sonication probe were used. In this method, filter paper is cut into thin strips and placed in a funnel. The sonication probe is then placed over the filter strips in the funnel, and the vacuum pump is turned on. A solvent of choice is added, and the sonication probe is turned on to initiate extraction. Since the amount of solvent in the funnel decreases by vacuum filtration, additional solvent must be added. The solvent added will dilute the extract and allow efficient extraction. The membrane filter used is usually a Teflon filter with 0.2–0.5 μm pore size. It is necessary to use a filter that is stable to the solvent employed. This technique accomplishes extraction and filtration at the same time. The filtration time depends on the rate of filtration and normally ranges from 0.5–1.5 hr.

Soxhlet extraction is considered a standard extraction method, although it takes a long time and requires a large amount of solvent. Ultrasonic extraction does not require a special apparatus, nor does it require a long time for extraction. Therefore it would be a method of choice for a large number of samples.

3. Pressurized Solvent Extraction

In pressurized solvent extraction (PSE) the sample is first loaded into the extraction cell, which is then filled with a solvent, followed by heating (100–150°C) and pressurizing (1200–2500 psi) to effect efficient and rapid extraction [66] (Fig. 5). The diffusion rate and mobility of a solvent increases at high temperatures, leading to an increased solubility of sample materials. Generally when the temperature of a solvent is raised, it transforms into a gas. Under high pressures, however, the solvent can remain liquid and can incorporate even volatiles. In comparison to Soxhlet extraction or ultrasonic extraction, PSE can dramatically reduce the amount of solvent and time required for extraction. Further, all the operations are performed in a completely closed system and no solvent vapor is released outside of the apparatus. It is fully automated and is a reliable extraction technique.

The operational procedures for PSE are as follows. If the sample is a solid mass, it is powderized for increased surface area. If the sample contains more than 10% moisture, some drying agent (e.g., Varian Hydromatrix, Na_2SO_4) is added so that solvent-sample contact area is increased. Some adsorbent may be used to

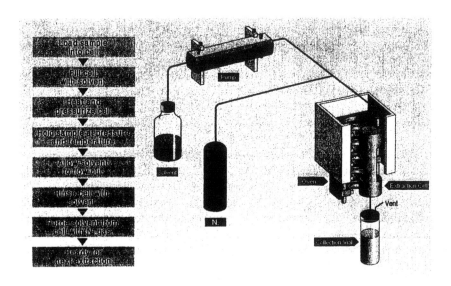

FIGURE 5 Schematic illustration of PSE (pressurized solvent extraction).

prevent loss of volatile components. The sample is placed in a stainless steel pressure-resistant extraction vessel, and an appropriate amount of solvent is added. The cell is loaded into the PSE. The cell is heated, pressurized, and held at a specified temperature and pressure for about 5 min. The cell is then cooled, and the pressure is released. The extract is transferred into a collection vial while the cell is rinsed with solvent. The cell is purged with N_2 to completely remove the remaining solvent. Generally it takes only 15-20 min for complete extraction of a sample. Solvents that would be used in Soxhlet extraction can be used for PSE.

There have been no reports describing nitro-PAH extraction by PSE. However, recovery of PAHs using PSE has been reported, and it was found that there was no significant difference in extraction efficiency between PSE and Soxhlet extraction. PSE is described here in the expectation that this technique will replace Soxhlet and ultrasonic extraction in the future.

4. Supercritical Fluid Extraction

In contrast to Soxhlet extraction, in which organic solvents are used for extraction, supercritical fluid extraction (SFE) employs CO_2 instead [67–69]. Supercritical fluid can be obtained by applying high pressure and temperature on CO_2 gas beyond the critical point of CO_2. CO_2 under this condition is not a gas, nor is it a liquid (nor a solid, of course), and is called supercritical fluid. Supercritical fluid is characterized by having viscosity and diffusion characteristics similar to gases which exert a high osmotic pressure and a high solvent power that is a characteristic for solvents.

Generally there is a correlation between the solvent power of supercritical fluid and its density; therefore by controlling the pressure and temperature the solvent power can be altered, allowing selective extraction of compounds of interest in a short period of time. Further, by addition of a small amount of a solvent called "modifier," the range of components extracted can be varied. Because of the structural requirements, pressure-resistant parts are required; however, procedures can be automated to process continuously a number of samples at one time. In addition, SFE from some manufacturers allows the extract to be directly introduced into the injection port of GC or LC. At present SFE can only handle relatively dry solid samples, and other forms of samples cannot be used with this instrument. However, it has been reported that SFE was used for selective extractions of diesel exhaust particulates and other particulate samples, and the rest of the cleanup procedures could be eliminated.

In summary, the advantages of SFE are (1) it shortens the time required for extraction; (2) selective extraction eliminated further pretreatments; and (3) it can be fully automated for reliability. Because of these advantages SFE will be widely used in the coming years, and further development will continue. In Table 3, the extraction techniques discussed above are compared.

Table 3   Extraction Performances Comparison

Extraction method	Average solvent consumption[1] (ml)	Average extraction[2] time (min)	Laboratory contamination by solvent	Automation
PSE	15–40	12–18	No	Good
Soxhlet	200–500	240–2880	Yes	Fair
Automated Soxhlet	50–100	60–240	Yes	Good
Ultrasonication	100–300	30–60	Yes	Poor
SFE	8–50	30–120	No	Good

[1,2]Solvent consumption per sample (ml) and extraction time (min).

C. Cleanup

Particulates generated by the combustion of fossil fuels contain organics with molecular weight up to 800 and also contain a trace amount of PAHs and nitro-PAHs as basic and acidic fractions. Diesel particulate emissions are known to contain over 120 nitro-PAHs; therefore nitro-PAHs that can be synthesized in a laboratory should also be found in diesel emissions. Extracts from airborne or diesel particulates described in the previous chapter contain, in addition to species of interest, a variety of chemicals at low concentrations that can act as interfering species in analysis. This is because a fractionation process is required to analyze PAHs and nitro-PAHs from extracts with high accuracy.

1.   Liquid-Liquid Partition

Fractionation by liquid-liquid partition has been established by Funkenbush et al. [70]. This is a technique that separates extracts into three fractions, i.e., acidic, basic, and neutral (Fig. 6). At present the first fractionation is performed according to this technique. The purpose of the procedure is to extract selectively a neutral fraction that contains PAHs and nitro-PAHs from the rough extract. The procedure is as follows. The solvent of the extract from particulates is replaced with benzene or ethyl ether. Aqueous basic solution is added to the extract and the mixture is shaken, and then the aqueous layer is removed. The same process is then repeated with acidic solution. Washing the organic layer with distilled water gives a neutral fraction. Species soluble in basic and acidic fractions are removed by this fractionation. The neutral fraction contains not only nitro-PAHs but also various hydrocarbons. Therefore the neutral fraction obtained by liquid-liquid partition can hardly be analyzed reliably by GC or HPLC, because this cleanup process is usually not adequate and the fraction still contains a number of interfering compounds that would show up as background noises. To increase the S/N ratio and accuracy in the chromatographic analysis, further cleanup treatments with column chromatography or HPLC are usually required.

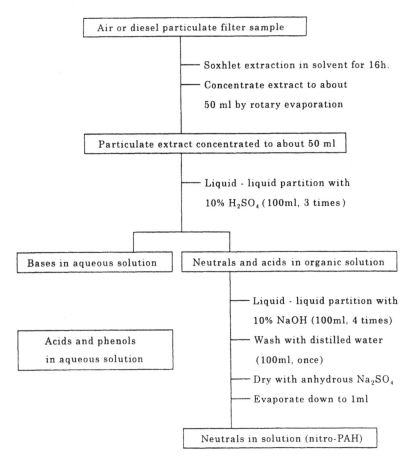

FIGURE 6 Scheme for liquid-liquid partition of organic materials in air and diesel particulates.

2. Column Chromatography and Solid Phase Extraction

Column chromatography is used to clean up the organic extracts from particulate matters [4,7,56,71–74]. Generally this process alone is not sufficient for a cleanup. Neutral fractions obtained by liquid-liquid partition which contains PAHs and nitro-PAHs are further fractionated by this method. Column chromatography does not require special expensive devices and can be prepared by packing an adsorbent material in a glass column. Adsorbent material commonly used is silica gel, and as shown in Fig. 7 a nitro-PAH-containing fraction is obtained by switching polarities of solvents. This affinity chromatography can be

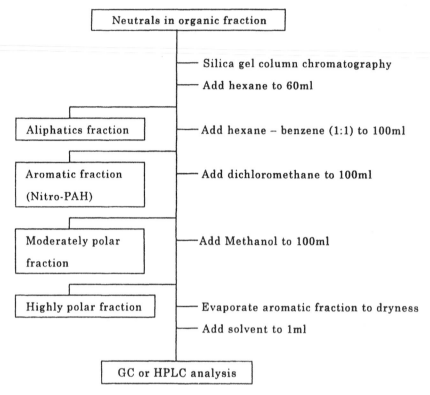

FIGURE 7 Scheme for column chromatographic fractionation of hydrocarbons.

prepared by the user in the laboratory, but it is not easy to prepare one with a good uniformity. In recent years the use of solid phase extraction cartridges has been reported. Solid phase extraction (SPE) cartridges are now commercially available. They are prepared columns and require only a few ml of solvent [75]. The extraction condition of SPE is similar to that of column chromatography.

3. HPLC Fractionation

HPLC is commonly used for separation and analysis of a mixture of compounds. Here, HPLC is used as a pretreatment to clean up samples with contamination from living bodies, foods, and the environment [20,27,63,73,75–79]. The conventional and simplest way to clean up samples is fractionation by TLC. The advantage of HPLC over TLC [80], column chromatography, and other pretreatment methods is its high resolution.

Currently a variety of preparatory columns are commercially available.

Columns that are used for pretreatment of airborne and diesel particulates analysis are ones packed with silica gel. The neutral fraction of atmospheric and diesel exhaust particulates still contains a wide spectrum of compounds, ranging from low molecular weight to high molecular weight materials and compounds with low polarity to high polarity. HPLC columns used for the cleanup process are preparatory columns that can take a large quantity of sample at one time and give high resolution as well as high mechanical strength to withstand heavy usage.

A silica gel column separates compounds by its adsorbing capability, and silica gel, a packing material, offers both high mechanical strength under high pressures and a large number of theoretical plates. Therefore, silica gel columns meet the requirements described above.

In contrast to TLC and column chromatography, HPLC allows the use of a detector to monitor the fractions eluted, leading to accurate fractionation. This fractionation method is sometimes used in combination with the liquid-liquid partition method; however, it is used to selectively fractionate compounds of interest directly from the extracts without liquid-liquid partition.

HPLC fractionation affords facile fractions with high accuracy in a short period of time. The HPLC-cleaned samples are concentrated, followed by chromatographic analysis by GC and HPLC for quantification.

When HPLC is also used as an analytical instrument, the heart-cut method, in which a preparatory column, a flow switching valve, and an analytical column are connected in series, is often employed. This allows the fractions obtained by the preparatory column to be immediately analyzed by the analytical column by switching the flow valve, allowing the cleanup and analysis operations to occur at one time.

D. Chromatographic Analysis

Nitro-PAHs are significantly more difficult to analyze than PAHs, because (1) the vapor pressure of nitro-PAHs is smaller; (2) nitro-PAHs exhibit higher polarity and affinity; (3) environmental concentration is 1/10 to 1/100 that of the PAHs. The extraction and cleanup procedures must be done prior to conducting the chromatographic analysis on the samples for nitro-PAH using a detector with a high sensitivity and selectivity. Recommended GC and HPLC analytical methods are described below.

1. GC Analysis

Nitro-PAHs have a large number of positional isomers of the nitro group with varied toxicities. It is, therefore, important to be able to separate each of the isomers. Also it is desirable to have as many nitro-PAHs analyzed as possible in one analytical operation. Capillary columns are important in analyzing highly contaminated samples, because the number of theoretical plates is as high as several hundred thousand. In GC analysis it is important to select the best analytical column and high selectivity detector.

Among detectors, GC/MS, which utilizes mass spectrometer (MS) as a detector, has many functions and is considered a powerful analytical tool; therefore it is worthy of a separate chapter.

Analytical Columns

The GC columns used to separate nitro-PAHs have been nonpolar fused silica capillary columns. Currently the following four liquid phases are used.

1. Methyl silicone (Ultra-1, HP-1)
2. 5% phenyl silicone (DB-5, Ultra-2, SE-52, SE-54, HP-5, SPB-5)
3. 50% phenyl, 50% methyl-silicone (DB-17, OV-225)
4. 7% cyanopropyl, 7% phenyl, 86% methyl-silicone (DB-1701, SPB-1701)

1. and 2. above are of standard liquid phases and are used as nonpolar columns; 3. and 4. are medium polarity columns and are used to separate isomers that are difficult to separate with nonpolar columns. The columns and analytical conditions for nitro-PAHs reported up to 1984 have been reviewed by White et al. [94]. The analytical conditions reported thereafter are summarized in Table 4.

When capillary column-GC analysis is conducted, retention data that have been reported are very useful and should provide valuable indications [81,82].

Injection Methods

Capillary column chromatography requires as small as 1–2 μl of sample for analysis. When analyzing nitro-PAHs that exist at low concentrations in the

Table 4 Summary of GC Conditions Used for Nitro-PAH Analysis

| | | Analytical column | | | GC apparatus | |
| | | Dimension[b] | | | | |
Author	Liquid phase[a]	L (m)	i.d. (mm)	df (μm)	Injection technique	Detector
K. T. Menzies [83]		30	0.32	—	SPLS[c]	ECD
J. N. Pitts [84]	DB-5	60	0.25	—	on-column[d]	MS
W. E. Bechtold [85]	DB-5	25	0.25	—	—	MS
R. Williams [71]	—	30	0.30	—	on-column	MS (Q[h], NICI[j])
T. Ramdahl [73]	DB-5*	30	—	—	on-column	MS (HRMS[k])
T. Ramdahl [101]	SPB-5*	57	—	—	on-column	MS (Q)
W. M. Draper [76]	DB-5	30	0.32	0.25	on-column	ECD

Table 4 Continued

Author	Liquid phase[a]	L (m)	i.d. (mm)	df (μm)	Injection technique	Detector
A. Cecinato [20]	DB-5	12	0.32	1.0	PTV[e]	MS (SEC[l])
H. Lee [86]	DB-5	30	0.25	0.1	SPLS	MS (Q)
Korfmacher [63]	DB-5*	30	0.25	0.25	on-column	MS (Ap[m], NICI)
J. Arey [52]	HP-5	30	—	—	on-column	MS (Q)
J. M. Bayona [19]	HP-1	25	0.20	0.1	SPLS	MS (Q, NICI)
P. Ciccioli [51]	DB-5*	15	0.32	1.0	on-column	MS (SEC)
	DB-17*	15	0.25	0.5	on-column	MS (SEC)
T. Yoshikura [78]	DB-5	30	0.32	0.25	SPLS	MS (NICI)
J. Arey [18]	HP-5*	50	0.20	0.33	SPLS	MS (Q)
	DB-17*	30	0.25	0.25	SPLS	MS (Q)
	SB-Smectic*	25	0.20	0.15	SPLS	MS (Q)
P. Ciccioli [100]	DB-5*	15	0.32	1.0	on-column	MS (SEC)
	DB-17*	15	0.25	0.5	on-column	MS (SEC)
B. Zielinska [87]	DB-5	60	—	—	on-column	MS
R. Niles [95]	DB-5	30	0.25	0.25	on-column	MS (Q), TEA
Y. Nishikawa [98]	Ultra-1	25	0.31	0.52	SPLS	MS (SEC)
J. Schihabel [88]	DB-5*	30	0.32	0.25	SPLS	MS (Tam[n])
R. Kamens [89]	DB-1701	30	—	—	SPLS	MS (Q)
	DB-5	30	—	—	SPLS	MS (Q)
J. Feltes [49]	DB-17*	30	0.32	0.25	—	ECD, TEA
	OV-225*	30	0.32	0.25	—	ECD, TEA
	DB-5*	30	0.32	0.25	—	MS (EI, NICI)
D. Helmig [72]	DB-5*	60	0.25	0.25	on-column	MS
	DB-1701*	30	0.26	0.25	on-column	MS
	SB-Smectic*	25	0.20	0.15	on-column	MS
V. Lopez-Avila [81]	DB-5*	30	0.53	1.5	SPLS	ECD
	SPB-5*	30	0.53	1.5	SPLS	ECD
	DB-1701*	30	0.53	1.0	SPLS	ECD
J. Wienecke [90]	DB-5	60	—	0.25	SPLS	MS (Q)
T. Yamashita [91]	DB-17	15	0.53	1.0	TD[f]	MS (SEC)
K. Akiyama [92]	5% PMSlicon	25	0.32	—	SPLS, PTV	S(sulfur)ID
K. Buchholz [75]	PTE-5	30	0.25	0.25	SPME[g]	MS (IT[o])

[a]Liquid phases coated on capillary column. [b]Three capillary column dimensions are listed, L, i.d., and df, and each means length (m), internal diameter (mm), and film thickness (μm). [c]Splitless injection. [d]On-column injection. [e]Programmable temperature vaporization injection. [f]Thermal description injection. [g]Solid phase micro extraction injection. [h]Quadruple mass spectrometer. [j]Negative ion chemical ionization. [k]High resolution mass spectrometry. [l]Magnetic sector mass spectrometer. [m]Atmospheric pressure chemical ionization. [n]Tandem mass spectrometry. [o]Ion trap mass spectrometer.

*Asterisked liquid phases contain Retention Data.

atmosphere, the splitless injection, in which the whole sample volume is used, or on-column injection, in which the whole sample is directly loaded into the capillary column, are employed. However, the split injection, though popular, is not used. In analysis of nitro-PAHs both splitless and on-column injection methods are used equally (Table 4). Schuetzle et al. [6] and Herterieh [65] have reported that in the analysis of 1-NP 30–40% loss was observed with splitless injection compared to on-column injection.

Detectors

ECD (Electron Capture Detector):   The ECD is a detector that utilizes a phenomenon in which an electrophilic compound detects a decrease of the fixed current generated when electrons emitted from a β-ray source are captured; it shows a high sensitivity to certain compounds. Sometimes the ECD does not have enough selectivity against contamination, even when a high resolution GC column is used [81]. Since the ECD is inexpensive compared to an MS, it is one of the most popular detectors used with GC. It exhibits high detection sensitivity on halogenated compounds and is therefore a detector of choice in nitro-PAH analysis.

Draper [76] has detected 9 nitro-PAHs including 1-NP from diesel exhaust particulates collected on a Teflon-coated glass fiber filter. Samples were extracted into dichloromethane, followed by pretreatments with Sep-pak cartridge and HPLC fractionation. The detection limit for nitro-PAHs by ECD was 10–25 μg/g, and for dinitropyrenes, 2 μg/g. Since nitro-PAHs in airborne particulates exist at low concentrations, Morita et al. [93] converted the sample to amine derivatives by HFB for facile detection by ECD, then 7 nitro-PAHs including 1-NP were detected. This ECD detector is used in situations where MS is unavailable. Because retention time is the only information obtained from the ECD, care should be exercised when peaks overlap due to the presence of non-nitro-PAH in addition to many nitro-PAH isomers. It is recommended that when using ECD as a detector, analysis should be performed using at least two different columns to check the validity of the data obtained.

TID (Thermal Ionization Detector):   Like ECD, the TID is a detector utilizing an electrophilic compound capturing electrons. Unlike the ECD, it senses current generated when electrons are stripped off from the cesium layer on the surface of a heated ceramic. Using nitrogen as a detector gas, this detector has low sensitivities on halogenated and phosphorous compounds and does not respond to PAHs. Therefore in contrast to ECD it has extremely high selectivity against nitro-PAHs and also has a high sensitivity. White et al. [94] studied analytical conditions for 45 nitro-PAHs using SE-52 capillary column (30m × 0.2mm, 0.25 μm film) and suggested the possibility of analyzing at pg level.

TEA (Thermal Energy Analyzer):   In the TEA a C-NO$_2$ bond is cleaved by thermal energy, and the NO$_2$ radicals generated are excited by ozone to excited

NO_2. The light emitted during its decay is detected. This is therefore an NO_2 specific detector [95]. The TEA was originally developed for nitrosoamine analysis, but by increasing the decomposition temperature from 50°C to 800–900°C (the decomposition temperature of nitro-PAHs), this is applied to the analysis of nitro-PAHs. The TEA's selectivity is extremely high, since it detects only nitro groups.

Yu et al. [8] looked at diesel emissions extracts for 10 different nitro-PAHs including 1-NP and reported that the detection limit for 1-NP was 25 pg (S/N = 3) with dynamic ranges of thousands to tens of thousands.

Dennis et al. [29] examined 24 different foods, including dairy products, vegetables, meats, and teas and detected 1-NP (from teas) and 9-nitroanthracene (peated malt). The detection limit was 12 pg (equivalent to 0.02 mg/kg for a 50 g sample) for 1-NP.

GC/MS

The MS's greatest advantage is that it provides detection and identification of compounds at one time. Because of this feature, it is an indispensable detector in today's analytical technology. The MS is widely used in the analysis of nitro-PAHs from the environment. The MS spectrometer consists of an ion source, a mass analyzer, and a detector. Depending on the type of ionization method and mass analyzer, several types of MS are available. The analytical information obtained differs greatly according to the type of MS. When performing GC/MS analysis, it is recommended that internal standards be used for increased analytical accuracy. For internal standards, d_9-1-nitropyrene is used for quantification of mononitro-PAHs and d_5-dinitropyrene for dinitro-PAHs. Internal standards are added to the extracts.

Ionization Methods: EI (Electron Impact Ionization): MS can be operated in two types of ionization mode, i.e., EI and CI (chemical ionization). The information obtained varies with the mode. In EI the ionization of molecules is achieved by collision of high-energy electrons (70 eV, lower than 10^{-4} torr vacuum) with the target molecules. The ionization of molecules causes the molecules to cleave into small pieces called fragments. The mass patterns observed by EI follow some regularity, which depends on the type of compound. Currently the library contains mass patterns for about 200,000 compounds, and the number is increasing every year. Mass patterns of unknown compounds are compared to those of standards or fragment mass patterns listed in the library for identification. The fragmentation information obtained from the Mass Spectral Database library is now an essential part of the identification process.

However, if MS operated in EI (SCAN mode) is used as a detector for GC, high detection sensitivity cannot be expected. In EI (SIM mode), the most abundant ion in the mass fragmentation pattern is selected and used as the high sensitive monitor ion.

Although the detection sensitivity is low, a large amount of information can be obtained and coupled with operational ease and low instrumentation cost. Because of this, EI is considered the basic ionization method. This is the reason for the widespread use of this technique.

Ionization Methods: CI (Chemical Ionization): In CI, molecules introduced into MS are ionized by gentle collision with an ionized reagent gas (methane, isobutane, etc.) which has been generated in advance by the collision with thermoelectrons from the filament. Because the molecules are ionized under a mild condition, unlike in EI mode, molecules are not easily fragmented and show up as a single peak. CI produces positive and negative molecular ions at the same time. It is divided into two categories, depending on which ion is to be monitored: PCI (positive chemical ionization) and NCI (negative chemical ionization). In the case of nitro-PAHs, NCI produces several to thousands times more ions than PCI. CI provides high sensitivity detection, and is therefore a useful ionization method for nitro-PAHs present at low concentrations [63,96] (Fig. 8).

MS (Mass Analyzer): Quadrapole MS: The Quadrapole MS consists of two pairs of poles or four poles (Q-poles). High frequency voltages are applied between the two pairs of poles, and by sweeping the voltages, among ions introduced into the Q-pole, only the ions that can pass through the Q-pole are sequentially detected. The mass range obtained by Q-pole MS is not large, up to several thousand at one mass resolution. However, since the device is small in size, low in cost, and easily operated and maintained, this is the most widely used detector. The compounds from environmental pollutants are low molecular weight with masses up to 500 or less; therefore all the compounds that are chromatographed can be detected.

Okumura [97], who has studied analytical methods of aquatic compounds at trace levels, reported that the detection limit for Q-pole MS(EI) was 330 pg for 1-NP and 130 pg for 3-nitrofluoranthene. The concentration of 1-NP in airborne particulates was found to be within the detection limit of Q-pole MS(EI); however, it was reported that for analysis of dinitropyrenes that were at even lower concentrations, Q-pole MS(NCI) and HPLC/Flu, which provided higher detection sensitivity, were used [98].

Ramdahl and Urdal [96], and Iida and Daishima [110] have demonstrated that on Q-pole GC/MS, which is in relatively wide use, nitro-PAHs at low concentrations could be detected with NCI mode. Iida and Daishima [110] analyzed nitro-PAHs in the airborne and diesel particulates by ultrasonic extraction of particulates collected on filter paper, followed by HPLC fractionation and GC/MS(NCI). Quantification of 1,3-, 1,6-, and 1,8-dinitropyrene was made, and the detection limits for 36 different nitro-PAHs were reported, showing that the detection limit for 1-NP was 0.5 pg, while that for dinitropyrenes was 1–2 pg [110].

Ion trap MS is a technique based on Q-pole MS. In this instrument, ions produced are trapped in ring electrodes. By sweeping a high frequency voltage, ions are sequentially released from the trap electric field. Since ion trap MS

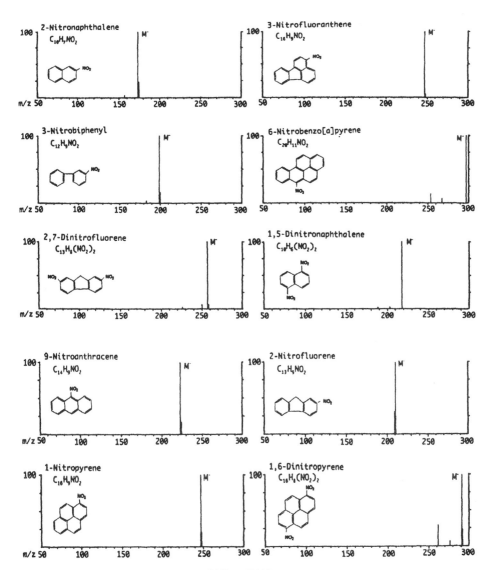

FIGURE 8 Ms(NCI) spectrum of Nitro-PAHs.

utilizes all the ions generated for detection, the spectral sensitivity is higher than normal Q-pole MS.

MS/MS: MS/MS consists of three sets of Q-poles (Q1, Q2, and Q3). Q2 is used as a collision cell which utilizes the electric field of QMS. Ions passing through the Q1 are fragmented in Q2 by collision with neutral molecules (inert gas), and the fragments are analyzed in Q3. Although the combination with Q-pole MS gives operational ease and is useful for routine work, the high cost of the instrument is a problem. In MS/MS it gives a high selectivity as well as a high S/N ratio. Therefore, it would be a strong analytical tool in determining nitro-PAHs in the environment, even with EI mode.

Schuetzle et al. [6] and Schilhabel and Levsen [99] have used this technique to analyze nitro-PAHs that contained dinitro groups in diesel particulate emissions.

HRMS (High Resolution Mass Spectrometry): HRMS, also called double-focus mass spectrometry, employs magnetic and electric fields to enhance its resolution. By HRMS's increased resolution (over 5,000), the exact mass of ions can be determined with accuracy to four decimal points (0.1 AMU). Therefore, HRMS can provide a capability for elemental analysis, i.e., determination of chemical compositions. The characteristics of this instrument are high sensitivity and high resolution, allowing nitro-PAHs at low concentrations to be analyzed even with EI mode. HRMS is particularly useful for analysis of isomers. Ramdahl and others reported studies on various nitro-PAH isomers [20,51,100] (Fig. 9). High cost is the only problem with this instrument.

2. HPLC Analysis

Although the compounds that can be analyzed are limited, HPLC can be used for determinations of low-level environmental nitro-PAHs, especially in situations when GC/MS is not readily available. In HPLC analysis, the selection of analytical column and detector is particularly important. Fluorescent detectors, electrochemical detectors, and chemiluminescence detectors offer high detection sensitivity when used with HPLC. Detectors such as UV or MS are sufficient for basic research but are insufficient for the analysis of environmental nitro-PAHs. The pretreatments required for HPLC analysis of nitro-PAHs are basically the same as for GC analysis, i.e., sampling, solvent extraction, and fractionation. The fractionation (cleanup) process should not be omitted, even if the detector offers high selectivity. In comparison to capillary columns used in GC, the number of theoretical plates for HPLC columns is small; therefore analyses of nitro-PAHs with many coexisting species would result in low S/N ratios and/or overlapping of peaks. The advantages of using HPLC include good reproducibility and an easy conditioning process for the instrument.

a) *Analytical Columns*

Reverse-phase columns are used in the analysis of nitro-PAHs, and the sizes are generally a few mm id × 20–30 cm [34,74,100–105]. The packing material for the columns is ODS, which is rigid porous silica gel (Si) sililated with an octadecyl group ($-C_{18}H_{37}$). Because of the high pressure resistance of C_{18} (ODS) columns, they can be used at a high column inlet pressure, leading to increased theoretical plates per unit time. In addition, porous packing materials have a large surface area, allowing a large amount of sample to be injected. Capillary columns are

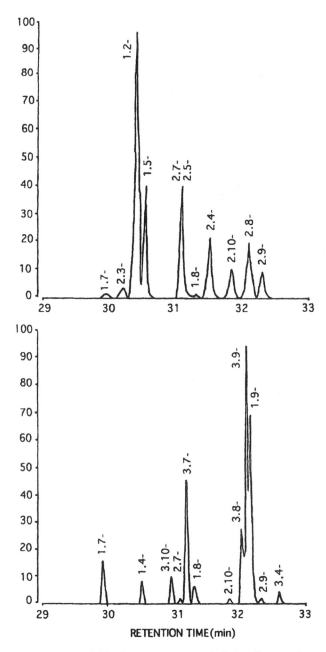

Figure 9 HRMS chromatogram of dinitrofluoranthene isomers (Ramdahl). The electron impact–multiple ion detection traces of the m/z molecular ion showing the gas chromatographic separation of two sets of dinitrofluoranthene isomers on a 57 m DB-5 capillary column. The numbers given indicate the specific isomers.

useful in analyzing highly contaminated samples. The mobile phase used is methanol-water or acetonitrile-water.

There have been many reports on the use of general analytical columns, but Jin and Rappaport [107] used a micropore column to analyze 10 different nitro-PAHs, e.g., 1,3-dinitropyrene, and demonstrated the high resolution capability of the column.

Detectors

Among detectors used with HPLC, fluorescent detectors, electrochemical detectors, and chemiluminescence detectors have a high detection sensitivity. The fluorescent detector is particularly important in the analysis of nitro-PAHs. When using this detector, nitro groups must be reduced to amino groups, because nonfluorescent nitro-PAHs cannot be detected. On-line reduction column and NaSH reduction methods are important in reducing amino groups.

Fluorescent Detector: On-line Reduction Column: The fluorescent detector itself is a high-selectivity detector. If this is used with a low-resolution HPLC column, nitro-PAHs can be selectively analyzed without influence from coexisting species. In the reduction column method, a reduction column that has been packed with Pt-Rh catalyst is connected on-line after the analytical column, and compounds reduced *in situ* are detected by a fluorescent detector [105,106].

The reduction column is prepared with 1% (9 + 1) Pt-Rh on alumina or silica, which is dry-packed in a short column of 4.6 mm $\phi \times 70$ mm. This is used at 80°C or higher for increased reduction efficiency. If the reduction efficiency is not adequate, the catalyst must be reduced by heating it to 350°C under a flow of purified air and then under hydrogen gas for 3 h to complete the reduction. For the mobile phase 80% methanol is used; however, acetonitrile cannot be used since it inhibits the reduction.

The on-line reduction method allows direct analysis without any pretreatments on the sample, when analyzing compounds existing at relatively high concentrations in the environment, such as 1-nitropyrenes. However, some compounds and isomers found at low concentrations, such as dinitropyrenes, require cleanup operations. The reduction columns last a long time.

Tejada et al. [105] have used this method. They suggested reducibility of 19 nitro-PAHs, and reported the detection limits for 16 compounds among them (Table 5). If this method were applied to diesel particulate samples, it would be impossible to analyze all of the compounds at one time. For all practical purposes, a clean-up process is required for all environmental compounds. The packing material for the reduction column can be replaced with the reducing materials used in the catalytic converter for automobile exhaust gas [34]. Figure 10 shows a chromatogram of nitropyrenes.

Fluorescent Detector: NaSH Reduction In this method, nitro-PAHs are reduced to amino-PAHs with a reducing agent prior to injection into an HPLC.

The reducing agents used are $NaBH_4$-$CuCl_2$, Zn-HCl, reducing catalysts, NaSH, etc. Described here is the NaSH reduction which Tanabe et al. [79] have reported. It is easy to perform and has a good reproducibility. The solvent of the nitro-PAH fraction is first evaporated to dryness by gently blowing nitrogen gas. Then 0.5 ml of 10% NaSH (70% purity)/H_2 is added to it. The reduction of nitro-PAHs is performed at 90°C for 90 min under reflux. At room temperature the reduction has been confirmed to be complete in 24 h. 2 ml of 0.15 N NaOH and 0.5 ml of benzene are added to the resultant amino-PAH, and the mixture is shaken vigorously. Upon centrifugation, the benzene layer is directly injected into an HPLC and analyzed by a fluorescent detector.

Tanabe et al. [79] have analyzed six different nitro-PAHs, i.e., 1-nitropyrene, 1,3-, 1,6- and 1,8-dinitropyrenes, 2-nitrofluoranthene, and 2-nitrofluorene. Nakajima et al. [104] have employed this method to reduce 1-NP in azalea leaves

Table 5   Relative Retention Time (RRT) of Nitro-PAHs on ODS-Catalyst Column, Wavelength Pairs, and On-Column Detection Limit for Fluorescence Detection of Corresponding Amino-PAH

Compound	RRT[a]	Wavelength[b], nm	DL[c], pg
1,5-dinitronaphthalene	0.26	231/390	30
1-nitronaphthalene	0.33	243/429	4
2,-nitrophthalene	0.36	234/403	14
1,3,6,8-tetranitropyrene	0.56	397/465	
1-nitrofluorene	0.56	285/370	14
9-nitroanthracene	0.58	263/505	34
2-nitroanthracene	0.70	260/495	16
1,3,6-trinitropyrene	0.77	396/460	
1,6-dinitropyrene	0.79	369/442	10
1,8-dinitropyrene	0.89	395/454	50
8-nitrofluoranthene	0.99	300/550	
1-nitropyrene	1.00	360/430	10
3-nitrofluoranthene	1.05	300/530	80
1,3-dinitropyrene	1.08	395/445	
6-nitrochrysene	1.53	273/437	47
6-nitrobenzo[a]pyrene	2.35	420/475	7

[a]Conditions: column, 15 cm × 4.6 mm i.d. Du pont Zorbax ODS + 5 cm × 4.6 mm i.d. Pt/Rh column: mobile phase, 80% methanol in water; flow, 1 ml/min; detector, Parkin-Elmer Model 650-S fluorescence spectrophotometer.
[b]Wavelengths: excitation/emission.
[c]DL: detection limit at S/N between 7 to 20. Absence of value indicates that only qualitative standards were available.

of roadside trees both in urban and rural areas, and found that the concentration was 1.17–2.50 mg/g in urban areas and 0.14–0.23 mg/g in rural areas. The advantages of this method are simple reduction with NaSH and readily available reagent.

Detection limits for GC and HPLC detectors are listed in Table 6, and analysis examples are shown in Table 7.

Electrochemical and Chemiluminescence Detectors: An electrochemical detector is a detector which measures the electrode reaction current generated when electrochemical active species on the surface of the electrode are oxidized or reduced [103,109]. It has a high selectivity and sensitivity and is used in trace amount analyses of various species.

Jin and Rappaport [107] attached this detector to an HPLC and analyzed diesel soot, and detected 12 different nitro-PAHs. The measurable range was found to be 10–100 pg per compound. Murahashi et al. [74] have used HPLC-chemiluminescence detectors and determined the nitropyrene concentrations in airborne particulates. The chemiluminescence reagent used was an acetonitrile solution containing 20 µM of bis(2,4,6-trichlorophenyl) oxalate and 15 mM of hydrogen peroxide. The detection limit for 1-NP was 1.5 fmol (10–100 fmol standard) and for 1,3-, 1,6- and 1,8-dinitropyrenes, it was 0.25–0.35 fmol (1–100 fmol standard) [108].

Table 6   Detection Limit of GC and HPLC Detector (pg)

	GC analysis			HPLC analysis		
	MS(EI) [97]	MS(NCI) [110]	TEA [8]	Chemilu [106]	Flu [105]	Flu [79]
S/N	—	—	3	—	7–20	2
1-nitronaphthalene		0.08	8	—	4	—
2-nitronaphthalene		0.1	7	—	14	—
1,5-dinitronaphthalene		0.1	9	—	30	—
1-nitrofluorene		—	—	—	14	—
2-nitrorfluorene		0.5	8	48	—	4
2-nitroanthracene		—	—	—	16	—
9-nitroanthracene		0.4	10	80	34	—
1-nitropyrene	330	0.5	25	60	10	2
1,3-dinitropyrene		1	—	0.05 [108]	—	2
1,6-dinitropyrene		1	—	0.025 [108]	10	1
1,8-dinitropyrene		2	—	0.025 [108]	50	2
6-nitrochrysene		—	—	24	47	—
6-nitrobenzo[a]pyrene		8	—	28	7	—
3-nitrofluoranthene	180	0.7	—	66	80	40

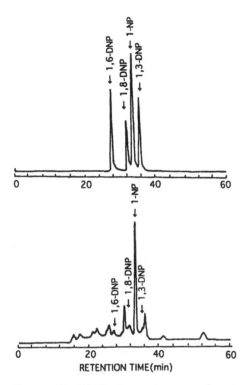

FIGURE 10 HPLC chromatogram of a standard mixture of nitropyrenes and airborne particulate exhaust. The standard solution of authentic nitroarenes was as follows: concentration of 1,3-dinitropyrene(DNP), 1,8-DNP, and 1-nitropyrene (1-NP) were 0.01 µg/ml, and the concentration of 1,6-DNP was 0.005 µg/ml. Twenty microliters of the standard solution was subjected to HPLC.

VI. CONCLUSION

Volatile nitro-PAHs in air and water can be directly analyzed by GC/MS upon sampling and concentration. On the other hand, analytical procedures for particulate nitro-PAHs are long and time-consuming, i.e., first step: sampling by HV air sampling or dilution tunnel sampling; second step: solvent extraction; third step: fractionation; fourth step: GC or HPLC analysis. Some of the procedures here are rather complicated, while others are simple but time-consuming.

In this chapter options are provided for each of the procedures, but which one is best for a specific purpose may be a question. In reality the question boils down to how well-equipped a laboratory is with leading-edge instruments. However, once target compounds are known, upon Soxhlet or ultrasonic extraction,

Table 7 Analysis Examples of Nitro-PAHs

1. GC analysis

Samples	Analytical method	Compound	Concentration
Air particulates (Ciccioli, 1989, Ref. 100)	HV sampling (quartz fiber filter); Soxhlet extraction (benzene-MeOH, 4+1, 6h); alumina and silica cleanup; HPLC fractionation; GC (15m DB-5, 15 m DB-17)/MS(EI)	1-nitrofluoranthene	:0.02 µg/g
		2-nitrofluoranthene	:0.03–0.30 µg/g
		3-nitrofluoranthene	:0.05–0.41 µg/g
		7-nitrofluoranthene	:0.02–0.2 µg/g
		8-nitrofluoranthene	:0.05–0.20 µg/g
		1-nitropyrene	:0.20–1.00 µg/g
		2-nitropyrene	:0.20–1.10 µg/g
Foods (Dennis, 1984, Ref. 29)	Homogenization; extraction (acetonitrile–lead acetate trihydrate–acetic acid); filtration; extraction (water, hexane, MeOH–water, hexane, dimethylforamide–water, hexane); rotary evaporation; Sep-pak cleanup; Concentration; GC (25m Cptm Sil 5CB)/TEA	9-nitroanthracene	:<0.06–0.9 µg/g
		1-nitropyrene	:<0.03–1.7 µg/kg
Diesel exhaust particulates (Draper, 1986, Ref. 76)	Sampling (Teflon-coated glass fiber filter, 47 mm); DCM extraction (250 ml, 16 h); silica Sep-pak cleanup; HPLC fractionation; GC (30 m Spelco Bellefonte PA)/ECD	1-nitronaphthalene	:47–0.77 µg/g
		2-nitronaphthalene	:0.87–0.94 µg/g
		9-nitronaphthalene	:0.34–1.4 µg/g
		2-nitrofluorene	:0.63–8.8 µg/g
		9-nitronaphthalene	:0.34–1.4 µg/g
		3-nitrofluoranthene	:<0.16 µg/g
		1-nitropyrene	:<0.12–5.0 µg/g
		1,3-dinitropyrene	:0.52–1.60 µg/g
		1,6-dinitropyrene	:<0.23 µg/g
		1,8-dinitropyrene	:<0.29 µg/g

Ponds and river (Feltes, 1990, Ref. 49)	DCM extraction or adsorption on Amberlite XAD resins and elution with DCM; GC (30 m DB-17, 30 m OV-17)/ECD and TEA	2-nitrotoluene :<0.01–22.0 µg/g
		3-nitrotoluene :<0.01–1.5 µg/g
		4-nitrotoluene :<0.01–4.8 µg/g
		2,6-dinitrotoluene :0.02–0.3 µg/g
		2,4-dinitrotoluene :0.02–1.2 µg/g
		3,4-dinitrotoluene :<0.01–0.1 µg/g
		2,4,6-trinitrotoluene :0.02–0.5 µg/g
		5-methyl-2-nitroaniline :<0.01–0.8 µg/g
		2-methyl-3-nitroaniline :<0.01–0.7 µg/g
		2-methyl-5-nitroaniline :<0.01–0.2 µg/g
		2-methyl-4-nitroaniline :<0.01 µg/g
Air and diesel particulates (Niles, 1989, Ref. 95)	Soxhlet extraction (methylene chloride, 48 h); Filtration: Rotary evaporation: gel filteration chromatography (Sephadex LH20); column fractionation (µ Bondpack-NH2); GC (30 m DB-5)/FID, TEA and MS	[air]
		9-nitroanthracene :0.16 µg/g
		9-methyl-10-nitroanthracene :0.02 µg/g
		3-nitrofluoranthene :0.24 µg/g
		1-nitropyrene :0.07 µg/g
		7-nitrobenz[a]anthracene :0.06 µg/g
		[diesel particulates]
		9-nitroanthracen :6 µg/g
		9-methyl-10-nitroanthracene :0.02 µg/g
		3-nitrofluoranthene :<1 µg/g
		1-nitropyrene :12 µg/g
		7-nitrobenz[a]anthracene :<1 µg/g

Table 7 Continued

1. GC analysis (cont.)

Samples	Analytical method	Compound	Concentration
Air and diesel particulates (Yamaki, 1986, Ref. 77)	HV sampling; ultrasonic extraction (benzene-MeOH,3+1); Liquid-liquid partition; HPLC fractionation; rotary evaporation; GC/MS(NCI)	nitrobiphenyl	:15–140 pg/m^3
		nitroacenaphthene	:<1–75 pg/m^3
		2-nitrofluorene	:24–290 pg/m^3
		9-nitroanthracen	:120–890 pg/m^3
		nitrofluoranthene isomer	:31–650 pg/m^3
		1-nitropyrene	:<7–500 pg/m^3
		1,3-dinitropyrene	:1.0–5.3 pg/m^3
		1,6-dinitropyrene	:<0.1–9.26 pg/m^3
		1,8-dinitropyrene	:<0.2–11 pg/m^3
Air particulates (Ramdahl, 1986, Ref. 101)	Ultra-HV sampling; Soxhlet extraction (DCM); silica column chromatography; HPLC fractionation; HRGC (30 m DB-5)/MS(EI)	2-nitrofluoranthene	:0.56–2.8 µg/g
		1-nitropyrene	:0.15–0.36 µg/g
		2-nitropyrene	:0.05–0.17 µg/g
		benzo[e]pyrene	:3.4–44 µg/g

2. HPLC analysis

Samples	Analytical method	Compound	Concentration
Gasoline and diesel engine particulates (Hayakawa, 1994, Ref. 103)	LV sampling (glass fiber filter, 0.2–2 h); ultrasonic extraction (benzene-MeOH, 3+1, twice); filtration (HLC-DISK, 0.45 µm) and washing; sodium hydrosulfide reduction; extraction (benzene); HPLC (Cosmosil 5C18)/Chemiluminescence Det.	1,3-dinitropyrene	:0.077–0.54 pmol/mg
		1,6-dinitropyrene	:0.046–1.3 pmol/mg
		1,8-dinitropyrene	:0.067–0.60 pmol/mg
		1-nitropyrene	:0.7–170 pmol/mg
		benzo(a)pyrene	:1.1–46 pmol/mg

Sample (Reference)	Method	Compound	Concentration
Azalea leaves (Nakajima, 1994, Ref. 104)	Ultrasonic extraction (ethyl acetate); silica gel chromatography; evaporation and dryness; ultrasonic dissolution (MeOH); catalytic reduction (NaSH); HPLC (Lichrospher 100RP-18, MERCK)/Flu	1-nitropyrene	:0.14–2.50 ng/g
Air particulates (Saitoh, 1990, Ref. 34)	HV sampling (glass filter): Soxhler extraction (DCM, 24 h); liquid-liquid partition, column chromatography (SiO_2 400 mg); HPLC (Du Pont Zorbax ODS)/Flu with catalytic reduction column	1-nitropyrene 1,3-dinitropyrene 1,6-dinitropyrene 1,8-dinitropyrene benzo[a]pyrene	:38–1210 pg/m^3 n.d. n.d. n.d. 0.77–3.68 pg/m^3
Air particulates (Tanabe, 1986, Ref. 79)	HV sampling (quartz fiber filter); ultrasonic extraction (ethanol); liquid-liquid partition; HPLC fractionation; catalytic reduction (NaSH); HPLC (Nucleosil 7C18)/Flu	1-nitropyrene 1,3-dinitropyrene 1,6-dinitropyrene 1,8-dinitropyrene 2-nitrofluorene	:14.8–134 pg/m^3 n.d.–4.7 pg/m^3 0.6–8.7 pg/m^3 n.d.–6.6 pg/m^3 n.d.
Gasoline and diesel exhaust particulates (Tejada, 1986, Ref. 105)	Dilution tunnel sampling (Teflon-coated glass fiber filter); extraction (methylene chloride); drying with nitrogen blow: dissolution (MeOH-methylene chloride, 1+1); HPLC (Du Pont Zorbax ODS)/Flu by on-column catalytic reduction (Pt/Rh)	1-nitropyrene	:3.7–173 µg/g
Diesel paticulates (Veigl, 1994, Ref. 114)	Sampling on a Teflon-coated glass fiber filter; reflux extraction (toluene); fractionation with silica gel column chromatography; multidimensional HPLC (Seibersdorf RP8)/Flu by on-column catalytic reduction (Pt/Rh)	1-nitropyrene	:0.25–7.26 ng/mg

followed by fractionation with column chromatography, samples can be analyzed by GC or HPLC, when an expensive GC/MS is unavailable.

Since the analytical procedures are long and tedious, there may be human and experimental errors. Skill and care are required for analysts. The first step may not be modified, but the second step may be done with PSE or SFE, the third step with HPLC fractionation, and the fourth step with GC/MS or HPLC/Flu analysis. Comparing GC analysis and HPLC analysis, and judging by resolution and identification capability, GC/NCIMS(Q-pole), GC/MS-MS, and GC/HRMS are important techniques for simultaneous analysis of nitro-PAHs that contain various isomers.

At a laboratory equipped with a GC/MS which gives a high detection sensitivity, the most reliable data that today's technology can provide can be obtained in various analytical situations, including analysis of isomers. However, the instrument is expensive, and its maintenance and operation require some skill. On the other hand, GC/NCIMS(Q-pole) provides a high sensitivity and selectivity detection and ease of use, and is becoming widely used. This instrument would currently be the most useful analytical tool for multicomponent analysis of nitro-PAH-containing isomers.

On HPLC/Flu it would be a difficult task to analyze 10 or more nitro-PAHs at one time as judged from the column resolution. However, if compounds are grouped into categories such as nitropyrenes, nitrofluoranthenes, etc., then because of the reasons listed below, HPLC/Flu should be recommended for routine analytical work.

1. Relatively inexpensive and general purpose chromatography
2. Easy conditioning and excellent analytical accuracy
3. High sensitivity analysis with a fluorescence detector

ACKNOWLEDGMENTS

The authors are indebted to Professors A. Koizumi and S. Kamiyama (Professor Emeritus) of Akita University School of Medicine and Dr. S. Daishima of Yokogawa Analytical Systems Inc. for invaluable assistance and suggestions in the preparation of this manuscript. We also gratefully acknowledge Dr. Y. Ichinowatari, president of Iwate Prefectural Institute of Public Health, who gave us the opportunity to prepare this review article. Finally, we would like to thank Ms. Jennifer Beck of the Iwate Prefectural Government Culture and International Relations Division for editing the manuscript.

REFERENCES

1. P. H. Howard, J. Santodonato, J. Saxena, J. E. Malling, and D. Greninger, *EPA-560/2-76-010*, Center for Chemical Hazard Assessment, Syracuse Research Corp., New York, 1976, p. 47.

2. M. Windholz, *The Merck Index*, 10th ed., 1983.
3. J. N. Pitts, Jr., K. A. Van Cauwenberghe, D. Grosjean, J. P. Schmid, D. R. Fitz, W. L. Belser, Jr., G. B. Knudson, and P. M. Hynds, *Science 202*: 515 (1978).
4. H. S. Rosenkranz, E. C. McCoy, D. R. Sanders, M. Butler, D. K. Kiriazides, and R. Mermelstein, *Science 209*: 1039 (1980).
5. M. C. Paputa-Peck, R. S. Marano, D. Schuetzle, T. L. Riley, C. V. Hampton, T. J. Prater, L. M. Skewes, T. E. Jensen, P. H. Ruehle, L. C. Bosch, and W. P. Duncan, *Anal. Chem. 55*: 1946 (1983).
6. D. Schuetzle, T. L. Riley, and T. J. Prater, *Anal. Chem. 54*: 265 (1982).
7. D. Schuetzle, *Environ. Health Perspectives 47*: 65 (1983).
8. W. C. Yu, D. H. Fine, C. K. Chiu, and K. Biemann, *Anal. Chem. 56*: 1158 (1984).
9. I. Alfheim, G. Becher, J. K. Hongslo, and T. Ramdahl, *Environ. Mutagen. 6*: 91 (1984).
10. D. Schuetzle and J. M. Dasey, *Environ. Sci. Res. 39*: 11 (1990).
11. H. Tokiwa, R. Nakagawa, and K. Horikawa, *Mutat. Res. 157*: 39 (1985).
12. M. N. Kayali, S. Rubio-Barroso, and M. Polo-Diezl, *J. Liq. Chromatogr. 17*: 3623 (1994).
13. P. Baudot, M. L. Viriot, J. C. Ander, J. Y. Jezequel, and M. Lafontaine, *Analusis(Fra) 19*: 85 (1991).
14. S. A. Wise, L. C. Sander, and W. E. May, *J. Chromatogr. 642*: 329 (1993).
15. J. H. Johnson, S. T. Baglet, L. D. Gratz, and D. G. Leddy, *SAE Technical Paper Series*, 940233: International Congress & Exposition, Detroit, MI, February 28–March 3, 1994, p. 1.
16. J. Arey, B. Zielinska, R. Atkinson, A. M. Winer, T. Ramdahl, and J. N. Pitts, Jr., *Atmos. Environ. 20*: 2339 (1986).
17. D. Barcel, *Anal. Chim. Acta 263*: 1 (1992).
18. J. Arey and B. Zielinska, *HRCCC 12*: 101 (1989).
19. J. M. Bayona, D. Barcel, and S. J. Albaig, *Biomed. Environ. Mass Spectrom. 16*: 461 (1988).
20. A. Cecinato, E. Brancaleoni, C. Di Palo, R. Draischi, and P. Ciccioli, *Phys. Behav. Atmos. Pollut. 1986*: 58 (1986).
21. S. K. Hammond, T. J. Smith, S. R. Woskie, B. P. Leaderer, and N. Bettinger, *Am. Ind. Hyg. Assoc. J. 49*: 516 (1988).
22. E. Garshick, M. B. Schenker, A. Munoz, M. Segal, T. J. Smith, S. R. Woskie, S. K. Hammond, and F. E. Speizer, *Am. Rev. Respir. Dis. 135*: 1242 (1987).
23. E. Garshick, M. B. Schenker, A. Munoz, M. Segal, T. J. Smith, S. R. Woskie, S. K. Hammond, and F. E. Speizer, *Am. Rev. Respir. Dis. 137*: 820 (1988).
24. J. L. Mumford, X. Z. He, R. S. Chapman, S. R. Cao, D. B. Harris, X. M. Li, Y. L. Xian, W. Z. Jiang, W. C. Xu, J. C. Chuang, W. E. Wilson, and M. Cooke, *Science 235*: 217 (1987).
25. H. Tokiwa, *J. Japan Soc. Air Pollut. 27*: 73 (1992).
26. H. Vainio, K. Hemmini, and J. Wilbourn, *Carcinogenesis 6*: 1653 (1985).
27. T. Kinouchi, K. Nishifuji, H. Tsutsui, S. L. Hoare, and Y. Ohnishi, *Jpn. J. Cancer Res. 79*: 32 (1988).
28. J. Jacob, W. Karcher, R. Dumler, J. J. Belliardo, and A. Boenke, *Fresenius J. Anal. Chem. 340*: 755 (1991).
29. M. J. Dennis, R. C. Massey, D. L. McWeeny, and M. E. Knowles, *Food Addit. Contam 1*: 29 (1984).

30. T. E. Jensen, J. F. O. Richert, A. C. Clearly, D. L. LaCourse, and R. A. Gorse, Jr., *J. Air Pollut. Control Assoc. 36*: 1255 (1986).
31. M. A. McCartney, B. F. Chatterjee, E. C. McCoy, E. A. Mortimer, Jr., and S. Rosenkranz, *Mutat. Res. 171*, 99 (1986).
32. H. Klingenberg and H. Winneke, *Sci. Total Environ. 93*: 95 (1990).
33. R. B. Zweidinger, *Dep. Toxicolo. Environ. Sci. 10*: 83 (1982).
34. N. Saitoh, Y. Wada, A. Koizimi, and S. Kamiyama, *Jpn, J. Hyg. 45*: 873 (1990).
35. R. R. Beard, J. T. Noe, *Patty's Industrial Hygiene and Toxicology, Vol. II A, 3d rev. ed.* (G. D. Clayton and E. E. Clayton, eds.), Wiley-Interscience (1981), p. 2413.
36. H. S. Rosenkranz, *Mutation Res. 101*: 1 (1982).
37. J. N. Pitts, Jr., D. M. Lokensgard, W. Harger, T. S. Fisher, V. Mejia, J. J. Schuler, G. M. Scorziell, and Y. A. Katzenstein, *Mutation Res. 103*: 241 (1982).
38. Y. Manabe, T. Kinouchi, and Y. Ohnishi, *Mutation Res. 158*: 3 (1985).
39. K. El-Bayoumy, A. Rivenson, B. Johnson, J. DiBello, P. Little, and S. S. Hecht, *Cancer Res. 48*: 4256 (1988).
40. T. Kinouchi, K. Nishifuji, and Y. Ohnishi, *Carcinogenesis 11*: 1381 (1990).
41. A. K. Roy, P. Upadhyaya, F. E. Evans, and K. El-Bayoumy, *Carcinogenesis 12*: 577 (1991).
42. P. P. Fu, D. W. Miller, L. S. Von Tungeln, M. S. Bryant, J. O. Lay, Jr, K. Huang, L. Jones, and F. E. Evans, *Carcinogenesis 12*: 609 (1991).
43. K. El-Bayoumy, G. H. Suey, and S. S. Hecht, *Chem. Res. Toxicol. 1*: 243 (1988).
44. J. A. Bond, J. D. Sun, M. A. Medinsky, and R. K. Jones, *Toxicol. Appl Pharmacol. 85*: 102 (1986).
45. J. A. Bond, R. K. Wolff, J. R. Harkema, J. L. Mauderly, R. F. Henderson, W. C. Griffith, and R. O. McClellan, *Toxicol. Appl. Pharmacol. 96*: 336 (1988).
46. F. S. C. Lee, T. J. Prater, and F. Feris, *Polynuclear Aromatic Hydrocarbons* (Jones P. W., Lever P. Ed.), Ann Arbor Science Publishers, Ann Arbor, MI, 1979, p. 83.
47. E. D. Morgan, and N. Bradlet, *J. Chromatogr. 468*: 339 (1989).
48. S. F. Patil, and S. T. Lonkar, *J. Chromatogr. 600*: 344 (1992).
49. J. Feltes, K. Levsen, D. Volmer, and M. Spiekermann, *J. Chromatogr. 518*: 21 (1990).
50. V. Librando and S. D. Fazzino, *Chemosphere 27*: 1649 (1993).
51. P. Ciccioli, A. Cecinato, E. Brancaleoni, and A. Liberti, *HRCCC. 11*: 306 (1988).
52. J. Arey, B. Zielinska, W. P. Harger, R. Atkinson, and A. M. Winer, *Mutat. Res. 207*: 45 (1988).
53. P. T. Scheepers, H. J. Thuis, M. H. Martens, and R. P. Bos, *Toxicol. Lett. 72*: 191 (1994).
54. D. Grosjean, K. Fung, and J. Harrison, *Environ. Sci. Technol. 17*, 673 (1983).
55. G. Stark, J. Stauff, H. G. Miltenburger, and I. S. Fischer, *Mutat. Res. 155*: 27 (1985).
56. A. Koizumi, N. Saitoh, T. Suzuki, and S. Kamiyama, *Arc. Environ. Health 49*: 87 (1994).
57. M. P. Holloway, M. C. Biaglow, E. C. McCoy, M. Anders, H. S. Rosenkranz, and P. C. Howard, *Mutat. Res. 187*: 199 (1987).
58. P. S. Lee, R. A. Gorski, J. T. Johnson, and S. C. Soderholm, *J. Aerosol. Sci. 20*: 627 (1989).
59. G. K.-C. Low, B. E. Batley, and C. I. Brockbank, *J. Chromatogr. 392*: 199 (1987).

60. J. Arey, B. Zielinska, R. Atkinson, and A. M. Winer, *Environ. Sci. Technol.* 22: 457 (1988).

61. R. Barbella, A. Ciajolo, A. D'anna, and C. Bertoli, *Combust. Flame 77*: 267 (1989).

62. S. Goto, A. Kawai, T. Yonekawa, Y. Hisamatsu, and H. Matsushita, *J. Japan Soc. Air Pollut. 16*: 18 (1981).

63. W. A. Korfmacher, B. Rushing, and J. N. Pitts, Jr. *HRCCC 10*: 641 (1987).

64. S. Goto, A. Kawai, T. Yonekawa, and H. Matsushita, *J. Japan Soc. Air Pollut. 17*: 53 (1982).

65. R. Herterich, *J. Chromatogr. 549*: 313 (1991).

66. B. E. Richter, J. L. Ezzell, D. Felix, K. A. Roberts, and D. W. Later, *Am. Lab. 27*: 24 (1995).

67. K. D. Bartle, T. Boddington, A. A. Clifford, N. J. Cotton, and C. J. Dowle, *Anal. Chem. 63*: 2371 (1991).

68. N. Husers and W. Kleibohmer, *J. Chromatogr. A. 697*: 107 (1995).

69. H. B. Lee and T. E. Peart, *J. Chromatogr. A. 663*: 87 (1994).

70. E. F. Funkenbush, D. G. Leddy, and J. H. Johnson, *SAE Technical Paper Series 790418*, Society of Automotive Engineers, 1979.

71. R. Williams, C. Sparacino, B. Petersen, J. Bumgarner, R. H. Jungers, and J. Lewtas, *Int. J. Environ. Anal. Chem. 26*: 27 (1986).

72. D. Helmig and J. Arey, *Int. J. Environ. Anal. Chem. 43*: 219 (1991).

73. T. Ramdahl, B. Zielinska, J. Arey, R. Atkinson, A. M. Winer, and J. N. Pitts, Jr., *Nature 321*(6068): 425 (1986).

74. T. Murahashi, K. Hayakawa, Y. Imamoto, and Y. Miyazaki, *Bunseki Kagaku 43*: 1017 (1994).

75. K. D. Buchholz and J. Pawliszyn, *Anal. Chem. 66*: 160 (1994).

76. W. M. Draper, *Chemosphere 15*: 437 (1986).

77. N. Yamaki, T. Kohno, S. Ishiwata, H. Matsushita, K. Yoshihara, Y. Iida, T. Mizoguchi, S. Okuzawa, K. Sakamoto, H. Kachi, S. Goto, T. Sakamoto, and S. Daishima, *Dev. Toxicol. Environ. Sci. 13*: 17 (1986).

78. T. Yoshikura, T. Kamiura, A. Okamoto, M. Tanaka, K. Masumoto, M. Fukushima, and K. Kuroda, *Seikatsu Eisei 32*: 231 (1988).

79. K. Tanabe, H. Matsushita, C.-T. Kuo, and S. Imamiya, *J. Japan Air Pollut. 21*: 535 (1986).

80. W. E. Bechtold, J. S. Dutcher, A. L. Brooks, and T. R. Henderson, *J. Appl. Toxicol. 5*: 295 (1995).

81. V. Lopez-Avila, J. Benedicto, E. Baldin, and W. F. Beckert, *HRCCC 14*: 601 (1991).

82. A. Robbat, Jr., N. P. Corso, P. J. Doherty, and M. H. Wolf, *Anal. Chem. 58*: 2078 (1986).

83. K. T. Menzies, C. M. Wong, H. Magnil, and M. A. Rangel, *Development on a Routine Technique for Analysis of Nitro-PAH in Diesel Particulate*, Arthur D. Little, 1985.

84. J. N. Pitts, Jr., J. A. Sweetman, B. Zielinska, A. M. Winer, and R. Atkinson, *Atmos. Environ. 19*: 1601 (1985).

85. W. E. Bechtold, T. R. Henderson, and A. L. Brooks, *Mutat. Res. 173*: 105 (1986).

86. H. B. Lee, G. Dookhran, and A. S. Y. Chau, *Analyst 112*: 31 (1987).

87. B. Zielinska, J. Arey, R. Atkinson, and A. M. Winer, *Atmos. Environ. 23*: 223 (1989).

88. J. Schilhabel and K. Levsen, *Fresenius' Z. Anal. Chem. 333*: 800 (1989).
89. R. M. Kamens, J. Guo, J. M. Perry, H. Karam, and L. Stockburger, *Environ. Sci. Technol. 23*: 801 (1989).
90. J. Wienecke, H. Kruse, and O. Wassermann, *Chemosphere 25*: 1889 (1992).
91. T. Yamashita, Y. Yasuda, K. Haraguchi, S. Sueda, and K. Kido, *J. Japan Soc. Air Pollut. 27*: 65 (1992).
92. K. Akiyama and S. Sakamoto, *J. Environ. Chem. 4*: 647 (1994).
93. K. Morita, K. Fukamachi, and H. Tokiwa, *Bunseki Kagaku 31*: 255 (1982).
94. C. M. White, A. Robbat Jr., and R. M. Hoes, *Anal. Chem. 56*: 232 (1984).
95. R. Niles and Y. L. Tan, *Anal. Chim. Acta. 221*: 53 (1989).
96. T. Ramdahl and K. Urdal, *Anal. Chem. 54*: 2256 (1982).
97. T. Okumura, *J. Environ. Chem. 5*: 597 (1995).
98. Y. Nishikawa, K. Taguchi, T. Okumura, K. Imamura, and S. Asada, *Anual Report of Environ. Pollution Control Center, Osaka Prefectural Government, Jpn. 11*: 109 (1989).
99. J. Schilhabel and K. Levsen, *Fresenius' Z. Anal. Chem. 333*: 800 (1989).
100. P. Ciccioli, A. Cecinato, E. Brancaleoni, R. Draisci, and A. Liberti, *Aerosol Sci. Technol. 10*: 296 (1989).
101. T. H. Ramdahl, J. Arey, B. Zielinska, R. Atkinson, and A. M. Winer, *HRCCC 9*: 515 (1986).
102. T. Ramdahl, B. Zielinska, J. Arey, and R. W. Kondrat, *Biomed. Environ. Mass Spectrom. 17*: 55 (1988).
103. K. Hayakawa, M. Butoh, Y. Hirabayashi, and M. Miyazaki, *Jpn. J. Toxicol. Environ. Health 40*: 20 (1994).
104. D. Nakajima, T. Teshima, M. Ochiai, M. Tabata, J. Suzuki, and S. Suzuki, *Environ. Contam. Toxicol. 53*: 888 (1994).
105. S. B. Tejada, R. B. Zweidinger, and J. E. Sigsby, Jr., *Anal. Chem. 58*: 1827 (1986).
106. W. A. Maccrehan, W. E. May, S. D. Yange, and B. A. Benner, Jr., *Anal. Chem. 60*: 194 (1988).
107. Z. Jin and S. M. Rappaport, *Anal. Chem. 55*: 1778 (1983).
108. M. Maeda, K. Tsukagoshi, M. Murata, M. Takagi, and A. Yamashita, *Anal. Sci. 10*: 583 (1994).
109. M. T. Galceran and E. Moyano, *Talanta 40*: 615 (1993).
110. Y. Iida and S. Daishima, *A Review of Air Pollution Compounds, Nitroarene* (H. Matsushita, ed.), The Japan Information Center of Science and Technology, Tokyo, 1988, p. 133.
111. F. S.-C. Lee and D. Schuetzle, *Handbook of Polycyclic Aromatic Hydrocarbons*, Vol. 1 (A. Björseth, ed.), Marcel Dekker, New York and Basel, 1983, p. 27

7

Chromatographic Analysis of 2,3,7,8-Tetrachlorodibenzo-*p*-dioxin (2,3,7,8-TCDD)

Takayuki Shibamoto
University of California, Davis, California

I. INTRODUCTION

Dioxin, in particular 2,3,7,8-tetrachlorodibenzo-*p*-dioxin (2,3,7,8-TCDD), is the most potent man-made carcinogen. Acute lethality of 2,3,7,8-TCDD varies significantly among experimental animals. The guinea pig seems to be the most sensitive animal with an oral LD_{50} under 1 mg/kg. Hamsters are the most resistant species tested, requiring some 10,000 times greater doses than that of the guinea pig to achieve a similar level of lethality. There are no reports to provide information on human sensitivity toward 2,3,7,8-TCDD. However, based on accidental exposures, some researchers estimate that humans are more resistant to polychlorinated aromatic compounds than animals [1,2].

Dioxin contamination in the environment has received much attention, not only from environmental chemists but also from the public, because of its potent carcinogenicity. The presence of dioxins in the ecosystem and their adverse effects on animals and humans have been of concern because it has been spread through the use of herbicides [3]. In particular, since the July 10, 1976, accident at a chemical plant for the synthesis of 2,4,5-trichlorophenol in Seveso, Italy, public concern about dioxin contamination has been raised. Analytical techniques used to monitor dioxin contamination in various matrices in the Seveso area were intensively reviewed [4].

The major sources of TCDD contamination are chemical manufacturing, high-temperature processes such as the paper industry, and reservoirs such as chemical waste dumps. Recently, incineration of waste materials has been added to the major sources. Early public concern about TCDDs in the environment arose in the late 1960s when TCDDs were reported as contaminants in phenoxy herbicides. Later, ppb to ppm levels of 2,3,7,8-TCDD were found in a herbicide, 2,4,5-trichloro phenoxy acetic acid [5]. Because of the extreme toxicity of 2,3,7,8-TCDD, investigation of the environmental impact of this contaminate became a pressing need. Sufficient information on the fate of 2,3,7,8-TCDD to determine whether low concentrations reaching plants, soils, and water posed any threat to humans and to the ecosystem was urgently required. In order to satisfy this demand, the development of analytical methodology for TCDD became important. Since the mid 1960s, many scientists have worked to develop highly sensitive and selective determination methods for the presence of 2,3,7,8-TCDD. Meanwhile, the limit of TCDD detection has improved from ppm in 1960 to ppq (parts per quadrillion) in the mid 1980s [6]. Also, specificity has improved from 20 isomers being included in the chromatographic peak used to measure 2,3,7,8-TCDD to only one of all the 22 TCDD isomers in 1979 [7].

There have been numerous reports on analytical methodology for detecting TCDD. For example, the National Research Council of Canada published comprehensive reports on TCDD, which consisted of a summary of the available documents up to 1993. The report contains analytical methodology for trace residue in various matrices, bioanalyses, and criteria for safety in the laboratory [8]. In *Chlorinated Dioxins and Dibenzofurans in Perspective*, 10 of the 37 chapters cover analytical methods for dioxins and dibenzofurans, including HPLC, CC cleanup, GC/MS, and GC/MS/MS [9]. Details of procedures for detection of dioxin in flue gases and ash from combustion processes are available in a monograph, *Dioxin Emissions from Combustion Sources* [10]. The proceedings of an international symposium and workshop entitled Dioxins in the Environment, held at Michigan State University December 6–9, 1983, devoted one chapter to Sampling and Analytical Techniques [11]. Following increased public concern about environmental contamination by TCDD, the demand to lower the detection limit for TCDD increased. In order to increase the detection limit, cleanup or sample preparation procedures became extremely important.

II. USE OF COLUMN CHROMATOGRAPHY FOR CLEANUP OF COMPLEX SAMPLE MATRICES

Because TCDDs are present in air, water, soil, sediment, and ash as well as in foods, sampling requires diverse techniques. Once samples are collected, the next critical step is cleanup. In order to obtain high quality quantitative data, the cleanup process is extremely important. Complex matrices of environmental and food samples often contain substantial amounts of materials that interfere with the analysis of TCDDs. In order to obtain highly sensitive analytical results, it is essential to remove interfering materials as much as possible. At the present time, once adequate cleanup is accomplished, high-resolution and highly sensitive gas chromatography (GC) or gas chromatography/mass spectrometry (GC/MS) can achieve satisfactory analysis of TCDDs.

Since Tswett invented column chromatography (CC) in 1906, it has been widely used to separate organic materials form complex matrices. Development of GC and high-performance liquid chromatography (HPLC) seems to decrease the importance of CC in analytical chemistry. However, currently CC still remains one of the most efficient methods to clean up a compound of interest.

There are several excellent review articles containing CC cleanup methods. For example, removal of interferences from environmental and food samples has been well reviewed by Clement and Tosine [12]. Bulk matrix removal included the use of magnesia/Celite 545, size-exclusion chromatography, and modified silica gel. Separation of chlorinated organic compounds such as DDT, PCBs, and TCDDs involves the use of silica gel and florisil, alumina, and carbon. Table 1 summarizes conventional column chromatography used to clean up complex samples. Silica gel and alumina are still most widely used for the cleanup of chlorinated organic samples. In many cases, silica gel is combined with silica gel with concentrated H_2SO_4. For example, selective analysis of 2,3,7,8-TCDD was performed using Alumina B-Super I® (ICN Biomedicals, Eschwege, Germany). Ten μL of standard mixture containing 0.7–2 ng/μL each of eight polychlorinated dioxins (PCDDs) including TCDDs was applied on a column (1.0×15 cm) filled with 5 g of Alumina B-Super I and 3 g of Na_2SO_4, and prewashed with 100 mL of hexane. Within the first eluate with 60 mL of hexane/dichloromethane (80/20), all PCDDs including TCDDs were recovered with 80–100% recovery efficiencies [13]. A column combined with two alumina and silica gel has been used widely for PCDD cleanup. For example, samples of oil extract from water leachate of a hazardous waste landfill (200 g) was first cleaned up using a column of 20 g Alumina B-Super I and 10 g of Na_2SO_4 [14]. The sample was prewashed with 400 mL hexane and then eluted with 100 mL of hexane/dichloromethane (1/1). The concentrated eluate from the above sample was applied to a column, filled with 10 g of silica gel, 20 g of silica gel/44% conc. H_2SO_4, and 10 g Na_2SO_4 and developed

Table 1 Column Chromatography Used for Cleanup Samples in TCDD Analysis

Sample	Packing material	Solvent	Literature
Oil extracts from water leachate of a hazardous waste landfill	ICN Alumina B-super + Na_2SO_4 ICN Silica (63/200 Active 60A) + conc. H_2SO_4 + Bio-beads S-X3	Benzene, hexane/dichloromethane Hexane Hexane/ethyl acetate	13 14
Nonfat and fat tissue, milk, cream, grain, dry plant material, wet plant material, soil, and blood	Silica gel (100–200 mesh) Alumina (A-540)	20% benzene in hexane (w/w) 20% dichloromethane in hexane	15
Tissue and sediment, paper products	Silica gel 60 (70/230 mesh) Acid alumina (AG4) Activated carbon (Amoco PX-21) Sulfuric-acid-impregnated-silica gel (40%, w/w)	Toluene, methanol, cyclohexane/ dichloromethane dichloromethane (50/50) Dichloromethane/methanol/benzene (75/20/5)	20 21
Paper mill waste water samples	Silica gel (acid, base, neutral) basic alumina, Px-21 carbon/celite	8% dichloromethane in hexane, 50% dichloromethane in hexane	16
Fish	18% Carbopack C (80/100 mesh) on Celite 545	Toluene, dichloromethane/methanol/ toluene (75/20/5)	51
Waste water	Silica gel (100/200 mesh), basic alumina (100/120 mesh), 44% sulfuric acid on silica, 33% 1 M sodium hydroxide on silica	Hexane, hexane/dichloromethane (50/50)	22
Soil, vegetation samples	H_2SO_4 treated silica gel, alumina	Dichloromethane, hexane/acetone	4
Bottom-feeding fish, game fish	Activated florisil, carbon on silica gel	Hexane, dichloromethane/hexane (2/98), benzene/dichloromethane (1/3)	13
Bird eggs	Deactivated basic alumina	Acetone	19
Air samples from an incinerator	Silica gel, acid- & base-alumina silver nitrate/ silica gel	Dichloromethane/hexane (2/98 & 50/50)	24
Paper products	Silica gel, basic alumina, charcoal: Charbopak C on Celite 545	Hexane	40
Plants, soil	Na_2SO_4, silica gel, Celite-H_2SO_4	20% dichloromethane in hexane	54

with 150 mL of hexane. The concentrate was chromatographed on a column of Bio-Beads S-X3® (Bio-Rad, Munich, Germany), equilibrated with 400 mL of cyclohexane/ethylacetate (1/1). Finally, the concentrate from the Bio-Beads S-X3 column was treated with a column packed with 5 g of Alumina B-Super I and 3 g of Na_2SO_4. The cleanup sample was analyzed using GC/MS and 70.5 ppb $\pm$ 1.2 (mean $\pm$ SD, $n = 3$) of 2,3,7,8-TCDD was found. Using a similar CC method, over 70% recovery was achieved from samples of beef fat, soil, and rice spiked with ppt levels of TCDD [15].

In chemically complex aqueous (ppq) and solid sample (ppt) matrices, 2,3,7,8-TCDD was successfully isolated using a series of liquid chromatographic cleanup sequences, which include a combination silica gel column containing acid-modified, base-modified, and neutral silica gel, a basic alumina column, and a PX-21 carbon/Celite column [16]. A unique carbon absorbent consisting of 18% Carbopack C® (80/100 mesh, Supelco, Inc., Bellefonte, PA), on Celite 545® (Fisher Scientific, Fair Lawn, NJ) was applied to clean up 2,3,7,8-TCDD in a fish homogenate. The method achieved over 90% recovery of TCDD [17]. Cattabeni et al. [4] reviewed the analytical procedures used for the samples collected at Seveso, Milan, Italy after the factory synthesizing 2,4,5-trichlorophenol accidentally dispersed chemicals including 2,3,7,8-TCDD. The CC cleanup methods included an extrelut column (biological substrate), silica gel (soil, vegetation), florisil column (most substrata), and an alumina column (most substrata).

A series of CC techniques was also used to clean up fish samples for TCDD [18]. After ground fish samples were extracted with hexane/dichloromethane (1/1) using a soxhlet extractor, concentrated extracts were cleaned up using CC packed with activated florisil. The column was eluted with dichloromethane and dichloromethane/hexane (2/98). The eluates were passed directly onto a column containing carbon on silica gel, which was developed with benzene/dichloromethane (1/3), toluene. The eluates were then analyzed for TCDD by GC/MS.

A bird-egg sample was cleaned using an adsorbent chromatography column consisting of sodium sulfate and deactivated basic alumina. Lipid materials were satisfactorily removed using this column [19]. The column used in this study was one of the typical cleanup columns containing several adsorbents [20–22].

Typical cleanup columns are shown in Figure 1. These columns (I, II, and III) were connected in series [20]. Column I consists of sodium sulfate, potassium sulfate, and silica gel as shown in Figure 1. The amount of each adsorbent varies in different sample matrices. The key step in using different adsorbents is a column of carbon dispersed on glass fibers (Column II). This column adsorbs most planar polynuclear polychlorinated aromatics and allows the major portion of biological coextractants to pass through.

Figure 2 shows the column prepared for further cleanup of the samples. The sample was first passed through a strongly basic adsorbent, cesium silicate, and a strongly acidic adsorbent, 40% sulfuric acid impregnated silica gel with hexane,

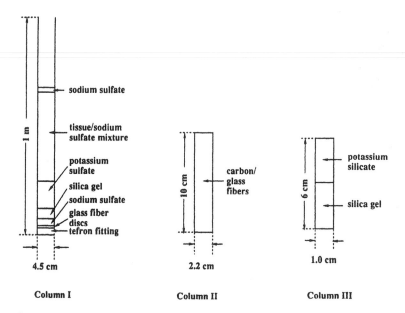

FIGURE 1 Cleanup columns used for dioxin analysis. (Reconstructed from Smith et al., Ref. 20.)

and then subjected to chromatography on acid alumina with dichloromethane. Sometimes a column of carbon dispersed on glass fibers can be used alone. The PCDDs are then removed from the column by reverse elution with toluene [23].

Most commonly used solvents for CC cleanup are hexane and dichloromethane. Benzene was also commonly used, but it is currently not recommended due to its toxicity. More polar solvents such as ethyl acetate, acetone, and methanol are also used. Solvent choice is also essential to obtain sufficient cleanup. However, mixtures of hexane and dichloromethane in various ratios seem to give satisfactory results (Table 1).

III. GAS CHROMATOGRAPHY (GC) AND GAS CHROMATOGRAPHY/MASS SPECTROMETRY (GC/MS)

Since the invention of gas chromatography (GC) in the early 1950s, determination of trace contaminants in various matrices became possible. Currently, it is possible to detect ppt (sometimes ppq) levels of organic chemicals using highly specific and sensitive GC detectors such as a nitrogen-phosphorus detector (NPD), a flame photometric detector (FPD), and an electron capture detector (EC).

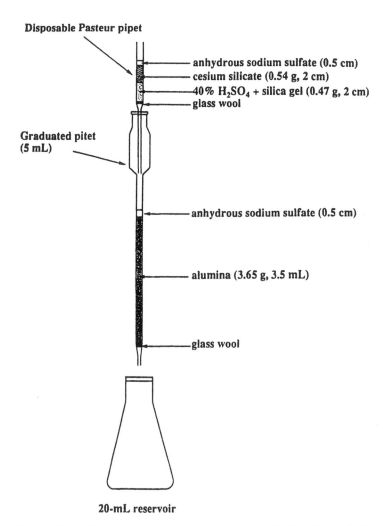

FIGURE 2 A further cleanup column prepared by Smith et al. (Reconstructed from Smith et al., Ref. 20.)

A total of 75 different chlorinated dibenzo-*p*-dioxins (PCDDs) are possible, having from 1 to 8 chlorine substituents. Among 22 possible TCDDs, 2,3,7,8-TCDD (Fig. 3) has received the most attention because of its toxicity. Analytical methodology developed is always focused on this particular dioxin. Because dioxin has numerous isomers, high-resolution GC columns are required to determine specific isomers such as 2,3,7,8-TCDD. Until early 1975, gas chromato-

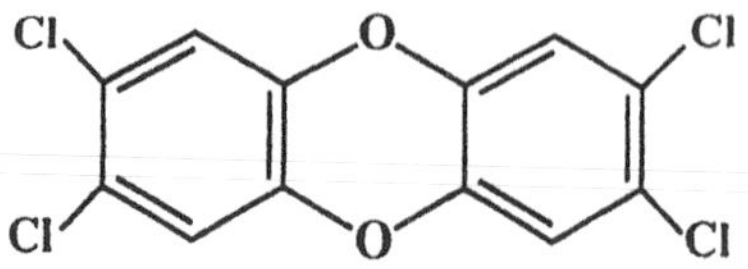

FIGURE 3   Structure of 2,3,7,8-tetrachlorodibenzo-*p*-dioxin.

graphic columns used for dioxin analysis were mainly packed columns. The stationary phase was prepared in individual laboratories. Commonly used stationary phases for packed columns were nonpolar and seminonpolar columns such as OV-1, OV-3, OV-7, OV-17, OV-101, OV-105, OV-210, SE-30, SE-52, SP-2100, and XE-60 [12]. Polar columns, such as Carbowax 20M, were also used. For example, fish ash samples were analyzed using a 1.8 m × 2 mm i.d. glass column packed with 0.2% Carbowax 20M on 120/140 mesh Chromosorb W interfaced to a mass spectrometer with a selected ion monitoring mode (SIM) [24]. Complete separation of all dioxin isomers using a low-resolution packed column is almost impossible. However, determination of 2,3,7,8-TCDD in a mixture of all 22 TCDDs was performed using a low-resolution packed column after the mixture was prefractionated by HPLC [25,26].

Column choice is one of the most important tasks to achieve satisfactory analysis of complex organic mixtures because the column performs the separation. Currently, high-resolution capillary columns are commercially available. Table 2 shows typical GC columns used for TCDD analysis. Almost all TCDD analysis was performed using commercial fused silica capillary columns. There are only a few reports that involve a noncommercial capillary column for TCDD analysis. For example, a column coated with a Silar 10c, which is deactivated Carbowax 20M, was used to analyze 2,3,7,8-substituted dioxin isomers (2,3,7,8-tetra-, 1,2,3,7,8-penta, 1,2,3,4,7,8-, 1,2,3,6,7,8-, and 1,2,3,7,8,9-hexachloro-*p*-dioxin) in fly ash from municipal incinerators [27]. A fused silica capillary column (25 m × 0.32 mm i.d.) coated with a newly developed liquid crystalline polysiloxane stationary phase showed unsurpassed selectivity for the separation of 2,3,7,8-TCDD from other tetra isomers [28]. A noncommercial column was developed to resolve both 2,3,7,8-TCDD and 2,3,7,8-tetrachloro dibenzofuran (TCDF) from all of their other respective tetrachlorinated isomers [29]. This study involved computer model predictions of the isomer separation based on data obtained from commercial columns, including DB-5, DB-225, DB-WAX, and SP2250. A column developed according to the results of the computer model predictions was successfully applied in the analysis of a wide variety of samples, including animal tissues, pulp/paper mill samples, and aqueous and other chemical wastes. However, high-resolution fused silica capillary columns, which give satisfactory resolution to TCDD mixtures, are commercially available at reasonable prices today.

Table 2 Gas Chromatographic Capillary Columns Used for TCDD Analysis

Stationary phase	Length × i.d.	Application	Literature
DB-5	30 m × 0.25 mm	Tissue, sediment	15
Silar 10c	55 m × 0.27	Fish	51
DB-1	30 m × 0.3 mm (0.25 μm)	Wastewater	22
DB-5	30 m × ?	Fish	18
SP2330	60 m × ?	Fish	18
DB-5	30 m × 0.25 mm	Oil extracts	14
SP2331	40 m × 0.25 mm		
DB-5	60 m × ?	Paper mill wastewater	61
SP2330	60 m × ?		
DB-225	30 m × ?		
SP2331	40 m × 0.25 mm	Water leachates from	13
		waste landfill	52
DB-5	30 m × 0.25 mm	Bird eggs	19
	(0.25 μm)		
SP2330	60 m × ?	Soft-shell clam	35
SP2331	60 m × 0.32 mm	Sediment samples	36
DB-5	60 m × 0.32 mm		
Polysiloxane	25 m × 0.32 mm	Dioxin standard mixture	28
DB-5	30 m × 0.32 mm	Air/smoke samples	39
	(0.25 μm)		
DB-5	30 m × 0.2 mm	Pulp, paper mills	12
SP2331	60 m × ?		
DB-5	60 m × 0.25 mm	Car exhaust samples	37
SP2330	60 m × 0.25 mm		
OV-17	50 m × 0.2 mm	Fish, sediment	53
OV-17, OV-101	20–30 m × 0.35–0.37 mm	Plant, soil	54
OV-1	25–50 m × 0.3 mm	Air samples from	38
	(0.25 μm)	incinerators	
SP2330	60 m × 0.25 mm (0.2 μm)		
Silar 10c (Carbowax deactivated)	55 m × 0.26 mm	Fish, herring gull	27

Therefore, it is not necessary to prepare noncommercial columns unless one intends to study the performance of a liquid phase toward organic chemicals including TCDDs.

As with commercially available high-resolution fused silica capillary columns in the mid 1980s, reasonably priced bench-top gas chromatograph/mass spectrometers (GC/MS) have been developed by commercial manufacturers, so that the determination of 2,3,7,8-TCDD is now heavily dependent on GC/MS, because it can perform identification and quantitation simultaneously [30]. Today most studies on TCDD analysis are performed using a GC/MS. In only a few cases, ECD has been used to study a column performance in TCDD analysis. For example, ECD was used to study the performance of different stationary phases (OV-101, OV-17, Silar 10c) in separation of PCDD [31]. Later, the same author studied resolutions of 22 TCDD isomers on OV-101, OV-17, and Silar 10c using a GC/MS. Though 2,3,7,8-TCDD could be separated by using the Silar 10c column, this isomer was coeluting with 1,2,7,9-TCDD and is only partially separated from 1,4,6,9-TCDD on OV-17. On OV-101, 2,3,7,8-TCDD was coeluting with 1,2,7,8-TCDD. The separation of 2,3,7,8-TCDD from all the other TCDD isomers required a rather long (55 m) and narrow (0.25 mm i.d.) Silar 10c column [32]. Penta-chlorinated dibenzo-*p*-dioxins eluted on the Silar 10c column in the TCDD elution range. However, they were easily distinguished from TCDDs by the presence of more intense molecular ions (M^+) at m/z 354.

As Table 2 shows, the most widely used columns for 2,3,7,8-TCDD analysis are DB-5, which is (5% phenyl)-methylpolysiloxane, and SP2331. Stationary phases similar to DB-5 are HP-5, Ultra-2, SPB-5, CP-Sil 8CB, RSL-200, Rtx-5, BP-5, CB-5, OV-5, PE-5, 007-2(MPS-5), SE-52, SE-54, XTI-5, PTE-5, and HP-5MS [33].

Application of a 60 m SP2331 column seems to be the most useful to perform isomer-specific determinations of 2,3,7,8-TCDD. Isomer-specific determinations of 2,3,7,8-TCDD in samples from various matrices, including paper mill waste water [34], soft-shell clam [35], sediment samples [36], car exhaust samples [37], and air samples from incinerators [38] were done using a 60 m SP2330 fused silica capillary column. Determination of total CDDs and separation of CDDs from CDFs were conducted using 30 m DB-5 fused silica capillary column [18,19,22,39,40]. For example, a 30 m DB-5 fused silica capillary column was used for the primary analysis of fish samples, and a 60 m SP2330 fused silica capillary column was used to verify results [18]. Figure 4 shows chromatograms of tetra- and penta-isomers of dioxins obtained by a newly developed bonded phase DB^{TM}-dioxin.

One of the most commonly used quantitation methods for 2,3,7,8-TCDD in various matrices with GC/MS is the use of labeled internal standards, $^{13}C_{12}$-2,3,7,8-TCDD and $^{37}Cl_4$-2,3,7,8-TCDD. For quantitation, the responses of the m/z at 320 and 322 for the 2,3,7,8-TCDD, m/z at 322 and 334 for $^{13}C_{12}$-2,3,7,8-TCDD internal standard, and m/z at 328 for the $^{37}Cl_4$-2,3,7,8-TCDD internal standard are measured. The concentration of 2,3,7,8-TCDD in a sample is determined by comparing the ratio of the m/z at 322 and 334 peaks or the m/z at 334 and 328 to the calibration curve, which is established using the standard solutions [41].

Generally, determination of 2,3,7,8-TCDD in various matrices has been mostly conducted as a combination of CC clean up followed by GC/MS analysis with a $^{13}C_{12}$-labeled internal standard. Almost 99% of GC/MS analysis has been conducted using the EI mode of mass spectra. The mass spectrum of 2,3,7,8-TCDD

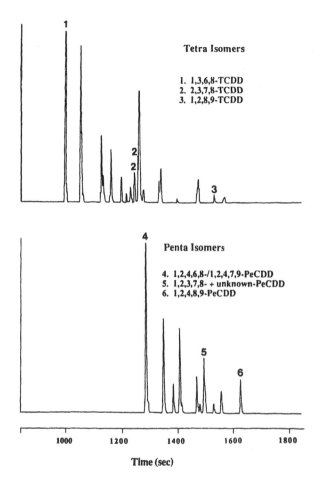

FIGURE 4 Gas chromatograms of tetra and penta isomers of dioxins obtained with a 60 m × 0.25 mm (d_f = 0.15 μm) fused silica capillary column bonded phase DB™-dioxin. The oven temperature was held at 180°C for 1 min and then programmed to 270°C at 2.5°C/min and held at 270°C for 40 min. (Courtesy of Dr. Jennings, J & W Scientific, Folsom, CA.)

shows the molecular ion and its isotope peaks at M⁺ 320 (78.5), m/z 322 (100), and 324 (47.9). Major ions are at m/z 257 (32.9, M⁺—COCl), 259 (34.3, M⁺—COCl), 194 (31.6, M⁺—2COCl), 196 (19.6, 2COCl⁺), 109 (19.0, $C_6H_2Cl^+$), 97 (36.0, $C_5H_2Cl^+$), and 74 (37.4, $C_6H_2^+$) [42]. 2,3,7,8-TCDD shows a significant peak at m/z 74; the spectrum of other tetra-TCDDs—1,2,3,8-TCDD and 1,2,3,4-TCDD—shows corresponding peaks at m/z 75 ($C_6H_3^+$) and 76 ($C_6H_4^+$), respectively. These ions may allow a determination of the number of chlorine substituents in each ring of the dioxin system [43]. There are only a few applications of negative ion chemical ionization (NICI) MS to TCDD analysis. The sensitivity of methane NICI was reported as comparable to that of EI ionization for the (M-2Cl)-ions [44,45], but Oehme and Kirschmer [38] reported that the sensitivity for 2,3,7,8-TCDD in the NICI mode was at least 100 times lower than that of EI.

After sample cleanup using a CC technique, concentrations of 2,3,7,8-TCDD ranging from 1.0 ng/kg in Lake Superior lake trout to 48.9 ng/kg in lake trout from Lake Ontario were determined using GC/MS [23]. Subsequent to intensive cleanup processes using CC techniques described above, quantitation of ppt levels of 2,3,7,8-TCDD in 12 species of freshwater fish [46,47], crayfish, mussels, muscle and eggs of birds [48], and commercial fish feeds [49] was obtained [20]. Using a VG 70-70S double-focusing high-resolution mass spectrometer interfaced to a Hewlett-Packard Model 5898 gas chromatograph equipped with a 60 m DB-5 or a SPB-5 fused silica capillary column, 0.07 pg/g pulp and 0.99 pg/g pulp of 2,3,7,8-TCDD were detected in unbleached paper pulp and bleached paper pulp, respectively, suggesting that bleaching with a chlorinated compound increased the presence of 2,3,7,8-TCDD in pulp [40].

Electron-impact (EI) ionization was preferred for TCDDs (sensitivity 1–10 pg) but showed decreasing sensitivity for the higher chlorinated species such as penta- to octachlorinated (10–50 pg). The ion used to monitor TCDD in this study was $M^+ = 320$ [43].

Using $^{13}C_{12}$-2,3,7,8-TCDD as a GC/MS internal standard, levels of 2,3,7,8-TCDD in the sediment from Newark Bay have been monitored. A significant decrease of 2,3,7,8-TCDD levels from 1100 ppt in the 1960s to 230 ppt in the 1980s in the sediment was observed [36]. Soft-shell clams (*Mya arenaria*) in the same bay contained levels of 11–13 ppt in 1986 [35]. Fish and sediment samples from one lake (Lake Winthrop) out of six Massachusetts lakes contained 71 and 5.9 ppt respectively of 2,3,7,8-TCDD in 1983 [50].

There are only a few reports on the analysis of TCDDs using high-performance liquid chromatography. This may be due to its low resolution and low sensitivity compared with GC. Thus, HPLC has been commonly used for one of the sample preparation steps. For example, a normal-phase (silica gel) HPLC was used to fractionate complex mixtures into several simple fractions, and then further analysis of TCDDs was conducted using GC/MS [25]. Automated isolation procedures for the 22 authentic TCDD isomers were achieved using a reverse phase HPLC [9]. However, the results exhibited lower resolution than GC.

REFERENCES

1. F. Coulston and F. Pocchiari, *Accidental Exposure to Dioxins: Human Health Aspects*, Academic Press, New York, 1983.

2. M. Kuratsune, M. Takesumi, J. Matsuzaka, and A. Yamaguchi, Epidemiologic study on yusho, a poisoning caused by ingestion of rice oil contaminated by a commercial brand of polychlorinated biphenyl, *Environ. Health Prospect 1*: 119–128 (1972).

3. W. W. Muelder and L. A. Shadoff, The preparation of uniformly labeled ^{14}C-2,7-dichlorodibenzo-*p*-dioxin and ^{14}C-2,3,7,8-tetrachlorodibenzo-p-dioxin, in: *Chlorodioxins—Origin and Fate* (Blair, E. H., ed.), American Chemical Society, Washington, D.C., 1973, pp. 1–6.

4. F. Cattabeni, A. DiDomenico, and F. Merli, Analytical procedures to detect 2,3,7,8-TCDD at Seveso after the industrial accident on July 10, 1976, *Ecotoxicol. Environ. Safety 12*: 35–52 (1986).

5. M. P. Esposito, T. O. Tiernan, and F. E. Dryden, *Dioxins*, EPA-600/2-80-197 (1980).

6. W. B. Crummett, T. J. Nestrick, and L. L. Lamparski, Analytical methodology for the determination of PCDDs in environmental samples: an overview and critique, in: *Dioxins in the Environment* (M. A. Kamrin and P. W. Rodgers, eds.), McGraw-Hill, New York, 1995, pp. 57–83.

7. T. J. Nestrick, L. L. Lamparski, and D. I. Townsend, Identification of tetrachlorodibenzo-*p*-dioxin isomers at the 1-ng level by photolytic degradation and pattern recognition techniques, *Anal. Chem. 52*: 1865–1874 (1980).

8. National Research Council Canada. Polychlorinated dibenzo-*p*-dioxins: limitations to the current analytical techniques. Subcommittee on pesticides and industrial organic chemicals, NRCC/CNRC, Ottawa, Canada, 1981.

9. C. Rappe, G. Choudhary, and L. H. Keith, eds., *Chlorinated Dioxins and Dibenzofurans in Perspective*, Lewis, Chelsea, MI, 1986, pp. 305–365.

10. J. C. Harris, R. C. Anderson, B. E. Goodwin, and C. E. Rechsteiner, *Dioxin Emissions from Combustion Sources: A Review of the Current State of Knowledge*, Authur D. Little, New York, 1981.

11. M. A. Kamrin and P. W. Rodgers, *Dioxins in the Environment*, Hemisphere, New York, 1985.

12. R. E. Clement and H. M. Tosine, Analysis of chlorinated dibenzo-*p*-dioxins and dibenzofurans in the aquatic environment, in: *Analysis of Trace Organics in the Aquatic Environment* (B. K. Afghan and A. S. Y. Chan, eds.), CRC Press, Boca Raton, FL, 1989, pp. 165–168.

13. C. Forst, L. Stieglitz, and G. Zwick, Development of a method for the isomer-specific determination of PCDDs/PCDFs in leachate-oil extracts of a waste landfill, in: *National Technical Information Service, No. KFK-4327*, 1987, p. 9.

14. C. Forst, L. Stieglitz, and G. Zwick, Isomer-specific determination of PCDD/PCDF in water leachate of a waste landfill, in: *Organic Micropollutants in the Aquatic Environment: Proceedings of the Fiftieth European Symposium, Rome, Italy, 1987*, Kluwer, Boston, 1988, pp. 52–58.

15. R. A. Hummel, Cleanup techniques for the determination of parts per trillion residue levels of 2,3,7,8-tetrachlorodibenzo-*p*-dioxin (TCDD), *J. Agric. Food Chem. 25*: 1049–1053 (1977).

16. T. O. Teirnan, J. H. Garrett, J. G. Solch, D. J. Wagel, G. F. VanNess, and M. L. Taylor, Improved separations procedures for isolating 2,3,7,8-TCDD and 2,3,7,8-TCDF from chemically complex aqueous and solid sample matrices and for definitive quantitation of these isomers at ppq to ppt concentrations, *Chemosphere 18*: 93–100 (1989).

17. T. Zacharewski, L. Safe, S. Safe, B. Chittim, D. DeVault, K. Wiberg, P.-A. Bergqvist, and C. Rappe, Comparative analysis of polychlorinated dibenzo-*p*-dioxin and dibenzofuran congeners in Great Lakes fish extracts by gas chromatography–mass spectrometry and in vitro enzyme induction activities, *Environ. Sci. Technol. 23*: 730–735 (1989).

18. P. J. Marquis, M. Hackett, L. G. Holland, M. L. Larsen, B. Butterworth, and D. W. Kuehl, Analytical methods for a national study of chemical residues in fish, *Chemosphere 29*: 495–508 (1994).

19. L. L. Williams, J. P. Giesy, D. A. Verbrugge, S. Jurzysta, and K. Stromborg, Polychlorinated biphenyls and 2,3,7,8-tetrachlorodibenzo-*p*-dioxin equivalents in eggs of double-crested cormorants from a colony near Green Bay, Wisconsin, USA, *Arch. Environ. Contam. Toxicol. 29*: 327–333 (1995).

20. L. M. Smith, D. L. Stalling, and J. L. Johnson, Determination of parts-per-trillion levels of polychlorinated dibenzofurans and dioxins in environmental samples, *Anal. Chem. 56*: 1830–1842 (1984).

21. H. Beck, A. Bross, K. Eckart, W. Mathar, and R. Wittkowski, PCDDs, PCDFs and related compounds in paper products, *Chemosphere 19*: 655–660 (1989).

22. T. L. Peters, T. J. Nestrick, and L. L. Lamparski, The determination of 2,3,7,8-tetrachlorodibenzo-*p*-dioxin in treated waste water, *Water Res. 18*: 1021–1024 (1984).

23. D. D. Vault, W. Dunn, P.-A. Bergqvist, K. Wiberg, and C. Rappe, Polychlorinated dibenzofurans and polychlorinated dibenzo-*p*-dioxins in Great Lakes fish: a baseline and interlake comparison, *Environ. Toxicol. Chem. 8*: 1013–1022 (1989).

24. O. Hutzinger, K. Olie, J. W. A. Lustenhouwer, A. B. Okey, S. Bandiera, and S. Safe, Polychlorinated dibenzo-*p*-dioxins and dibenzofurans: a bioanalytical approach, *Chemosphere 10*: 19–25 (1981).

25. L. L. Lamparski and T. J. Nestrick, Synthesis and identification of the 10 hexachlorodibenzo-*p*-dioxin isomers by high performance liquid and packed column gas chromatography, *Chemosphere 10*: 3–10 (1981).

26. L. L. Lamparski and T. J. Nestrick, The isomer-specific determination of tetrachlorodibenzo-*p*-dioxin at part per trillion concentrations, in: *Chlorinated Dioxins and Related Compounds: Impact on the Environment* (O. Hutzinger, R. W. Frei, E. Merian, and F. Pacchiari, eds.), Pergamon Press, Oxford, 1982, p. 1.

27. H. R. Buser and C. Rappe, Isomer-specific separation of 2378-substituted polychlorinated dibenzo-*p*-dioxins by high-resolution gas chromatography/mass spectrometry, *Anal. Chem. 56*: 442–448 (1984).

28. K. P. Naikwadi and F. W. Karasek, Development of GC capillary column for isomer-specific separation of toxic isomers of PCDDs and PCDFs in environment samples, *Chemosphere 20*: 1379–1384 (1990).

29. T. O. Tiernan,, J. H. Garrett, J. G. Solch, and L. A. Harden, New capillary gas chromatography column for the simultaneous isomer-specific analysis of 2,3,7,8-TCDD and 2,3,7,8-TCDF, *Chemosphere 20*: 1371–1378 (1990).

30. R. E. Finnigan, M. S. Story, and D. F. Hunt, Overcoming bottlenecks in environmental sample analysis, in: *Advances in the Identification and Analysis of Organic Pollutants in Water* (L. H. Keith, ed.), Ann Arbor Science Publishers, Ann Arbor, MI, Vol. 2, 1981, pp. 555–570.

31. H. R. Buser, High-resolution gas chromatography of polychlorinated dibenzo-*p*-dioxins and dibenzofurans, *Anal. Chem. 48*: 1553–1557 (1976).

32. H. R. Buser, C. Rappe, and P.-A. Bergqvist, Analysis of polychlorinated dibenzofurans, dioxins and related compounds in environmental samples, *Environ. Health Persp. 60*: 293–302 (1985).

33. J & W Scientific 1996/97 Catalog and Technical Reference, J & W Scientific, Folsom, CA, 1996, p. 22.

34. T. O. Tiernan, J. H. Garrett, J. G. Solch, D. J. Wagel, G. F. VanNess, and N. L. Taylor, Improved separations procedures for isolating 2,3,7,8-TCDD and 2,3,7,8-TCDF from chemically-complex aqueous and solid sample matrices and for definitive quantitation of these isomers at ppq to ppt concentrations, *Chemosphere 18*: 93–100 (1989).

35. R. P. Brown, K. R. Cooper, A. Cristini, C. Rappe, and P.-A. Bergqvist, Polychlorinated dibenzo-*p*-dioxins and dibenzofurans in *Mya arenaria* in the Newark/Raritan Bay Estuary, *Environ. Toxicol. Chem. 13*: 523–528 (1994).

36. H. Y. Tong, S. J. Monson, M. L. Gross, R. F. Bopp, H. J. Simpson, B. L. Deck, and F. C. Moser, Analysis of dated sediment samples from the Newark Bay area for selected PCDD/Fs, *Chemosphere 20*: 1497–1502 (1990).

37. A. G. Bingham, C. J. Edmunds, B. W. L. Graham, and M. T. Jones, Determination of PCDDs and PCDFs in car exhaust, *Chemosphere 19*: 669–673 (1989).

38. M. Oehme and P. Kirschmer, Isomer-selective determination of tetrachlorodibenzo-*p*-dioxins using hydroxyl negative ion chemical ionization mass spectrometry combined with high-resolution gas chromatography, *Anal. Chem. 56*: 2754–2759 (1984).

39. R. D. Kleopfer, R. L. Greenall, T. S. Viswanathan, C. J. Kirchmer, A. Gier, and J. Muse, Determination of polychlorinated dibenzo-*p*-dioxins and dibenzofurans in environmental samples using high resolution mass spectrometry, *Chemosphere 18*: 109–118 (1989).

40. K. Wiberg, K. Lundstrom, G. Glas, and C. Rappe, PCDDs and PCDFs in consumers' paper products, *Chemosphere 19*: 735–740 (1989).

41. C. Rappe and H. R. Buser, Chemical properties and analytical methods, in: *Halogenated Biphenyls, Terphenyls, Naphthalenes, Dibenzo-p-dioxins, and Related Products* (R. D. Kimbrough, ed.), Elsevier/North Holland Biomedical Press, New York, 1980, pp. 48–68.

42. B. S. Middleditch, S. R. Missler, and H. B. Hines, *Mass Spectrometry of Priority Pollutants*. Plenum Press, New York, 1981, pp. 210–211.

43. H. R. Buser, Analysis of TCDD's by gas chromatography–mass spectrometry using glass capillary columns, in: *Dioxin: Toxicological and Chemical Aspects* (F. Cattabeni, A. Cavallero, and G. Galli, eds.), SP Medical Science Books, New York, 1978, pp. 27–41.

44. J. R. Hass, M. D. Friesen, D. J. Harvan, and C. E. Parker, Determination of polychlorinated dibenzo-*p*-dioxin in biological samples by negative chemical ionization mass spectrometry, *Anal. Chem. 50*: 1447–1479 (1978).

45. A. Cavallaro, L. Luciani, G. Ceroni, I. Rocchi, G. Invernizzi, and A. Gorni, Summary of results of PCDDs analyses from incinerator effluents, *Chemosphere 9*: 859–868 (1982).

46. R. D. Kleopfer and J. Zirschky, TCDD distribution in the Spring River, South Western Missouri, *Environ. Int. 9*: 249–253 (1983).

47. D. L. Stalling, L. M. Smith, J. D. Petty, J. W. Hogan, J. L. Johnson, C. Rappe, and H. R. Buser, Residues of polychlorinated dibenzo-*p*-dioxins and dibenzofurans in Laurentian Great Lakes fish, in: *Human and Environmental Risks of Chlorinated Dioxins and Related Compounds* (R. E. Tucker, A. L. Yong, and A. P. Gray, eds.), Plenum Press, New York, 1983, pp. 221–239.

48. C. Rappe, H. R. Buser, D. L. Stalling, L. M. Smith, and R. C. Dougherty, *Nature 292*: 524–526 (1981).

49. J. D. Petty, L. M. Smith, P. Bergqvist, J. L. Johnson, D. L. Stalling, and C. Rappe, Composition of polychlorinated dibenzofuran and dibenzo-*p*-dioxin residues in sediments of the Hudson and Housatonic rivers, in: *Chlorinated Dioxins and Dibenzofurans in the Total Environment* (G. Choudhary, L. H. Keith, and C. Rappe, eds.), Ann Arbor Science Publishers, Ann Arbor, MI, 1983, pp. 203–207.

50. J. J. Jonasch, Six ponds dioxin survey, *Northeastern Environ. Sci. 3*: 80–89 (1984).

51. B. Chittim, D. DeVault, K. Wiberg, P.-A. Bergqvist, and C. Rappe, Comparative analysis of polychlorinated dibenzo-*p*-dioxin and dibenzofuran congeners in Great Lakes fish extracts by gas chromatography–mass spectrometry and in vitro enzyme induction activities, *Environ. Sci. Technol. 23*: 730–735 (1989).

52. C. Forst, L. Stieglitz, and G. Zwick, Isomer-specific determination of PCDD/PCDF in oil extracts from water leachates of a waste landfill, *Chemosphere 17*: 1935–1944 (1988).

53. J. Lawrence, F. Onuska, R. Wilkinson, and B. K. Afghan, Methods research: determination of dioxins in fish and sediment, *Chemosphere 15*: 1085–1090 (1986).

54. H. R. Buser, Determination of 2,3,7,8-tetrachlorodibenzo-*p*-dioxin in environmental samples by high-resolution gas chromatography and low resolution mass spectrometry, *Anal. Chem. 49*: 918–922 (1977).

8

Organophosphorus Esters

JAMES N. SEIBER, JAMES E. WOODROW,
AND MICHAEL D. DAVID
University of Nevada, Reno, Nevada

I. INTRODUCTION

In 1820, the French chemist Lassaigne reacted alcohol with phosphoric acid, producing the first organophosphate esters and launching the field of organophosphorus chemistry. There are several types of organophosphorus chemicals—phosphines, phosphonium salts, phosphoric anhydrides, etc.—but particular utility has been found in the organophosphate (OP) esters, which are the focus of this review. These are esters primarily of phosphoric acid, but they include esters of phosphonic and phosphinic acids, thiono (sulfur) analogues, and derivatives containing, occasionally, amino and other substituent groups in place of the more common alkoxy and aryloxy substituents.

A. OP Pesticides

The rapid development of OP pesticide chemistry is often ascribed to the search for potential warfare agents—nerve poisons—during World War II, primarily in Germany and England. Simultaneously, however, there was recognition of the potential utility of many OPs as insecticides. In 1941 the insecticide TEPP—tetraethyl pyrophosphate—was synthesized by G. Schrader, and in 1944 the insecticide now known as parathion was synthesized, also by Schrader. Many other commercially important OP pesticides, primarily insecticides but including a few herbicides (such as glyphosate and the defoliant DEF) and a few fungicides, were developed thereafter and occupy a major niche in the pesticide market to this day (Fest and Schmidt, 1973) (Fig. 1).

OP insecticides are characterized by good contact and systemic activity in a variety of arthropods, and moderate to high toxicity in nontarget insects, aquatic species, mammals, and man. They vary considerably in physicochemical properties, from fairly high persistence as in chlorpyrifos and azinphosmethyl to low persistence materials such as TEPP and acephate. Most OPs are developed for use on food crops prior to harvest, but uses also exist in the home and home garden market, and for use on turf, ornamentals, and forestry. Because of their toxicity, use on foods, home usage, and the concern regarding contact with wildlife, very sensitive residue analytical methods for both foodstuffs and environmental media have been developed.

In addition to parent OPs, methods are also required for environmental and metabolic conversion products, at least for those of significant toxicity. As the general (and abbreviated) conversion pathway for parathion in Figure 2 shows, paraoxon may require analysis along with the parent. In some cases, acid breakdown products—DETP, DEP and *p*-nitrophenol in the case of parathion—are also targets of analysis, particularly when conducting human exposure studies where these metabolites may represent the chemical evidence of prior exposure to the parent.

Residue methods for OP pesticides have been reviewed in detail elsewhere. Bowman (1981) has traced method development from the early colorimetric, paper, and TLC-based methods to those that were in use at the time of that writing—primarily gas chromatographic methods using element selective detectors. The FDA Pesticide Analytical Manual (PAM, 1993), recently revised, provides considerable detail on analytical methods for OPs, primarily in foods, including extraction, cleanup, and determination steps with extensive instructions for gas chromatography and high-performance liquid chromatography methods. Multiresidue methods, in which most OPs may be analyzed (along with organochlorine, carbamate, and other pesticide classes) in a single integrated method, are presented in PAM Volume I. Volume II of PAM, and other sources such as the series Analytical Methods for Pesticides (Zweig and Sherma, 1972), provide details for individual OPs and, generally, their metabolites.

B. OP Hydraulic Fluids, Fuel Additives, and Fire Retardants

While much of the emphasis on OP analytical method development derives from the pesticide group, other classes of OPs are of interest as well (Fig. 3). Tri-

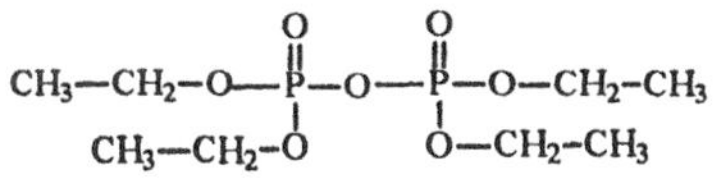

tetraethylpyrophysphate (TEPP)

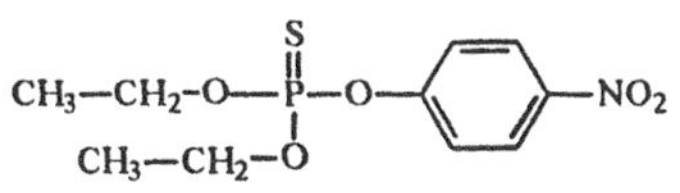

parathion

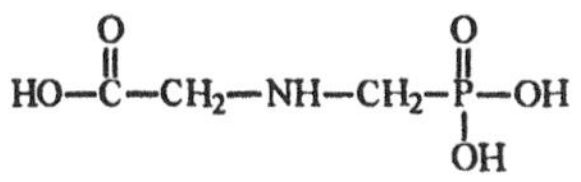

glyphosate

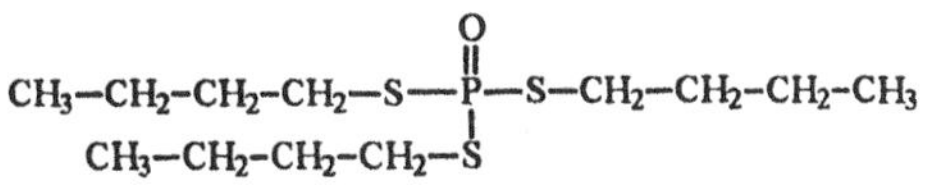

DEF (Butifos)

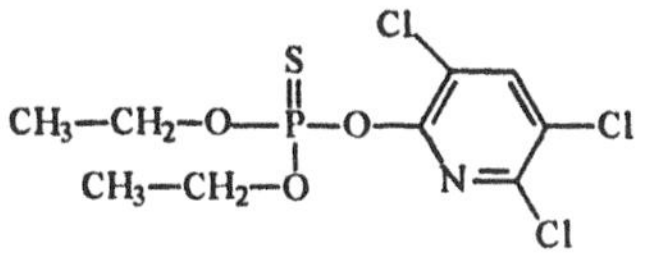

chlorpyrifos

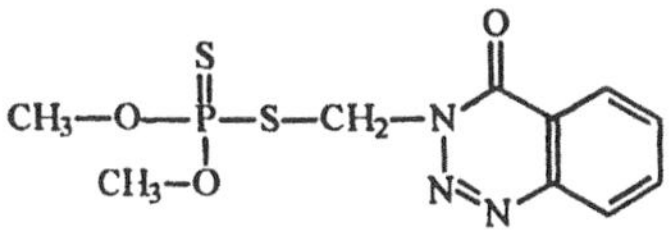

azinphos methyl

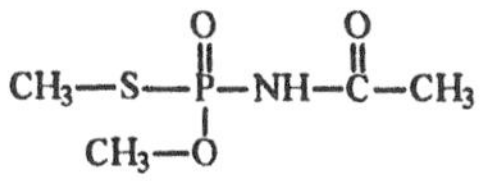

acephate

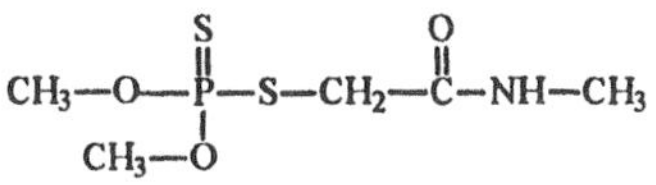

dimethoate

Figure 1 Structures of some common organophosphorus pesticides.

cresylphosphate (TCP), along with other aryl and mixed alkyl-aryl OPs, have been used extensively as fuel and lubricant additives, and as hydraulic fluids. These compounds possess high stability and generally low to moderate polarity, and have the significant property, owing to the presence of phosphorus in their structures, of flame retardancy. This is particularly important in such applications as aircraft hydraulic fluids. It also provides for an ongoing market; it has been estimated that 70–80% of the annual usage of OP ester hydraulic fluids in the

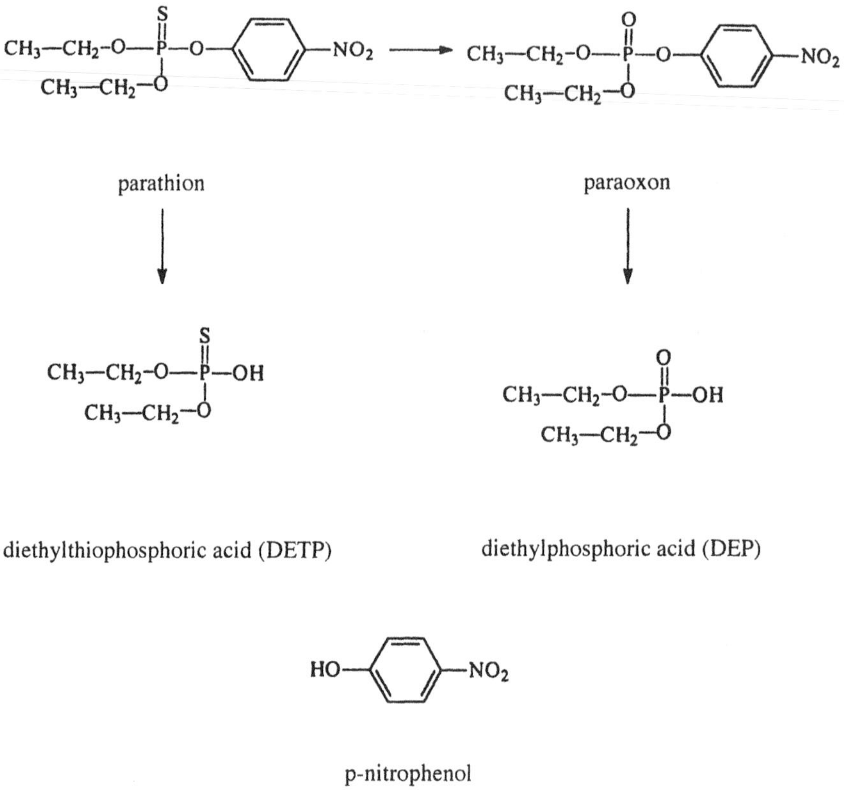

FIGURE 2 Major environmental and metabolic breakdown products of parathion.

1970s was for replacement of losses due to leakage, most of which resulted in direct release to the environment (Boethling and Cooper, 1985; Federal Register, Vol. 48; Heitcamp et al., 1984). A recent survey of soils from U.S. Air Force bases showed the presence of OP hydraulic fluids in soils and sediments around runways and maintenance operations in the range of 2–150 ppm (David and Seiber, 1996).

The acute toxicities of most nonpesticidal OPs are relatively low, and concerns over TCP have decreased since the neurotoxic ortho isomer was removed from commercial formulations. Still, the relatively long environmental persistence of TCP makes it a contaminant of concern. Tributylphosphate is a ubiquitous environmental and laboratory contaminant largely because of its use as an extractant for metal complexes including uranium processing (Thomas and Macaskie, 1996). A number of halogenated OP esters find use in such areas as

tricresylphosphate (TCP) 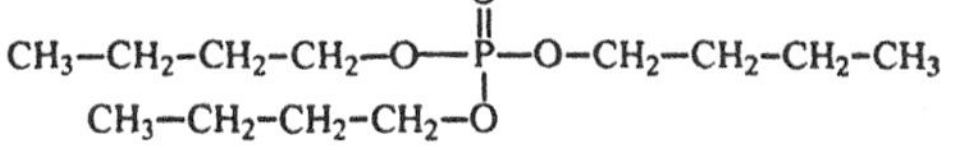   tributylphosphate (TBP)

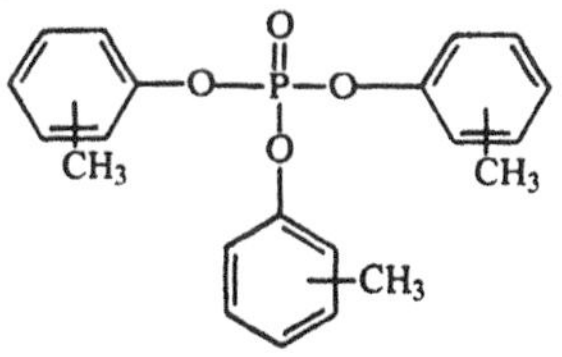

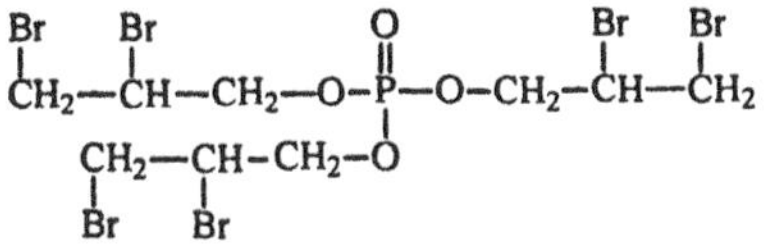

tri(2,3-dibromopropyl)phosphate (TRIS-BP) 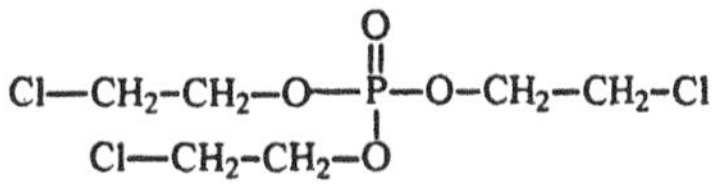   trichloroethylphosphate (TCEP)

FIGURE 3 Structures of some common organophosphate hydraulic fluids, additives, and fire retardants.

polyurethane and polyisocyanurate foams, other plastic products, and some garments, where high fire retardancy is required. One such compound, Tris-BP, was banned for use in clothing because of its genotoxocity. Tris (2-chloroethyl) phosphate and related chloroalkyl phosphates continue in widespread use (Hutzinger et al., 1976; Green, 1993; Aston et al., 1996).

Analytical methods for the nonpesticidal OPs have not received the attention of the pesticidal OPs, partly because they are of lower toxicity and are not used in or on foods or in other high-exposure situations. Because their physicochemical properties are similar in many respects to the pesticidal OPs, however, the same approaches to extraction, cleanup, and determination are applicable.

C. Goal of This Chapter

The purpose of this chapter is to review chromatographic approaches, primarily using GC and HPLC, to the determination of OPs in both groups. GC will receive more of the attention simply because it is in greater use for OPs than HPLC. This is due to the stability and volatility of most OPs under GC conditions and, particularly, the availability of two routine element-selective detectors that give high selectivity toward OPs, namely, the flame photometric detector and the alkali flame ionization (AFID) detector. Mass spectrometry–based detection is increasing in popularity in GC and, to a lesser extent, in HPLC detection. HPLC finds use when the more polar breakdown products of OP esters are to be analyzed. It suffers from lack of a commercial phosphorus-selective detection system and the fact that nonaromatic OPs do not absorb at longer UV wavelengths. TLC finds some specialty use for screening purposes, particularly in locations where GC is not available. This review will focus on determination methods, with less attention to extraction and cleanup details, which for the most part can be found in such methods compilations as the PAM (1993) and the AOAC Official Methods of Analysis (1990).

II. GAS CHROMATOGRAPHIC METHODS

GC is clearly the method of choice for OP esters because of its ability to resolve individual members of this chemical class, and to resolve individual analytes in suitably prepared extracts containing potential interferences. This latter characteristic is enhanced considerably because of the availability of selective detectors for OPs. Two element-selective detectors are in general use, and mass selective detectors can be used for OPs as well as most other chemical classes. OP acids and other polar metabolites may also be analyzed by GC after derivatization, generally back to an ester form.

A. Detectors

Many of the element-selective detectors are variations of the flame ionization detector, which is well described in standard references in instrumental analysis (see, for example, Willard et al., 1989) and gas chromatography (see, for example, Grob, 1985). Two such detectors are used routinely for OP analysis, namely the alkali-flame ionization detector (AFID), also known as the nitrogen-phosphorus thermionic selective detector (NP-TSD), and the flame photometric detector (FPD). Other selective detectors, such as the electrolytic conductivity and photo-ionization detectors, may be used for OPs that contain other heteroatoms such as sulfur, halogen, or nitrogen in addition to phosphorus. The atomic emission detector (AED) may be used for a variety of heteroatoms, including phosphorus, but it is not in general service because of its complexity and cost. The mass selective detector is useful for individual analytes but does not have the multi-residue screening capability of the AFID and the FPD.

     The alkali flame ionization detector (AFID) is essentially an FID with the addition of an alkali salt (potassium bromide, cesium bromide, rubidium sulfate, etc.) to the flame base (Willard et al., 1989; Grob, 1985). It was discovered by chance, when a salt deposit was inadvertently left on the detector base during a cleaning of the FID (Giuffrida, 1964). The contaminated detector showed a much enhanced response and selectivity toward OPs. The detector may be prepared by inserting a salt pellet on top of the FID flame base or by placing a metal cup containing the compressed salt as a substitute for the plain metal or quartz base. More commonly, the AFID may be purchased, with the advantage that the electrode and gas flow geometries are adapted to the AFID mode of operation. Although the exact mechanism of the enhanced response of the AFID to P-containing compounds is unknown, it is believed to involve an electron transfer between the alkali metal, formed in the reducing edge of the relatively cool AFID flame, and P-containing radicals produced from partial combustion of the OP (see Fig. 4). The vapor phase redox reaction results in a flux of cations and P-containing anions that lower the resistance between the electrodes and set up a current flow in the external circuit to the electrometer. A somewhat lower, but still useful, enhancement of response to organic nitrogen compounds in the AFID similarly involves a vapor phase redox chemistry, between the alkali metal reducing agent and cyano radicals, producing alkali metal cations and cyanide ions. The response factors (F), and mass detection limits, for typical organophosphorus and organonitrogen compounds are: OPs, 5000 relative to hydrocarbons with a detection limit of 1–10 pg; ONs, 50 relative to hydrocarbons with a detection limit of 1–10 ng.

Because of difficulty in reproducing the response characteristics of the AFID during routine operation, Kolb and Bischoff (1974) introduced a major modification that is now the standard for commercial thermionic detectors. In the modified detector (NP-TSD), the flame is replaced with a hydrogen-air plasma

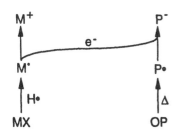

$M^{\cdot}$ = Alkali metal (Rb, Cs, K)

X = Anion (Cl^-, Br^-, SO_4^{2-})

$P^{\cdot}$ = Phosphorus-containing radical (PH_2, PO, PO_2)

$H^{\cdot}$ = Hydrogen atom from hydrogen-rich flame

FIGURE 4 Postulated mechanism for enhanced response of alkali-flame ionization detector to organophosphorus compounds.

created around the surface of an alkali-salt bead. The latter is heated by means of a resistance wire. This electrical heating of the salt bead can be more carefully controlled, so that the NP-TSD is much more stable and reproducible than the AFID. The lower temperature of the plasma also appears to increase the yield of cyano radicals from organonitrogen compounds. The response factors (F) for the NP-TSD are of the same magnitude as those listed above for the AFID, except that the F for organonitrogen to hydrocarbon is more favorable, on the order of 500 or more. The NP-TSD is a fairly inexpensive detector with favorable detection limits for trace analysis of OP, as well as carbamate, urea, triazine, etc., classes of pesticides in relatively clean extracts. Application to soil and biological matrices may require additional cleanup, and confirmation is essential because of possible interferences from high levels of non-P and non-N compounds that are also detected by NP-TSD.

The flame photometric detector (FPD) is based on the principle of thermal excitation of outer electrons to higher energy levels, with reversion to the ground state accompanied by emission of light of specific energies or wavelengths. For OP compounds, combustion produces electronically excited HPO radicals, which emit as an envelope of bands centered on 525 nm in returning to the ground state. For organosulfur compounds, excited S_2 is produced, which emits at 394 nm in returning to the ground state. The emitted light is focussed through an optical interference filter selected to transmit either 525 nm (P mode) or 394 nm (S mode) light to a photomultiplier tube that converts the incident light to an electric current. The detector is highly selective, with F values of close to 10,000 for P- and S-containing compounds relative to hydrocarbons. The useable limits of detection are roughly 10–100 pg for OPs and 0.1–10 ng for organosulfur compounds.

The expense of the FPD is well worth the investment. Much time is saved by bypassing the need for the extensive cleanup required for soil and biological extracts when using the less selective AFID or electron capture detectors, and in assigning identities to peaks. Also, the response of the FPD is very stable, making it the detector of choice for monitoring OP pesticides in foods and environmental samples. The linear range (10^5 for OPs) and baseline stabilities are other attributes of the FPD. Identification of OP pesticides can be materially aided by observing response characteristics on a dual P- and S-mode FPD; response on both modes, for example, indicates presence of both P and S, as might be expected for a thion OP pesticide, while response only to the P-mode might indicate the presence of an OP oxon or nonpesticidal OP ester (Bowman, 1981).

Mass spectrometry offers compound-specific detection capability as well as "absolute" confirmation of identity. In GC-MS, the mass spectrometer receives GC effluent (analyte plus carrier gas) delivered via transfer tubing to the MS ionization chamber. Electron impact ionization (EI) and the softer chemical ionization (CI) operate on the analyte to produce, usually, a parent or molecular ion peak and a series of fragment peaks. OP esters vary from providing very strong molecular ions (chlorpyrifos, triphenyl phosphate) with fairly simple fragmentation under electron impact ionization conditions to those which produce little or no molecular ion peak but extensive fragmentation (dimethoate, malathion). Having both EI and CI capabilities is particularly advantageous for this class of chemicals. Full scan spectra may be obtained at residue levels, with a detection limit of about 1 ng to the MS, which varies with the molecule, instrument type, and state of tune. Selective ion monitoring (SIM) is particularly advantageous for low-level analysis. As long as the analyst knows what to look for, MS may be set for up to eight ions diagnostic for that compound, with detection limits in the 10–100 pg range for most OP analytes. SIM may also be set to program between ions diagnostic for several analytes in a single run, providing some capability for multianalyte detection. But under ideal conditions, NP-TSD and FPD will usually give lower detection limits for OPs, and a much broader capability for screening for multiple OPs in an extract, than MSD.

B. Columns

GC columns for OPs were almost exclusively packed glass columns until about 1980 when fused silica capillary columns came into common use. Packed column technology for OP pesticides is described in Zweig and Sherma (1972) and Bowman (1981). Of the range of polarities of liquid phases available for packed columns, just five or so were used commonly for OPs, namely Dexsil 300, OV-101 (SE-30), OV-17, OV-210, and OV-225 (Bowman, 1981).

Occasionally polar columns, such as DEGS, DEGA, or carbowax, were used, particularly for fairly polar and volatile OP pesticides such as dichlorvos (Vapona). And occasionally mixed phases, such as OV-17/OV-210 were advantageously used for special resolution applications. Capillary column technology represented a major advance for analysis of OPs, along with most other classes of pesticides. Capillary columns provided much enhanced efficiency and resolution, lower detection limits because of greater signal-to-noise ratios, and inertness to more labile compounds associated with the absence of solid support and the use of fused silica column walls. With the advent of bonded-phase capillary columns, and the larger megabore columns (0.53 mm id), the drawbacks associated with earlier capillary column technology (need for inlet-spitter, bleed of liquid phase from the column) largely disappeared, so that capillary columns are now used almost exclusively for gas chromatography of OP esters. The columns of greatest use are as follows:

	Packed Phase Equivalent	Polarity
DB-1	OV-1, OV-101, Dexsil 300	Low
DB-5	OV-17	Slight
DB-210	OV-210	Moderate
DB-225	OV-225	Moderate

Tables of retention times for pesticides, including OPs, are given on several phases in the PAM (1993). Most such compilations are based upon packed column technology, but they can be applied to capillary columns because of the rough equivalency referred to above. This relationship can be seen in Table 1, generated for various N- and P-containing pesticides on DB-1, compared with literature packed column (OV-1, OV-101) retention data.

In addition to the advances in column and detector technology, instrument manufacturers have added important labor-saving ancillaries such as programmable auto-injectors, improved temperature programmers, and data acquisition systems ranging from programmable integrators to fully computerized data stations that can acquire and process information from several GC systems. These latter have found much favor owing to the reporting and record-keeping requirements of Good Laboratory Practices (GLP); the data station produces report-ready records of the type required by GLP.

C. Applications to OP Pesticide Analysis

An array of different techniques for the detection and assay of OP pesticides in various contaminated matrices has developed. These different techniques include spectrophotometry (Blanco and Sanchez, 1990; Chohan and Shah, 1992), biosensors (Rogers et al., 1991; Skladal, 1991 and 1992; Kumaran and Tran-Minh, 1992; Imato and Ishibashi, 1995; Khavkin and Khavkin, 1996), immunosorbent assays (Skerritt et al., 1992; Thompson, 1995), gas chromatography (GC) (Hinckley, 1989; Weisskopf and Seiber 1989; Seiber et al., 1990; Wan, 1990; Belanger et al., 1991; Consalter and Guzzo, 1991; Drevenkar et al., 1991; Holstege et al., 1991; Hsu et al., 1991; Wilson et al., 1991; Leoni et al., 1992; Lino and Silveira, 1992; Mansour et al., 1992; Bourgeois et al., 1993; Kotcon and Wimmer, 1993; Pylypiw, 1993; Richardson and Seiber, 1993; Skopec et al., 1993; Zabik and Seiber, 1993; Grob et al., 1994; Holstege et al., 1994; Kennedy et al., 1994; McCurdy et al., 1994; Miliadis, 1994 and 1995; Pico et al., 1994; Psathaki et al., 1994; Schenck et al., 1994; Cai et al., 1995; Gillespie et al., 1995; Gomez-Gomez et al., 1995; Aston and Seiber, 1996; Lee et al., 1996), enzymology (Drevenkar et al., 1993; Vasilic et al., 1992 and 1993), and liquid chromatography (HPLC) (Seiber et al., 1990; Sharma et al., 1990; Barceló et al., 1991; Kawasaki et al., 1992; Coulibaly and Smith, 1993; Driss et al., 1993; Ioerger and Smith, 1993 and 1994; Cappiello et al., 1994; Smith and Coulibaly, 1994; Lacorte and Barceló, 1995 and 1996; Pinto et al., 1995). The more popular and widely used techniques include GC and HPLC. GC, especially when coupled with element-specific detectors, has been amply proven to give the sensitivity needed to assay trace levels of OPs, and when automated it is one of the more cost-effective techniques for performing routine assays of large numbers of samples. HPLC has more limited application, depending on the contaminated matrix.

Gas chromatographic techniques have been developed for the assay of OPs and OP-related residues for a wide variety of contaminated matrices (e.g., water,

Table 1 Retention Data for N- and P-Containing Pesticides and Related Compounds on Open-Tubular Column (DB-1, NPD, 190°C Isothermal, 1:124 Split Injection) and Comparison with Packed Column Retention Data

Compound	t^a (min)	t' (min)	k'	t'/t' par[b]	SE-30[c]
Solvent	1.45	—	—	—	—
p-Chlorotoluidine	1.93	.47	.32	.08	—
Mevinphos	2.11	.66	.46	.12	.13
Butylate	2.21	.76	.52	.14	—
p-Nitrophenol	2.41	.96	.66	.18	—
Oxamyl oxime	2.48	1.03	.71	.19	—
Methomyl	2.53	1.08	.74	.20	—
Molinate	2.67	1.22	.84	.22	—
Propoxur	2.82	1.37	.94	.25	—
Ethoprop	3.01	1.56	1.08	.28	.33
Azobenzene	3.06	1.61	1.11	.29	—
CIPC	3.08	1.63	1.12	.30	—
2,3,5-Landrin	3.10	1.65	1.14	.30	—
Chlordimeform	3.22	1.76	1.21	.32	—
Trifluralin	3.29	1.84	1.27	.34	—
Phorate	3.43	1.98	1.36	.36	.38
Dimethoate	3.53	2.08	1.43	.38	.41
Carbofuran	3.56	2.11	1.46	.38	—
Aminocarb	3.88	2.43	1.68	.44	—
Diazinon	4.25	2.80	1.93	.51	.53
3-Ketocarbofuran	4.33	2.88	1.99	.52	—
3-Hydroxycarbofuran	5.10	3.65	2.52	.67	—
Methyl parathion	5.27	3.82	2.63	.70	.70
Carbaryl	5.35	3.90	2.69	.71	—
Terbutol	5.46	4.01	2.76	.73	—
Paraoxon	5.61	4.16	2.87	.75	.78
Ronnel	5.92	4.47	3.08	.81	.82
Methiocarb	6.06	4.61	3.18	.84	—
Malathion	6.46	5.04	3.48	.92	.90
Parathion	6.94	5.49	3.79	1.00	1.00
Chlorpyrifos	7.05	5.60	3.86	1.02	1.02
Methidaoxon	7.08	5.63	3.88	1.03	—
Cruformate	7.28	5.83	4.02	1.06	1.10
Methidathion	9.26	7.81	5.39	1.42	1.43
Def	12.41	10.96	7.56	2.00	1.92
Methyl Trithion	14.08	12.63	8.71	2.30	2.20
Azinphosmethyl	32.20	30.85	21.28	5.62	5.20

[a]t = Retention time, $t' = t - t$ (solvent), $k' = t'/t$ (solvent).
[b]t'/t' = Retention time relative to parathion.
[c]Relative retention time (to parathion) in *Pesticide Analytical Manual* (FDA, 1979) for 10% DC-200 (SE-30) at 200°C.
Source: Wehner and Seiber, 1981.

air, urine, animal tissues, etc.). Since most OP parent compounds are volatile to semivolatile and are stable under gas chromatographic conditions (i.e., injection-port and column temperatures), all that is needed for gas chromatographic determination is their isolation from the contaminated matrix in a form free of significant interferences. This approach would also accommodate the oxon conversion products, since they are formed chemically/metabolically by the simple replacement of sulfur by oxygen and would have volatilities and stabilities similar to those for the parent compounds (see example for parathion in Fig. 2). However for the hydrolysis and metabolic breakdown products, such as the polar alkylthiophosphates and alkylphosphates (Fig. 2), additional isolation and derivatization techniques must be employed prior to gas chromatographic assay.

We will describe several present studies concerned with the determination of OP pesticides and their environmental and metabolic conversion products in air, water, food products, plant and animal tissues, bird excreta, and human urine. We will briefly mention the methods used to isolate the analytes but then focus primarily on the gas chromatographic conditions (e.g. columns, detectors, etc.) used to determine OP residue levels and comment on the detection limits achieved using different conditions.

Typically, OP parent residues in air are determined by cumulative sampling through polymeric adsorbents, followed by solvent extraction and analysis (Seiber et al., 1990; Wilson et al., 1991; Zabik and Seiber, 1993; Kennedy et al., 1994; Aston and Seiber, 1996). OP parent residues in water, other aqueous fluids, and fats and oils are either subjected to liquid-liquid extraction or are concentrated on solid phase extraction (SPE) cartridges and then eluted with organic solvents for analysis (Hinckley, 1989; Lino and Silveira, 1992; Bourgeois et al., 1993; Kotcon and Wimmer, 1993; Zabik and Seiber, 1993; Miliadis, 1994 and 1995; Pico et al., 1994; Psathaki et al., 1994; Schenck et al., 1994; Gillespie et al., 1995; Gomez-Gomez et al., 1995; Lee et al., 1996). OP residues in edible oils can be determined directly without special preparation (Grob et al., 1994). Water and urine containing acidic phosphates and thiophosphates are first adjusted in pH prior to SPE extraction. Excess ammonium sulfate is added to the aqueous samples to help "salt out" the analytes, and the SPE eluates containing the phosphates and thiophosphates are reacted with tetrabutylammonium hydroxide in the injection port of the gas chromatograph to form volatile alkylated derivatives (Weisskopf and Seiber, 1989; McCurdy et al., 1994). Bird excreta are handled in a similar manner by extracting an acidified and salt-saturated aqueous solvent prior to SPE cleanup. Animal tissues (e.g., kidney and liver) are homogenized in acidified solvent and then subjected to a multistep preparation with a final gel permeation chromatographic cleanup prior to determination (Holstege et al., 1991; Richardson and Seiber, 1993). Plant tissues (e.g., fruits and vegetables) that contain OP parent residues are blended with solvents to extract the residues (Wan, 1990; Consalter and Guzzo, 1991; Hsu et al., 1991; Chohan and Shah, 1992; Leoni et al.,

1992; Mansour et al., 1992; Pylypiw, 1993; Skopec et al., 1993; Holstege et al., 1994; Cai et al., 1995), and other tissues such as pine needles are treated in turn with a water wash, surfactant wash, chloroform extraction, and grinding in organic solvent, and each preparation is analyzed separately to determine OP distribution in the plant (Aston and Seiber, 1996).

Table 2 contains gas chromatographic conditions, contaminated matrix, and analyte detection limits for some representative examples taken from the literature for various matrices. In all cases cited, the detector of choice for the primary determination of OP residues was the flame photometric detector in the phosphorus mode (FPD-P, 525 nm filter) coupled to bonded-phase fused silica open tubular columns (FSOTs). The FSOT column temperature programs varied somewhat depending on the particular liquid phase. Analyte confirmation was accomplished by comparing GC elution patterns with standards on FSOT columns with different liquid phases and/or by employing electron-impact mass spectrometry (EI-MS) with single-ion monitoring and looking for characteristic fragments. The limit of quantitation (LOQ) using the FPD-P varied depending on the matrix, with air having the lowest LOQ of $\sim 10^{-7}$ ppb and animal tissues having the highest of 20–50 ppb. While EI-MS is very specific, it is usually not as sensitive as the FPD-P (Table 2, water and animal tissues), and so EI-MS has been used in these citations to only confirm the presence of OP analytes by comparing ion fragment abundance ratios characteristic for each analyte. Regardless of the contaminated matrix, the OP analyte gas chromatograms are similar once the analytes have been isolated and interferences have been eliminated or at least minimized.

D. Application to Nonpesticidal OPs

Table 3 lists the representative nonpesticidal OPs along with their physical properties (Muir, 1984; Boethling and Cooper, 1985; Association of Official Analytical Chemists, 15th Edition, 1990). Residues of these compounds have been found in air, water, and soil (Aston and Seiber, 1996). Water solubility is a major controlling factor in whether the individual compounds are transported in surface waters or are concentrated in sediments or biota. Their generally low volatility and low water solubility dictate that sorption to soil and sediments will be a major environmental fate process. Field studies have observed rapid uptake of trialkyl/triaryl phosphates (TAPs) by bottom sediments (Muir et al., 1985).

Extraction of industrial OPs can be accomplished using similar strategies applied to OP pesticides. For extraction of soils and sediments, solvents utilized range in polarity from a 9:1 mixture of methanol and water (Anderson et al., 1993), used for triphenyl phosphate, to hexane (Saeger et al., 1979; Muir et al., 1983), used for a range of TAPs. Modern high-pressure extraction techniques, including supercritical fluid and high-pressure solvent (methanol), have also been used for these compounds (David and Seiber, 1996). Water samples can typically

Table 2 Examples of Gas Chromatographic Methods for Analysis of Organophosphorus Pesticides

Matrix	GC conditions (quantitation)		Analyte	LOQ[a]	GC conditions (confirmation)		LOQ[a]
	Column	Detector			Column	Detector	
Air	DB-210[b] (Zabik and Seiber, 1993; Aston and Seiber, 1996)	FPD-P	Diazinon Diazoxon Chlorpyrifos Parathion Paraoxon	0.7–1.4 pg/m^3	DB-5[c]	MSD[d]	NA[e]
Water	1. DB-210 (Zabik and Seiber, 1993) 2. DB-5[f] (LeNoir et al., 1997)	1. FPD-P 2. FPD-P	1. Same 2. Diazinon Chlorpyrifos	1. 1.3 ng/L 2. 0.6 ng/L	1. Same 2. DB-5[g]	1. MSD 2. MSD	1. NA 2. 1–2 ng/L
1. Urine 2. Excreta	1. DB-1[f] (Weisskopf and Seiber, 1989) 2. DB-1, DB-17[f] (Weisskopf and Seiber, 1989)	1. FPD-P 2. FPD-P	1. DMP, DEP, DMTP, DMDTP, DETP, DEDTP[h] 2. Same[h]	1. 20–30 ppb 2. 25 ppb	1. NA 2. DB-1701,[i] DB-1	1. NA 2. MSD	1. NA 2. NA

			DMP, DEP, DMTP, DMDTP, DETP, DEDTP[h]	20–50 ppb	DB-1701,[i] DB-1	MSD	50–80 ppb
Animal tissues	DB-1, DB-17 (Richardson and Seiber, 1993)	FPD-P	1. Diazinon, diazoxon, chlorpyrifos 2. Azinphosmethyl, chlorpyrifos, diazinon, dimethoate, ethion, isofenphos, malathion, (Me)parathion, phosmet	1. NA 2. 10–50 ppb	1. NA 2. SPB-1	1. NA 2. FPD-S	1. NA 2. 10–50 ppb
Plant tissues	1. DB-210[b] (Aston and Seiber, 1996) 2. SPB-1[j] (Pylypiw, 1993)	1. FPD-P 2. FPD-P					

[a]LOQ = limit of quantitation.
[b]30 m × 0.53 mm (id), 1.0 μm film thickness.
[c]40 m × 0.8 mm (id).
[d]MSD = mass-selective detector.
[e]NA = not available.
[f]30 m × 0.53 mm (id).
[g]15 m × 0.25 mm (id).
[h]DMP = dimethylphosphate; DEP = diethylphosphate; DMTP = dimethylthiophosphate; DMDTP = dimethyldithiophospate; DETP = diethylthiophosphate; DEDTP = diethyldithiophosphate.
[i]5 m × 0.53 mm (id).
[j]30 m × 0.53 mm (id), 0.5 μm film thickness. Alternatives: SPB-5, SPB-608, SPB-20.

Table 3   Physical Constants of Selected Tri-Aryl and Tri-Alkyl Phosphate Esters

OP	K_{ow} × 10^4	Water solubility (mg/L)[a,b]	Bpt. °C @ mm Hg	Vapor pressure (mmHg)/°C[a,b]	K^a	Henry's law constant[b] atm-m^3/mol
TPP	4.25/4.01	1.9	238@10	1.3/200	3100	1.8–3.6×10^{-7}
TCP	12.8/13.2	0.36	241–255@4	$1.4 \times 10^{-3}/30$	7200	1.1–2.8×10^{-6}
TXP	5.63[c]	0.9	248–265@4	$5.2 \times 10^{-8}/30$	4700	3.1×10^{-8}
						3.3×10^{-6}
NPDP	86.0	0.77	471@760	1.9×10^{-8} [d]	4900	1.4×10^{-8}
IPDP	20.2	2.2	220–230@1	1.1×10^{-6}	2800	2.4×10^{-7}
				2.8×10^{-7} [d]		6.2×10^{-8}
TBP	1.01	280	297@760	127/177		
T(2EH)P	1.68	>1000	216@5	0.23/150		

TPP = Triphenylphosphate. TCP = Tricresylphosphate (mixed isomers). TXP = Trixylylphosphate
NPDP = Nonylphenyldiphenylphosphate. IPDP = Isopropylphenylphosphate.
TBP = Tributylphosphate. T(2EH)P = Tri(2-ethylhexyl)phosphate
[a]K_{oc} estimated, (Muir, 1984).
[b]H_c calculated (Muir, 1984).
[c]TXP K_{ow} not 10^4 (Boethling and Cooper, 1985).
[d]VP estimated (Muir, 1984).

be extracted using either liquid-liquid extraction with ethyl ether or ethyl acetate, or using C-18 solid-phase extraction (unpublished methods of the authors).

Although water extracts are usually analyzed directly without cleanup, soil, sediment, and biological samples typically need to be cleaned up after extraction and before analysis. Cleanup procedures have been developed for TAPs and their breakdown products (Muir et al., 1981). Cleanup of TAP extracts can be accomplished with Florisil column chromatography under similar conditions to those used to purify OP pesticide residue extracts (unpublished methods of the authors). Extracts containing TAP breakdown products, such as diphenylphosphate, can be cleaned up on reverse phase (C18) silica cartridges (Muir et al., 1981).

The preferred method for analysis of nonpesticidal OP esters is capillary GC coupled with a phosphorous-specific detector, such as a FPD or NP-TSD. A flame-ionization detector (FID) may be employed for analyses that require simultaneous detection of OPs and non P-containing hydrocarbons, but sensitivity and selectivity are reduced. A benchtop mass spectrometer may also be used as a GC detector (Muir, 1984).

Typical phases for capillary GC of these compounds include DB-1 and DB-5 (David and Seiber, 1996). Figure 5 shows a GC/FPD chromatogram of several TAPs on a DB-1 column (30 m, 0.53 mm, 1 μm phase thickness). Oven tempera-

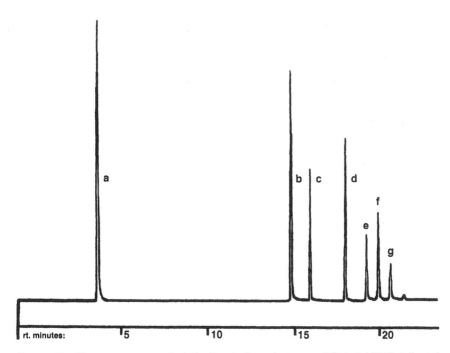

FIGURE 5 Chromatogram of trialkyl/aryl phosphates on DB-1: (a) TBP: tributyl phosphate; (b) TPP: triphenyl phosphate; (c) T(2EH)P: tri (2-ethylhexyl) phosphate; (d) ToCP: tri-ortho-cresylphosphate; (e) TmCP: tri-meta-cresylphosphate; (f) Bm, pCP: bis-meta, para-cresylphosphate; (g) Bp, mCP: bis-para, meta-cresylphosphate.

ture program began at 165°C for 1 minute, ramped at 5°/min to 245°C, where it remained for 7 minutes.

Volatility of the nonpesticidal OPs is typically lower than that of the pesticidal compounds, particularly for the aryl and mixed aryl/alkyl OPs. Oven temperatures are therefore usually higher than those used for the OP pesticides. Run times are also increased relative to the pesticides, exceeding 20 minutes in the case represented in Figure 5.

The primary factor that limits the resolution of components of a mixture of tri-aryl OPs is that, with the exception of TPP, they are all mixtures of closely related isomers. Alkyl or aryl substituents on the phenyl groups of the OP esters can be in the ortho, meta, or para positions, on any one of the three R groups. Tricresylphosphate, for example, is a mixture of tri-meta-cresylphosphate, bis-meta, para cresyl phosphate, bis-para, meta cresyl phosphate, and tri-meta cresyl phosphate. The ortho isomer is a potent delayed neurotoxicant (Heitcamp et al.,

1984), and was thus eliminated from commercial formulations by using only meta and para cresols in production feedstocks. Figure 5 shows GC separation of four isomers of TCP. Total TCP in a sample can be quantified as an area sum of the four primary peaks corresponding to the four isomers, or, with suitable separation, the isomers can be quantified independently.

III. HPLC METHODS

Although HPLC was practiced as early as the 1960s, advances in pumps, columns, and detectors made it a feasible method for routine use beginning in the mid-to-late 1970s and certainly in the 1980s. In about 1980, the sales of HPLC equipment for all uses surpassed that of GC equipment—a trend that has continued into the 1990s. Where GC can only be used for analytes capable of existing at some temperature in the gas phase, HPLC can be used for virtually any analyte regardless of volatility and stability. As a practical matter, most OP esters are amenable to GC, and that is the preferred method for their analysis because several phosphorus selective detectors exist for GC, while none exist for HPLC. For a few OPs, however, GC of the parent compound is not possible, because of lack of stability and/or because of high polarity. These compounds are illustrated by glyphosate, which is both nonvolatile and very polar (and ionizable). Also, OP partial esters, such as the dialkylphosphates and dialkylthiophosphates, are quite stable and fairly volatile as the free acids, but exist as salts in natural samples and are too polar (or ionized) to undergo GC as discrete peaks. For those types of compounds there are several choices:

> *GC after derivatization.* For glyphosate, a gas chromatographable derivative is the *N*-acetyl or *N*-trifluoroacetyl trimethyl ester (see, for example, Seiber et al., 1984). For the OP partial esters, alkylation produces the gas chromatographable triester (see, for example, Richardson and Seiber, 1993).
>
> *HPLC and derivatization.* For glyphosate and its primary metabolite, aminomethylphosphonic acid, Moye (1981) described a method based upon postcolumn derivatization to a fluorescent derivative by reaction with *o*-phthalaldehyde-2-mercaptoethanol.
>
> *HPLC without derivatization.* This can be used for virtually any neutral, acidic, or basic OP.

A. Detectors

UV-visible absorbance and fluorescence detectors probably constitute over 70% of HPLC detection systems in common use. They include fixed wavelength, variable wavelength, and scanning wavelength versions (Willard et al., 1989; Wheals, 1982; Yeung and Sinovec, 1986). The modern commercial detectors have much reduced baseline noise, higher signal-to-noise ratios, lower volume cells, and lower practical detection limits than their 1970s and 1980s predecessors. Needless to say, these advantages will be most reliable with UV-absorbing or fluorescing analytes. Aryl OP esters, such as parathion, have good UV absorption ranging to 300 nm, but alkyl OPs, such as the di- and tri-alkyl phosphates, absorb appreciably only at wavelengths below 250 nm—the same spectral region where many solvents, solvent impurities, and matrix-derived interferences absorb. Although analysis of such compounds by HPLC is still possible, it is generally only practiced in high concentration formulation analysis or with very clean environmental substrates such as water.

Fluorescence detectors have greater inherent sensitivity than UV-visible absorption detectors, but very few OPs are fluorescent. Azinphosmethyl (guthion) and azinphosethyl are fluorescent but also gas chromatographable, which represents the method of choice for residue analysis. As indicated above, for glyphosate and its metabolite one may prepare a fluorescent derivative (either pre- or post-column) to enhance detectability.

Dramatic improvements in HPLC–mass spectrometry instrumentation, particularly in the development of interfaces that do not overly compromise sensitivity or information output, have led to increasing use of HPLC/MS for detection of a variety of chemicals of environmental concern, including pesticides (Yergey et al., 1990; Brown, 1990; Voyksner and Cairns, 1989). Pesticide residue monitoring is being done increasingly using MS methods—GC-MS for the majority of chemicals but also LC-MS for those less amenable to GC.

B. Columns

Columns include the standard adsorption columns packed with silica or alumina, which commonly use an organic solvent as mobile phase; reversed phase partition columns using a C_2, C_8, C_{18}, cyclohexyl, or other organic ligand bonded to the stationary phase, usually with an aqueous mobile phase, and gel permeation or ion exchange specialty columns using organic solvents or aqueous buffers as mobile phases. The majority of HPLC of neutral, covalent chemicals is practiced on reversed phase columns, using water-methanol or water-acetonitrile as a solvent. HPLC retention data can be correlated with water solubilities and other physical properties (Kaliszan, 1987).

Normal-phase HPLC of a variety of pesticides and related environmental contaminants, including several OPs, has been practiced on a silica column with a hexane to methyl t-butylether gradient (Seiber et al., 1990). HPLC was used to clean up and fractionate extracts of air and water, prior to determination by GC. The combined HPLC fractionation–GC determination method provided very low (ppb) detection limits, clean GC chromatograms, and near-certain identifications.

New advances in HPLC include the use of microbore columns, supercritical fluids as mobile phases, and chiral columns for enantiomer separation. All are applicable to OPs, but few references specific to OPs have appeared.

C. Applications to OP Pesticide Analysis

Liquid chromatographic techniques serve as viable alternatives to gas chromatographic (GC) methods for the determination of OPs and OP-related compounds. HPLC is particularly useful for OP compounds that do not have the volatility and/ or thermal stability for GC methods. Typically, HPLC techniques applied to OP analysis utilize reverse-phase (i.e., C18) columns and protic mobile phases (e.g., methanol/water, acetonitrile/water, etc.). Detectors that have been commonly used for HPLC/OP applications include ultraviolet (Seiber et al., 1990; Kawasaki et al., 1992; Driss et al., 1993; Pinto et al., 1995), ultraviolet-diode array (Coulibaly and Smith, 1993; Ioerger and Smith, 1993; Lacorte and Barceló, 1995), and mass spectrometry using electrospray or thermospray interfaces (Barceló et al., 1991; Kawasaki et al., 1992; Ioerger and Smith, 1994; Smith and Coulibaly, 1994; Lacorte and Barceló, 1996). Some unusual applications include the use of electrochemical detectors (Pinto et al., 1995) and a particle beam interface (Cappiello et al., 1994), where the HPLC effluent is allowed to flow directly into the EI source of a mass spectrometer. HPLC techniques for OP analysis are usually automated, giving cost-effective ways of performing routine assays for large numbers of samples.

HPLC techniques have been applied to a number of OP-contaminated matrices, such as air, water, and human and animal tissues. While GC techniques typically require that highly polar and/or ionic analytes be derivatized to render them volatile, this is not a limitation for HPLC techniques, for which the mobile phase can be modified and the choice of column can be made to accommodate the physical and chemical properties of the analyte. Also, if the analyte lacks a suitable chromophore (e.g., alkylthiophosphates and alkylphosphates), a mass spectrometer is a reasonable choice as the HPLC detector.

In the following, we describe several studies that were concerned with the determination of OP pesticides and their environmental and metabolic conversion products. It is not within the scope of this discussion to present detailed information regarding sample preparation techniques. We will briefly mention the methods used to isolate the analytes, but then focus primarily on the HPLC conditions (e.g., columns, detectors, mobile phase) used to determine OP residue levels and comment on the detection limits achieved using different conditions.

A search of the literature has shown that the majority of HPLC/OP techniques have been applied to the determination of residues in surface, ground, and drinking water (Coulibaly and Smith, 1993; Driss et al., 1993; Cappiello et al., 1994; Lacorte and Barceló, 1995 and 1996; Pinto et al., 1995). This was followed by methods for determining OP residues in beef tissue (Coulibaly and Smith,

1993; Ioerger and Smith, 1993 and 1994; Smith and Coulibaly, 1994), human blood (Sharma et al., 1990; Kawasaki et al., 1992), lung (Sharma et al., 1990), and liver (Sharma et al., 1990), and in soils (Barceló et al., 1991). One method was used to determine OP residues in fogwater and the interstitial air (Seiber et al., 1990). Instead of using reverse-phase HPLC, employed in most of the methods cited, this method used silica gel normal-phase HPLC to fractionate fogwater and air samples into classes of pesticides, including OPs and OP conversion products, based on polarity. Determination and confirmation were accomplished using GC.

In preparation for determination, OP residues in water and blood serum are preconcentrated on solid phase extraction (SPE) precolumns (Kawasaki et al., 1992; Driss et al., 1993; Cappiello et al., 1994; Lacorte and Barceló, 1995 and 1996). The preconcentrated residues are then eluted and injected into an HPLC system either manually (Kawasaki et al., 1992; Cappiello et al., 1994) or by using automated switching systems (Driss et al., 1993; Lacorte and Barceló, 1995 and 1996). Instead of using precolumns, one novel approach adds a nonionic surfactant to aqueous OP samples and preconcentrates the OP residues in the surfactant phase that forms when the chilled dilute aqueous surfactant mixture is heated under controlled conditions to the cloud point (Pinto et al., 1995). OP residues in beef tissue are isolated by solvent extraction of the ground tissue in a blender, followed by filtration and concentration on SPE precolumns prior to determination by HPLC (Coulibaly and Smith, 1993; Ioerger and Smith, 1993 and 1994; Smith and Coulibaly, 1994). A method for OP residues on soil involves freeze-drying the soil prior to Soxhlet extraction with methanol, concentration, and Florisil cleanup (Barceló et al., 1991).

Table 4 contains liquid chromatographic conditions, contaminated matrix, and analyte quantitation limits for some representative examples taken from the literature for water, soil, animal tissue, and human blood. In some of the cases cited, ultraviolet and photodiode array detection were used as primary detectors to establish elution order, to evaluate chromatographic resolution, and for initial quantitation. Mass spectrometry was used for final determination and/or confirmation (Kawasaki et al., 1992; Coulibaly and Smith, 1993; Lacorte and Barceló, 1995 and 1996). In one case, electrochemical detection was used instead of mass spectrometry (Pinto et al., 1995), while in other cases single detectors were used, such as ultraviolet (Driss et al., 1993; Sharma et al., 1990), photodiode array (Ioerger and Smith, 1993), and mass spectrometry (Barceló et al., 1991; Ioerger and Smith, 1994; Smith and Coulibaly, 1994; Cappiello et al., 1994) for total determination of OP residues. One study expanded the capability of positive and negative ion thermospray mass spectrometry for identification/determination purposes by varying the composition of the reversed-phase eluting solvent to give characteristic base peaks (positive ion) and fragments (negative ion), depending on the chemical class (Barceló et al., 1991). Limit of quantitation (LOQ) for these different methods varied depending on the matrix, with water having the lowest

Table 4 Examples of HPLC Methods for Analysis of Organophosphorus Pesticides

Matrix	HPLC conditions		Analyte	LOQ[a]
	Column	Detector		
1. River water	1. Superspher C$_8$[b]	1. UV-diode array	1. Chlorpyrifos, diazinon, disulfoton, fenamiphos, fenthion, isofenphos, malathion, methidathion, pyridafenthion, temephos	1. 0.002–0.088 ppb[d]
2. Drinking water	2. Nucleosil C$_{18}$ ODS[c]	2. Variable UV	2. Methyl parathion, ethyl parathion, fenitrothion, diazinon, azinphos-et, azinphos-me, phosmet	2. 0.03–0.2 ppb
Soil	LiChrospher 100 RP-18[e]	Thermospray mass spectrometry	Fonofos, fensulfothion, trichlorfon, ethyl-parathion, phosmet, chlorpyrifos, fenitrothion	Parent: 20–50 ng (PI)[f]; 50–70 ng (NI)[f] Oxons: 1–2 ng (PI); 50–70 ng (NI)
Blood	μBondapak C$_{18}$[g]	Ultraviolet; APCI-MS[h], PI and NI modes	21 OPS	1–33 ppb (SIM)[i]; 133–667 ppb (TIC)[j]
Animal tissues	Bio-Sil C$_{18}$ HL90[j]	Thermospray mass spectrometry	17 OPs	1–5 ppm

[a]LOQ = limit of quantitation.
[b]Lacorte and Barceló, 1995.
[c]Driss et al., 1993.
[d]Confirmed using HPLC-thermospray mass spectrometry. All other cases cited in this table did not confirm.
[e]Barceló et al., 1991.
[f]PI = positive ion; NI = negative ion.
[g]Kawasaki et al., 1992.
[h]APCI-MS = atmospheric pressure chemical ionization mass spectrometry.
[i]SIM = single ion monitoring; TIC = total ion chromatogram.
[j]Loerger and Smith, 1994.

Table 5 HPLC Retention Times for TPP and
Breakdown Products

Compound	HPLC RT	Compound	HPLC RT
TPP	18.8	MPP	3.1
DPP	8.4	Phenol	7.1

TPP, DPP, and MPP are tri-, di-, and mono-phenylphosphate, respectively. UV/rad. detector. Column: Merck LiChrosorb RP 18, 5 μm, 250 × 4 mm, 40°C @ 1 ml/min. Solvent system, with gradient: solvent A: 95% pH 6.5 buffer (0.05M/l) with tetra-butylammoniumhydrogensulfate (0.005 M/l) 5% acetonitrile. Solvent B: acetonitrile. Gradient: initially 15% B in A; 40% B at 5 minutes; 70% B at 18 minutes; 15% B at 19 minutes; stop at 30 minutes.

LOQs in the range $2-88 \times 10^{-3}$ ppb and animal tissues having the highest in the range 1–5 ppm.

D. Applications to Nonpesticidal OPs

Liquid chromatography has been applied to analysis of these nonpesticidal OP esters when it is desirable to analyze an aqueous sample for both the parent OP and the OP-acid breakdown (typically hydrolysis) products. An appropriate HPLC detector is UV-Vis, especially for the aryl phosphates. A UV detector set at 260 nm can be used to detect the aryl OP parent, the P-containing diphenyl hydrolysis product, and the phenol or alkylphenol product (Anderson et al., 1993). Table 5 provides conditions for HPLC separation of TPP and its hydrolysis products.

REFERENCES

Anderson, C., D. Wischer, A. Schmeider, and M. Spiteller, Fate of triphenyl phosphate in soil, *Chemosphere 27*: 869–879 (1993).

Association of Official Analytical Chemists, *Official Methods of Analysis*, 15th ed., Arlington, VA, 1990, pp. 274–311.

Aston, L. S., and J. N. Seiber, Exchange of airborne organophosphorus pesticides with pine needles, *J. Environ. Sci. Health, B31*(4): 671–698 (1996).

Aston, L. S., J. Noda, J. N. Seiber, and C. A. Reece, Organophosphate flame retardants in needles of *Pinus ponderosa* in the Sierra Nevada foothills, *Bull. Environ. Contamin. Toxicol. 57*: 859–866 (1996).

Barceló, D., G. Durand, R. J. Vreeken, G. J. De Jong, H. Lingeman, and U. A. Brinkman, Evaluation of eluents in thermospray liquid chromatography-mass spectrometry for identification and determination of pesticides in environmental samples, *J. Chromatogr. 553*: 311–328 (1991).

Belanger, A., N. J. Bostanian, G. Boivin, and F. Boudreau, Azinphos-methyl residues in apples and spatial distribution of fluorescein in vase-shaped apple trees, *J. Environ. Sci. Health (B) 26*: 279–291 (1991).

Blanco, C. C., and F. G. Sanchez, A kinetic spectrophotometric method to determine the insecticide methyl parathion in commercial formulations and the aqueous environment, *Int. J. Anal. Chem. 38*: 513–523 (1990).

Boethling, R. S., and J. C. Cooper, Environmental fate and effects of triaryl and trialkyl/aryl phosphate esters, *Residue Reviews 94*: 49 (1985).

Bourgeois, D., J. Gaudet, P. Deveau, and V. N. Mallet, Microextraction of organophosphorus pesticides from environmental water and analysis by gas chromatography, *Bull. Environ. Contam. Toxicol. 50*: 433–440 (1993).

Bowman, M. C., Analysis of organophosphorus pesticides, in, *Analysis of Pesticide Residues* (H. A. Moye, ed.), John Wiley, New York, 1981, pp. 263–332.

Brown, P. R., High performance liquid chromatography. Past developments, present status, and future trends, *Anal. Chem. 62*: 995A–1008A (1990).

Cai, C. P., M. Liang, and R. R. Wen, Rapid multiresidue screening method for organophosphate pesticides in vegetables, *Chromatographia 40*: 417 (1995).

Cappiello, A., G. Famiglini, and F. Bruner. Determination of acidic and basic/neutral pesticides in water with a new microliter flow rate LC/MS particle beam interface, *Anal. Chem. 66*: 1416 (1994).

Chohan, Z. H., and A. I. Shah, Simple spectrophotometric screening method for the detection of organophosphate pesticides on plant material, *Analyst 117*: 1379 (1992).

Consalter, A., and V. Guzzo, Multiresidue analytical method for organophosphate pesticides in vegetables, *Fresenius' J. Anal. Chem. 339*: 390 (1991).

Coulibaly, K., and J. S. Smith, Thermostability of organophosphate pesticides and some of their major metabolites in water and beef muscle, *J. Agric. Food Chem. 41*: 1719–1723 (1993).

David, M. D., and J. N. Seiber, Comparison of extraction techniques, including supercritical fluid, high-pressure solvent, and Soxhlet, for organophosphorus hydraulic fluids from soil, *Anal. Chem. 68*: 3044 (1996).

Drevenkar, V., Z. Vasilic, B. Stengl, Z. Frobe, and V. Rumenjak, Chlorpyrifos metabolites in serum and urine of poisoned persons, *Chem. Biol. Interact. 87*: 315–322 (1993).

Drevenkar, V., Z. Radic, Z. Vasilic, and E. Reiner, Dialkylphosphorus metabolites in the urine and activities of esterases in the serum as biochemical indices for human absorption of organophosphorus pesticides, *Arch. Environ. Contam. Toxicol. 20*: 417–422 (1991).

Driss, M. R., M. C. Hennion, and M. L. Bouguerra, Determination of carbaryl and some organophosphorus pesticides in drinking water using on-line liquid chromatographic preconcentration techniques, *J. Chromatogr. 639*: 352–358 (1993).

Federal Register, Aryl Phosphates; Response to the Interagency Testing Committee, Vol. 48, No. 251, pp. 57452–60.

Fest, C., and K. J. Schmidt, *The Chemistry of Organophosphorus Pesticides*, New York: Springer-Verlag, 1973.

Gillespie, A. M., S. L. Daly, D. M. Gilvydis, F. Schneider, and S. M. Walters, Multicolumn solid-phase extraction cleanup of organophosphorus and organochlorine pesticide residues in vegetable oils and butterfat, *J. AOAC Int. (BKS) 78*: 431–437 (1995).

Giuffrida, L., A flame ionization detector highly selective and sensitive to phosphorus. A sodium thermoionic detector, *J. Assoc. Offic. Anal. Chem. 47*: 293–300 (1964).

Gomez-Gomez, C., M. I. Arufe-Martinez, J. L. Romero-Palanco, J. J. Gamero-Lucas, and M. A. Vizcaya-Rojas, Monitoring of organophosphorus insecticides in the Guadalete River (southern Spain), *Bull. Environ. Contam. Toxicol. (BFN), 55*: 431–438 (1995).

Green, J., A review of phosphorous-containing flame retardants, *J. Fire Science 10*: 470–487 (1993).

Grob, K., M. Biedermann, and A. M. Giuffre, Determination of organophosphorus insecticides in edible oils and fats by splitless injection of the oil into a gas chromatograph (injector-internal headspace analysis), *Z. Lebensm. unters Forsch. 198*: 325–328 (1994).

Grob, R. L., ed., *Modern Practice of Gas Chromatography*, 2nd ed., John Wiley, New York, 1985.

Heitcamp, M. A., J. N. Huckins, J. D. Petty, and J. L. Johnson, Fate and metabolism of isopropylphenyl diphenly phosphate in freshwater sediments, *Environ. Sci. Technol 18*: 434–439 (1984).

Hinckley, D. A., Analysis of pesticides in seawater after enrichment onto C_8 bonded-phase cartridges, *Environ. Sci. Technol. 23*: 995 (1989).

Holstege, D. M., D. L. Scharberg, E. R. Tor, L. C. Hart, and F. D. Galey, A rapid multiresidue screen for organophosphorus, organochlorine, and *N*-methyl carbamate insecticides in plant and animal tissues, *J. AOAC Int. (BKS) 77*: 1263–1274 (1994).

Holstege, D. M., D. L. Scharberg, E. R. Richardson, and G. Moller, Multiresidue screen for organophosphorus insecticides using gel permeation chromatography—silica gel cleanup, *J. Assoc. Off. Anal. Chem. 74*: 394–399 (1991).

Hsu, J. P., H. J. Schattenberg, and M. M. Garza, Fast turnaround multiresidue screen for pesticides in produce, *J. Assoc. Off. Anal. Chem. 74*: 886–892 (1991).

Hutzinger, O., G. Sundstrom, and S. Safe, Environmental chemistry of flame retardants. Part I. Introduction and Principles, *Chemosphere 1*: 3–10 (1976).

Imato, T., and N. Ishibashi, Potentiometric butyrylcholine sensor for organophosphate pesticides, *Biosensors & Bioelectronics, 10*: 435 (1995).

Ioerger, B. P., and J. S. Smith, Multiresidue method for the extraction and detection of organophosphate pesticides and their primary and secondary metabolites from beef tissue using HPLC, *J. Agric. Food Chem. 41*: 303–307 (1993).

Ioerger, B. P., and J. S. Smith, Determination and confirmation of organophosphate pesticides and their metabolites in beef tissue using thermospray/LC-MS, *J. Agric. Food Chem. 42*: 2619–2624 (1994).

Kaliszan, R., Quantitative structure-retention relationships in chromatography, *Chromatography*, June, pp. 19–29 (1987).

Kawasaki, S., H. Ueda, H. Itoh, and J. Tadano, Screening of organophosphorus pesticides using liquid chromatography-atmospheric pressure chemical ionization mass spectrometry, *J. Chromatogr. 595*: 193–202 (1992).

Kennedy, E. R., M. T. Abell, J. Reynolds, and D. Wickman, A sampling and analytical method for the simultaneous determination of multiple organophosphorus pesticides in air, *Am. Ind. Hyg. Assoc. J. 55*: 1172–1177 (1994).

Khavkin, M. J., and J. A. Khavkin, A ferromagnetic biosensor for simple assay of organophosphate pesticides, *Analytical Letters 29*: 1041 (1996).

Kolb, B., and J. Bischoff, A new design of a thermionic nitrogen and phosphorus detector for GC, *J. Chromatogr. Sci. 12*: 625–629 (1974).

Kotcon, J. B., and M. Wimmer, Monitoring movement of fenamiphos through soil water in peach orchards using quantitation by gas chromatography/mass spectrometry. *Bull. Environ. Contamin. Toxicol. 50*: 35–42 (1993).

Kumaran, S., and C. Tran-Minh, Determination of organophosphorus and carbamate insecticides by flow injection analysis, *Anal. Biochem. 200*: 187–194 (1992).

Lacorte, S., and D. Barceló, Determination of organophosphorus pesticides and their transformation products in river waters by automated on-line solid-phase extraction followed by thermospray liquid chromatography-mass spectrometry, *J. Chromatogr. 712*: 103–112 (1995).

Lacorte, S., and D. Barceló, Determination of parts per trillion levels of organophosphorus pesticides in groundwater by automated on-line liquid-solid extraction followed by liquid chromatography/atmospheric pressure chemical ionization mass spectrometry using positive and negative ion modes of operation, *Anal. Chem. 68*: 2464–2470 (1996).

LeBel, G. L., Williams, D. T., and Benoit, F. M., *J Assoc. Off. Anal. Chem. 64*: 991–998 (1981).

Lee, X.-P., T. Kumazawa, and O. Suzuki, Detection of organophosphate pesticides in human body fluids by headspace solid-phase microextraction (SPME) and capillary gas chromatography with nitrogen-phosphorus, *Chromatographia 42*: 135 (1996).

LeNoir, J., J. N. Seiber, and L. McConnell, Atmospheric deposition of pesticides to surface waters in southern Sierra Nevada mountains—Summer, 1996. Presented before the 213th ACS National Meeting, San Francisco, CA, April 13–17.

Leoni, V., A. M. Caricchia, and S. Chiavarini, Multiresidue method for quantitation of organophosphorus pesticides in vegetable and animal foods. *J. AOAC International 75*(3): 511 (1992).

Lino, C. M., and M. I. da Silveira, Organophosphorus pesticide residues in cow's milk: levels of *cis*-mevinfos, methyl-parathion, and paraoxon, *Bull. Environ, Contam. Toxicol. 49*: 211–216 (1992).

Mansour, M., D. Barceló, and J. Albaiges, Analytical methodology for screening organophosphorus pesticides in biota samples, *Sci. Total Environ. 123–124*: 45–56 (1992).

McCurdy, S. A., M. E. Hansen, C. P. Weisskopf, R. L. Lopez, F. Schneider, J. Spencer, J. R. Sanborn, R. I. Krieger, B. W. Wilson, and D. F. Goldsmith, Assessment of azinphosmethyl exposure in California peach harvest workers, *Arch. Environ. Health 49*: 289–296 (1994).

Miliadis, G. E., Organochlorine and organophosphorus pesticide residues in the water of the Pinios River, Greece, *Bull. Environ. Contam. Toxicol. (BFN) 54*: 837–840 (1995).

Miliadis, G. E., Determination of pesticide residues in natural waters of Greece by solid phase extraction and gas chromatography, *Bull. Environ. Contamin. Toxicol. 52*: 25–30 (1994).

Moye, H. A., High performance liquid chromatographic analysis of pesticide residues. In: *Analysis of Pesticide Residues* (H. A. Moye, ed.), John Wiley, New York, 1981, pp. 157–197.

Muir, D. G. C., N. P. Grift, and P. Solomon, Extraction and cleanup procedures for determination of diaryl phosphates in fish, sediment, and water samples, *J. Assoc. Off. Anal. Chem. 64*: 79–84 (1981).

Muir, D. G. C., A. L. Yarchewski, and N. P. Grift, Environmental dynamics of phosphate esters. III. Comparison of the bioconcentration of four triaryl phosphates by fish, *Chemosphere 12*: 155–166 (1983).

Muir, D. C. G., Phosphate esters, *Handbook of Environmental Chemistry 3C*: 41–46 (1984).

Muir, D. C., D. Lint, and N. P. Grift, Fate of three phosphate ester flame retardants in small ponds, *Environ. Toxicol. & Chem. 4*: 663–675 (1985).

Pesticide Analytical Manual, Department of Health and Human Services, Food and Drug Administration, Washington, D.C., 1979, 1993.

Pico, Y., A. J. Louter, J. J. Vreuls, and U. A. Brinkman, On-line trace-level enrichment gas chromatography of triazine herbicides, organophosphorus pesticides, and organosulfur compounds from drinking and surface waters, *Analyst 119*: 2025–2031 (1994).

Pinto, C. G., J. L. Pavon, and B. M. Cordero, Cloud point preconcentration and high-performance liquid chromatographic determination of organophosphorus pesticides with dual electrochemical detection, *Anal. Chem. 67*: 2606–2612 (1995).

Psathaki, M., E. Manoussaridou, and E. G. Stephanou, Determination of organophosphorus and triazine pesticides in ground- and drinking water by solid-phase extraction and gas chromatography with nitrogen-phosphorus or mass spectrometric detection, *J. Chromatogr. 667*: 241–248 (1994).

Pylypiw, H. M., Rapid gas chromatographic method for the multiresidue screening of fruits and vegetables for organochlorine and organophosphate pesticides, *J. AOAC International 76*: 1369 (1993).

Richardson, E. R., and J. N. Seiber, Gas chromatographic determination of organophosphorus insecticides and their dialkyl phosphate metabolites in liver and kidney samples, *J. Agric. Food Chem. 41*: 416–422 (1993).

Rogers, K. R., C. J. Cao, J. J. Valdes, A. T. Eldefrawi, and M. E. Eldefrawi, Acetylcholinesterase fiber-optic biosensor for detection of anticholinesterases, *Fundam. Appl. Toxicol. 16*: 810–820 (1991).

Saeger, V. W., O. Hicks, R. G. Kaley, P. R. Michael, J. P. Mieure, and E. S. Tucker, Environmental fate of selected phosphate esters, *Environ. Sci. Technol. 13*: 840–844 (1979).

Schenck, F. J., R.Wagner, M. K. Hennessy, and J. L. Okrasinski, Jr., Screening procedure for organochlorine and organophosphorus pesticide residues in eggs using a solid-phase extraction cleanup and gas chromatographic detection, *J. AOAC Int. (BKS) 77*: 1036–1040 (1994).

Seiber, J. N., D. E. Glotfelty, A. D. Lucas, M. M. McChesney, J. C. Sagebiel, and T. A. Wehner, A multiresidue method by high performance liquid chromatography–based fractionation and gas chromatographic determination of trace levels of pesticides in air and water, *Arch. Environ. Contam. Toxicol. 19*: 583–592 (1990).

Seiber, J. N., M. M. McChesney, R. Kan, and R. A. Leavitt, Analysis of glyphosate residues in kiwi fruit and asparagus using high-performance liquid chromatography of derivatized glyphosate as a cleanup step, *J. Agric. Food Chem. 32*: 678–681 (1984).

Sharma, V. K., R. K. Jadhav, G. J. Rao, A. K. Saraf, and H. Chandra, High performance liquid chromatographic method for the analysis of organophosphorus and carbamate pesticides, *Forensic Sci. Int. 48*: 21–25 (1990).

Skerritt, J. H., A. S. Hill, and H. L. Beasley, Enzyme-linked immunosorbent assay for quantitation of organophosphate pesticides: fenitrothion, chlorpyrifos-methyl, and pirimphosmethyl in wheat grain and flour-milling fractions, *J. AOAC International* 75: 519 (1992).

Skladal, P., Determination of organophosphate and carbamate pesticides using a cobalt phthalocyanine-modified carbon paste electrode and a cholinesterase enzyme membrane, *Analytica Chimica Acta,* 252: 11 (1991).

Skladal, P., Detection of organophosphate and carbamate pesticides using disposable biosensors based on chemically modified electrodes and immobilized cholinesterase, *Analytica Chimica Acta, 269*: 281 (1992).

Skopec, Z. V., R. Clark, P. M. Harvey, and R. J. Wells, Analysis of organophosphorus pesticides in rice by supercritical fluid extraction and quantitation using an atomic emission detector, *J. Chromatogr. Sci. 31*: 445–449 (1993).

Smith, J. S., and K. Coulibaly, Effect of pH and cooking temperature on the stability of organophosphate pesticides in beef muscle, *J. Agric. Food Chem. 42*: 2035–2039 (1994).

Thomas, R. A., and L. E. Macaskie, Biodegradation of tributyl phosphate by naturally occurring microbial isolates and coupling to the removal of uranium from aqueous solutions, *Environ. Sci. Technol. 30*: 2371–2375 (1996).

Thompson, H. M., Development of an immunoassay for avian serum butyrylcholinesterase and its use in assessing exposure to organophosphorus and carbamate pesticides, *Bull. Environ. Contam. Toxicol. (BFN) 54*: 237–244 (1995).

Vasilic, Z., V. Drevenkar, V. Rumenjak, B. Stengl, and Z. Frobe, Urinary excretion of diethylphosphorus metabolites in persons poisoned by quinalphos or chlorpyrifos, *Arch. Environ. Contam. Toxicol. 22*: 351–357 (1992).

Vasilic, Z., V. Drevenkar, B. Stengl, Z. Frobe, and V. Rumenjak, Diethylphosphorus metabolites in serum and urine of persons poisoned by phosalone, *Chem. Biol. Interact. 87*: 305–313 (1993).

Voyksner, R. D., and T. Cairns, Application of liquid chromatography–mass spectrometry to the determination of pesticides, In: *Analytical Methods for Pesticides and Plant Growth Regulators* (J. Sherma, ed.), Academic Press, New York, Volume XVII, 1989, pp. 119–166.

Wan, H. B., Small-scale method for the determination of organophosphorus insecticides in tea using sulfuric acid as clean-up reagent, *J. Chromatogr. 516*: 446–449 (1990).

Wehner, T. A., and J. N. Seiber, Analysis of *N*-methylcarbamate insecticides and related compounds by capillary gas chromatography, *J. High Resol. Chrom. and Chrom. Comm. 4*: 348–350 (1981).

Weisskopf, C., and J. N. Seiber, New approaches to analysis of organophosphate metabolites in the urine of field workers. In: *Biological Monitoring for Pesticide Exposure* (R. G. M. Wang, C. A. Franklin, R. C. Honeycutt, and J. C. Reinert, eds.), American Chemical Society, Washington, D.C., 1989.

Wheals, B. B., Detectors for HPLC. In: *Techniques in Liquid Chromatography* (G. F. Simpson, ed.), John Wiley, Chichester, 1982, pp. 121–140.

Willard, H. H., L. L. Merritt, Jr., J. A. Dean, and F. A. Settle, Jr., *Instrumental Methods of Analysis*, 7th ed., Wadsworth, Belmont, CA, 1989.

Wilson, B. W., M. J. Hooper, E. E. Littrell, P. J. Detrich, M. E. Hansen, C. P. Weisskopf, and J. N. Seiber, Orchard dormant sprays and exposure of red-tailed hawks to organophosphates, *Bull. Environ. Contam. Toxicol. 47*: :717–724 (1991).

Yergey, A. L., C. G. Edmonds, I. A. S. Lewis, and M. L. Vestal, *Liquid Chromatography/ Mass Spectrometry Techniques and Applications*, Plenum, New York, 1990.

Yeung, E. S., and R. E. Synovic, Detectors for liquid chromatography, *Anal. Chem. 58*: 1237A–1256A (1986).

Zabik, J. M., and J. N. Seiber, Atmospheric transport of organophosphate pesticides from California's Central Valley to the Sierra Nevada Mountains, *J. Environ. Qual. 22*: 80–90 (1993).

Zweig, G., and J. Sherma, *Analytical Methods for Pesticides and Plant Growth Regulators, Vol. VI, Gas Chromatographic Analysis*, Academic Press, New York, 1972.

9

Chromatographic Analysis of Insecticidal Carbamates

Jo A. Engebretson, Charles R. Mourer,
and Takayuki Shibamoto*
University of California, Davis, California

I. INTRODUCTION

Carbamate compounds are part of a diverse group of synthetic pesticides that have been developed, produced and used on a large scale during the past 50 years. The general formula for carbamates is shown in Figure 1. Carbamates used as pesticides include carbamate fungicides and herbicides, fungicidal dithiocarbamates, nematocides, and carbamate insecticides.

Carbamate insecticides are mostly represented by monomethyl carbamates, which are esters of carbamic acid where the N atom is substituted with a methyl group. An excellent history of the discovery of these compounds is provided by Kuhr and Dorough (1976). These compounds are most soluble in organic solvents. Other carbamates are more aliphatic in nature and may possess sufficient miscibility with water to act as effective plant systemic insecticides (e.g., aldicarb)

---

*IR-4 Western Region Leader Laboratory.

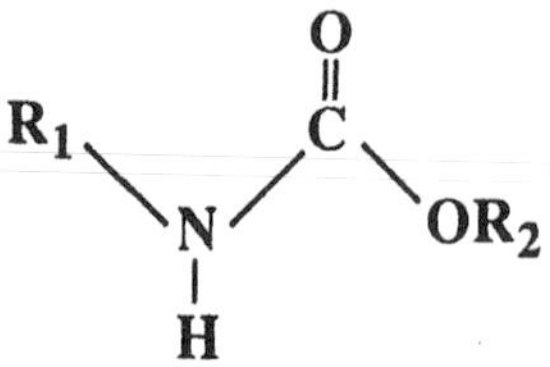

FIGURE 1   Common structure of carbamates where R$_1$ and R$_2$ are alkyl or aryl groups.

(World Health Organization, 1986). The synthesis and commercialization of the carbamate pesticides has been in progress since the 1950s.

Insecticidal carbamates have high acute toxicity. Their toxicity to insects, nematodes, and mammals is based on inhibition of acetylcholinesterase, which is the enzyme responsible for the hydrolysis of acetylcholine into choline and acetic acid. The cholinesterases become carbamylated. The resulting accumulation of acetylcholine in the parasympathetic nerve synapses (muscarine-like action), the motor end-plate (nicotine-like action), and in the central nervous system provokes symptoms similar to those caused by exposure to organophosphorous pesticides (Machemer and Pickel, 1994). See Figure 2 for a diagram of the mode of action of carbamate insecticides (Coats, 1982).

Figure 3 lists the main representative carbamate insecticides and their structures: aldicarb, aldoxycarb, bendiocarb, carbaryl, carbofuran, methomyl, oxamyl, pirimicarb, promecarb, propoxur, thiodicarb and thiofanox. Sensitive residue analytical methods for these compounds include the use of thin-layer chromatography (TLC), gas chromatography (GC), high performance liquid chro-

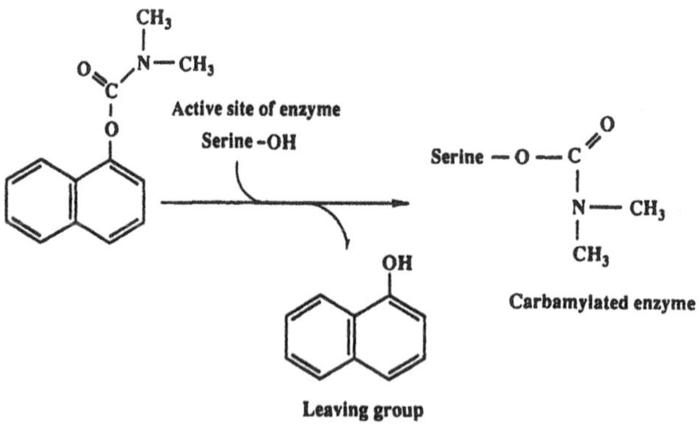

FIGURE 2   Mode of action of carbamates. (From Coats, 1982.)

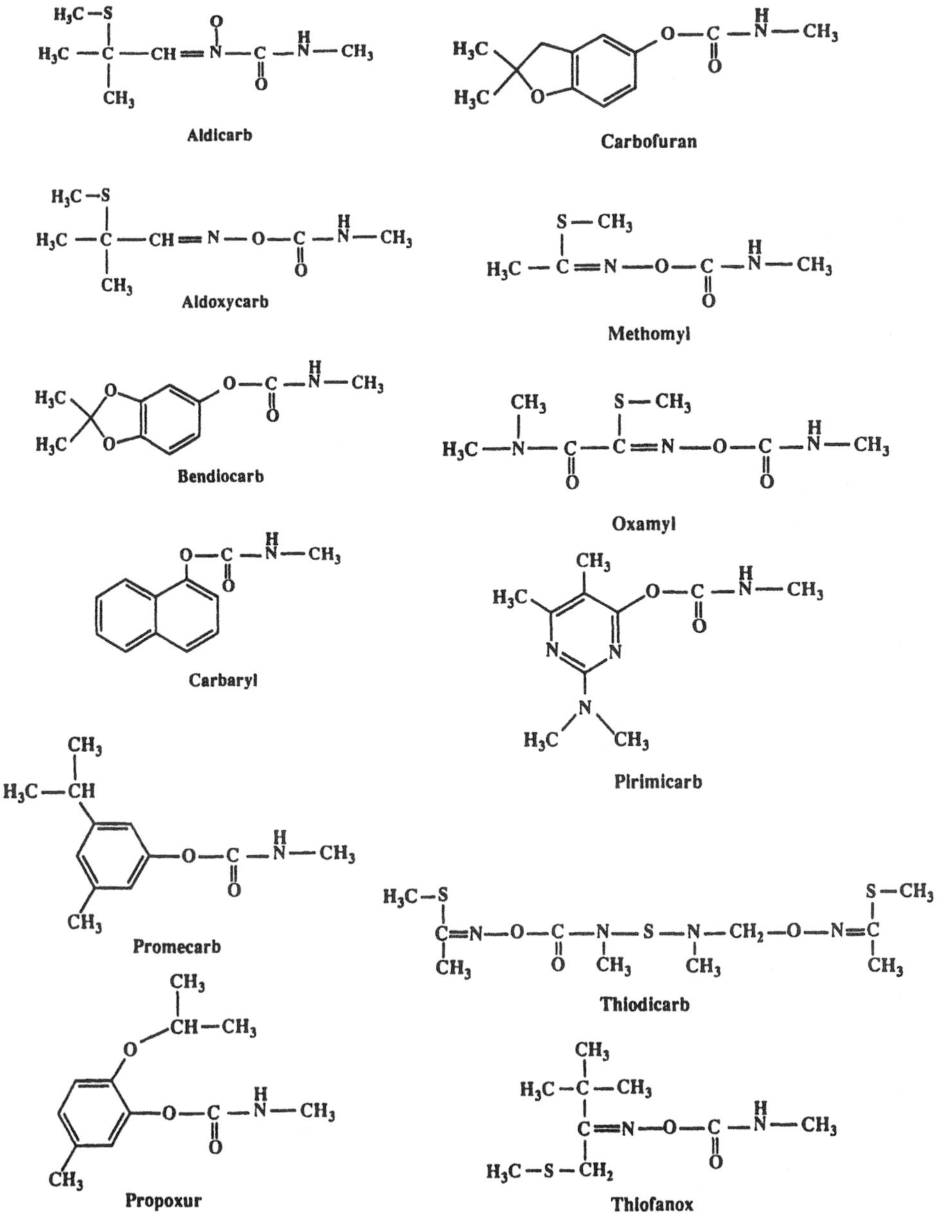

FIGURE 3 Common carbamate insecticides.

matography (HPLC), gas chromatography-mass spectrometry (GC/MS), and liquid chromatography-mass spectrometry (LC/MS). As mentioned in the chapter "Organophosphorus Esters," the FDA Pesticide Analytical Manual (PAM) (I, 1979, 1993) provides detail on analytical methods for carbamates, primarily in foods. Multiresidue methods often include carbamate analyses (Luke et al., 1981; PAM I, 1979, 1993). A general review of analytical methods is available in *Carbamate Pesticides: A General Introduction* (World Health Organization, 1986). The scope of this chapter is to provide a review of the most current chromatographic methods available for the analysis of carbamate insecticides.

II. THIN-LAYER CHROMATOGRAPHY

Thin-layer chromatography is a useful qualitative and quantitative tool for analysis of carbamate insecticides, which may not be as amenable to GC analysis due to lack of thermostability. Advantages of TLC include unrestricted access to the separation process; introducing magnetic, thermal, electrical, and other physical forces to improve resolution; high sample throughput; truly multidimensional separations; and the use of controlled multiple gradients (Sherma, 1986).

TLC utilizes various plate substances, solvent developments, and visualization techniques for identifying and quantifying carbamate compounds (Sherma, 1987). For example, Zoun and Spierenburg (1989) used various solvents with silica gel high performance TLC (HPTLC) plates to quantify carbamates in gizzard, gastrointestinal content, bait, food, and environmental samples. Plates were developed with bromine vapor, bovine liver suspension, and substrate solution to visualize cholinesterase-inhibiting pesticides. Table 1 shows hRf values for representative carbamate insecticides detected. McGinnis and Sherma (1994) used HPTLC plates with *p*-nitrobenzenediazonium fluoroborate visualization quantified by densitometric scanning. A more typical method is presented by Raju and Gupta (1991) using silica gel G plates and hexane-acetone solvent (4 + 1) and development with diazotized *p*-nitroaniline or *p*-amino acetophenone followed by sodium hydroxide at levels of 10 ppm carbaryl. This method has the distinct advantage of the matrix not being hydrolyzed prior to analysis. Patil and Shingare (1993) visualized carbaryl with phenylhydrazine hydrochloride in an alkaline medium at 0.1 μg. An extensive survey of TLC plates and solvents was conducted by Rathore and Begum (1993) on several carbamate pesticides with detection by potassium hydroxide solution and *p*-nitrobenzenediazonium tetrafluoroborate solution at 25–100 μg levels. This method provides a means to qualitatively detect carbamates in mixtures.

More recent advances in TLC include sequential thin-layer chromatography (S-TLC), automated multiple development (AMD), and analogies between TLC and HPLC for method development. Rathore and Sharma (1992) conducted S-TLC on carbaryl and related compounds. When a mixture contains compounds

Table 1 *h*R*f* Values of Cholinesterase-Inhibiting Carbamate Insecticides after HPTLC Analysis

	Solvents			
Compound	Xylene	Dibutyl ether	Butyl acetate	Methyl isobutyl ketone
Aldicarb	0	2	47	74
Bendiocarb	0	13	79	92
Carbaryl	2	18	80	92
Carbofuran	0	5	70	87
Methomyl	0	0	22	57
Oxamyl	0	0	0	4
Pirimicarb	0	1	36	69
Propoxur	0	10	77	93
Thiofanox	0	3	59	82

Values are means from at least three experiments.
Source: From Zoun and Spierenburg (1989).

that differ considerably in polarity, a single development with one mobile phase may not provide the desired separation. If the plate is developed successively with different mobile phases that have different polarity or strength, separation is possible. Common solvents such as benzene, carbon tetrachloride, chloroform, distilled water, 1-4-dioxane, and ethyl acetate are used for sequential development of the silica gel G plates. Detection was with potassium hydroxide and *p*-nitro-benzenediazonium tetrafluoroborate. Some important separations achieved were carbaryl from pheno, *o*-nitrophenol, α-naphthol, and carbofuran, in carbon tetra-chloride followed by distilled water; carbofuran from phenol, *o*-nitrophenol, α-naphthol, carbaryl, and β-naphthoxyacetic acid in distilled water followed by chloroform.

In the AMD technique introduced by Burger (1984), the plate is repeatedly developed in the same direction with either the same or a different solvent, with drying of the plate between each run. Butz and Stan (1995) screened 265 pesti-cides in water, using liquid-liquid extraction and solid-phase extraction (SPE), and with this technique 50 ng levels are achieved. Koeber and Niessner (1996) went a step further by using supercritical fluid extraction (SFE) coupled with HPTLC/ AMD for analysis of pesticides in soils. Detection limits were at the subnanogram level, proving the increased separation efficiency and spot reconcentration of HPTLC in combination with AMD. The simultaneous determination of various pesticides in real soil samples was made possible by use of SFE since soil coextractants could be eliminated. Correlations between TLC and HPLC have been attempted by several authors including Reuke and Hauck (1995). They propose TLC as a pilot technique for HPLC because it has the advantage of being simple; it uses various phase systems identical with or comparable to HPLC; the retention mechanisms that exist between the two techniques are similar; and it can be used to determine irreversible adsorption. They showed excellent transfer results from TLC to HPLC for other compounds, and although pesticides had a linear relationship, aldicarb and carbofuran transferability was not possible.

III. GAS CHROMATOGRAPHY

Because most methyl carbamate insecticides are thermally labile, gas chromatography (GC) of these compounds is challenging. Older methods usually derivatized the N-methylcarbamate hydrolysis product (Lawrence, 1979) with reagents such as heptafluorobutyric anhydride (Lawrence et al., 1977; Nagasawa et al., 1977). methanesulphonyl chloride (Maitlen and McDonough, 1980), and pentafluorobenzyl bromide (Tjan and Jansen, 1979; Cline et al., 1990). GC analysis was also facilitated by forming 2,4-dinitrophenyl ether derivatives (Cohen et al., 1970; Holden, 1976) or 2,4-dinitroaniline derivatives (Holden et al., 1969). Recent methods utilizing derivatizing reagents include trifluoroacetic anhydride for carbaryl in honeybees (Kendrick et al., 1991) for analysis on a nitrogen-phosphorus detector (NPD). A new analytical method has been published that combines on-line derivatization-extraction and gas-liquid chromatography for the determination of N-methylcarbamates. The phenols, hydrolysis products of the carbamates, are derivatized and extracted in a continuous fashion using penta-fluoropropionic anhydride. The thermal instability of these compounds is overcome by forming fluoroderivatives that are analyzed at the nanogram level using an electron capture detector (ECD) (Ballesteros et al., 1993a,b).

With advances in column configuration and detector sensitivity, more direct methods are available. Brauckhoff and Thier (1992) present a typical GC analysis of methyl carbamate insecticides where the residues are extracted with ethyl acetate, and an aliquot of the extract is evaporated to dryness. The residue is extracted with water and the solution passed through a Reverse Phase-18 disposable cartridge using an acetonitrile–water mixture as eluant. The eluate is shaken with dichloromethane, and the dichloromethane phase is evaporated. The methyl carbamates are determined by capillary GC with splitless injection using a thermionic nitrogen-specific detector. The key to the success of this method is the low injection temperature of 50°C and the capillary column, with which recoveries in real samples ranged from 0.01 to 0.1 ppm, or generally 70 to 115%. They suggest a SE-52, 0.1 μm column for adequate separations. See Figure 4 for typical methyl carbamate insecticide gas chromatogram showing ample separation between compounds. Consistent recoveries have also been accomplished with a cold injection system such as programmable cool on-column injection (Lartiges and Garrigues, 1995).

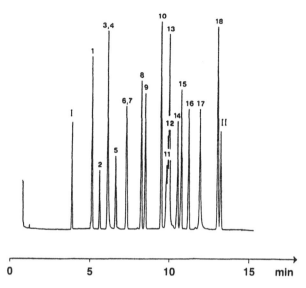

FIGURE 4 Standard mixture representing 2 ng of each methyl carbamate (4 ng each of 8,12,14,15,16,17 and II). 1, aldicarb; 2, butocarboxim; 3, methomyl; 4, propoxur; 5, thiofanox; 6, promecarb; 7, bendiocarb; 8, carbofuran; 9, aminocarb; 10, pirimicarb; 11, ethiofencar; 12, 3-keto-carbofuran; 13, desmethylpirimicarb; 14, dioxacarb; 15, carbaryl; 16, mercaptodimethur (methiocarb); 17, 3-hydroxy-carbofuran; 18, desmethylformamido-pirimicarb. I, lauronitrile; II, octadecanonitrile (internal standards).

Some single analyte methods utilize NPD detection when comparing various columns for carbofuran residues (Ling et al., 1990; Siebers et al., 1992) and packed columns for aldicarb analysis (Picó et al., 1990; Albelda et al., 1994). An interesting method is proposed by Bagheri and Creaser (1991) for determination of carbaryl in apples using GC combined with fluorescence spectrometry after derivatizing with trimethyl anilinium hydroxide.

Gas chromatography methods for multiresidues of pesticides often include carbamate insecticides such as carbaryl, carbofuran, and pirimicarb. All steps in the analysis of the carbamates by GC have been examined by various researchers. Extraction solvents have been evaluated by Barrio et al. (1994). Solid-phase extraction of various matrices for carbamates have been investigated (de la Colina et al., 1993; Holland et al., 1994). Packed column supercritical fluid chromatography was used with an NPD for carbamate pesticide analysis by Berger et al. (1994). Carbamate separations have been achieved on GC columns such as a BP-10 (Brooks et al., 1990); a DB-210 in coffee beans (Narita et al., 1994) and polished rice (Hirahara et al., 1994); and a SIL5-CB column (Kadenczki et al.,

1992). For carbamate analysis, most columns are microcapillary with a steady temperature ramp starting at about 50°C to protect carbamate integrity, and NPD detection is used. Other detectors used include flame thermionic detection (FTD) (Specht et al., 1995) and Hall detection in nitrogen mode (HECD-N) (Albanis and Hela, 1995). Finally, there is a method of analysis where a silica HPLC column cleanup prior to GC determination provides a connection between a liquid cleanup system and GC detection (Seiber et al., 1990). By collecting a specific fraction containing a carbaryl marker, they were able to separate carbaryl, carbofuran, and methomyl in air samples for quantitative analysis.

IV. HIGH-PERFORMANCE LIQUID CHROMATOGRAPHY

High-performance liquid chromatography methods for analysis of carbamate insecticides are often the preferred methods for these thermally labile compounds. An excellent review article about HPLC methods for determination of *N*-methylcarbamate pesticides in water, soil, plants, and air was published by McGarvey (1993). Advanced techniques in recent years include extraction/clean-up, column chromatography, detection, derivatization, and mathematical modeling.

Residues of carbamate insecticides in water are commonly extracted using liquid-liquid extraction or solid-phase extraction. Liquid-liquid extraction involves partitioning the matrix extract with an organic solvent such as dichloromethane, chloroform, toluene, benzene, hexane, or ethyl acetate. Since solubilities of carbamates can vary dramatically, recoveries of individual carbamates may vary for any one liquid-liquid extraction method. Solid phase extraction (SPE) has been surveyed by Odanaka et al. (1991) and has been used successfully in several multiresidue techniques for water (McGarvey, 1993; Parrilla, 1994). A further advancement, from the off-line SPE column, is an on-line trace enrichment technique (Marvin et al., 1990a; Marvin et al., 1990b; Liska et al., 1992; Marvin et al., 1991; Chiron and Barceló, 1993; Chiron et al., 1993; Driss et al., 1993; Slobodnik et al., 1993; Hiemstra and de Kok, 1994; Pichon and Hennion, 1994; Papadopoulou-Mourkidou, et al., 1996). These techniques usually involve passing a relatively small sample (3–50 mL) through a C_{18} or PLRP-S (styrene-divinylbenzene copolymer) precolumn for concentration and by using a series of switching valves, or manually, connecting the precolumn to the analytical column. This system automation allows for unattended operation, improvement of the limits of determination, and a lower likelihood of human error in sample handling affecting the reproducibility and precision of the measurements to be made. The SAMOS (System for Automated Measurement of Organic Micropollutants in Surface Waters) and the Prospekt module are the commercially available systems utilized in the methods described above. A typical LC chromatogram of *N*-methylcarbamate pesticides in environmental water samples is shown in Figure 5. Automated

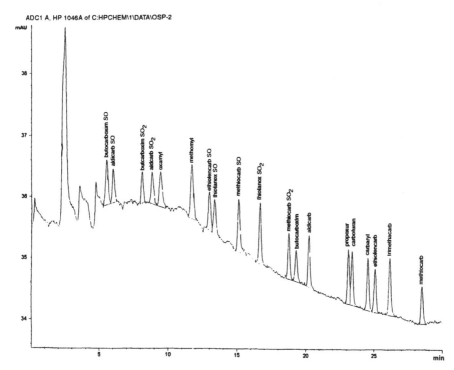

FIGURE 5 HPLC chromatogram of a 5 mL surface water sample, fortified with 18 *N*-methylcarbamates and metabolites at the 0.1 µg/L level, after on-line trace enrichment on a 10 × 4.0 mm i.d. C_{18}/OH precolumn, using the OSP-2 system. Trimethacarb was used as an internal standard.

on-line extraction has also been used for *N*-methylcarbamate pesticide determination in fruits and vegetables (de Kok and Hiemstra, 1992). Holstege et al. (1994) used automated gel permeation chromatography (GPC) with in-line silica gel minicolumns for analysis of *N*-methylcarbamate insecticides in plant and animal tissues. A LC chromatogram of their liver extract is shown in Figure 6. Supercritical fluid extraction (SFE) of pesticides prior to HPLC analysis has been studied by Alzaga et al. (1994).

Several other extraction/cleanup techniques have been used for the *N*-methylcarbamate pesticides. Graphitized carbon black cartridge extraction has been used successfully in multiresidue methods for pesticides in drinking water (DiCorcia and Marchetti, 1991, 1992; Guenu and Hennion, 1996). GPC has also been used in vegetables and fruits (Chaput, 1988) and liver extraction/cleanup (Ali, 1989). Carbofuran has been analyzed using resin extraction, which has the same

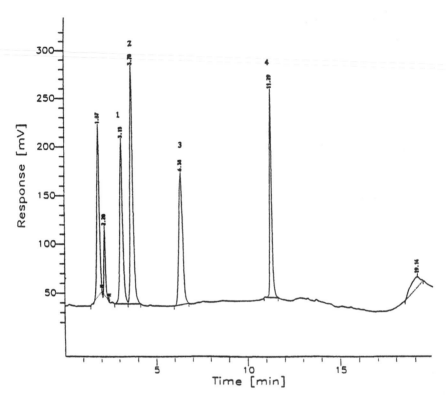

Figure 6 LC Chromatogram of bovine liver extract fortified at 0.5 ppm with MC standard mix B. A 10 μL volume of a 0.5 g/mL sample extract was injected under standard LC conditions. Peaks: (1) aldicarb sulfone, (2) methomyl, (3) 3-hydroxycarbofuran, and (4) propoxur.

degree of precision as SPE-C$_{18}$ and is less expensive (Basta and Olness, 1992). A novel technique that was used on very small samples of crop (0.5 g) was that of matrix solid-phase dispersion isolation (MSPD), which involves blending the matrix with C$_{18}$ followed by solvent washing and elution of pesticides (Stafford and Lin, 1992). Meat was extracted for propoxur and other carbamate analysis by partition with acetonitrile, blending the matrix with pelletized diatomaceous earth, and then using supercritical fluid extraction (SFE) with CO$_2$ (Argauer et al., 1995).

Chromatographic separation for carbamate pesticides usually employs reversed-phase chromatography with C$_{18}$ or C$_8$ columns and aqueous mobile phases. Retention and resolution of 14 *N*-methylcarbamates and metabolites in normal- and reversed-phase modes on a variety of columns has been reported

(Sparacino and Hines, 1976). Silica, cyanopropyl, and propylamine columns were studied in normal-phase mode with two mobile phase systems (isopropanol-heptane and dichloromethane-heptane). C_{18} and ether phase columns were studied in reversed-phase mode with three mobile phase systems (water-methanol, water-tetrahydrofuran, and water-acetonitrile). Although normal-phase mode was for the most part satisfactory, reversed-phase mode gave generally superior results. The C_{18} column and water-acetonitrile mobile phase gave overall best performance in terms of resolution of the pesticides and UV transparency of the mobile phase.

After HPLC separation, the most common means of detecting of carbamates are by UV absorbance, fluorescence, electrochemical detection, or mass spectrometry (summarized in a separate portion of this chapter, LC/MS). UV absorbance has been the most frequently used detection method because of its wide applicability and consequent presence in most HPLC systems (Beauchamp et al., 1989; Cabras et al., 1989; Page and French, 1992; Honing et al., 1994a; McGarvey, 1993; Dreyfuss et al., 1994; Eisert et al., 1995; Mora et al., 1995; Strait et al., 1991). A library of spectra for several carbamate pesticides was published by Bogusz and Erkens in 1994. However, UV is subject to interference from sample coextractives and also lacks sensitivity for some compounds. A derivatization technique using alkaline hydrolysis of carbamates followed by subsequent coupling with diazotized sulphanilic acid, the products of which can be monitored by UV, was published by Tena et al. (1992a,b). Derivatization provided stronger absorption at 280 nm and also allowed monitoring of the chromatogram at 506 nm, minimizing the possibility of interference from coextractives. This method employed a flow cell that was packed with C_{18} bonded silica, which served to retain and concentrate the derivative in the flow cell and allow determination at low concentration levels. In order to maximize sensitivity, a postcolumn pump was used to deliver water to the flow, downstream from the reactor, to dilute the aqueous acetonitrile mobile phase and favor retention of the derivative on the C_{18} solid-phase in the flow cell. An acidified ethanol reagent was delivered through a switching valve located just in front of the flow cell to elute the derivative from the solid phase after each peak had been completely integrated. Figure 7 shows absorbance of some carbamate compounds before and after derivatization with sulphanilic acid.

Most carbamate compounds do not possess native fluorescence. Early work used dansyl chloride for derivatization prior to injection, where detection limits were between 1 and 10 ng (Frei and Lawrence, 1973; Frei et al., 1974). A significant development for analysis of carbamate insecticides occurred in 1977 when Moye et al. introduced a postcolumn derivatization reaction of carbamates. Sodium hydroxide introduced by a postcolumn reagent delivery pump was used to hydrolyze the carbamates at 90°C and release methylamine. This methylamine

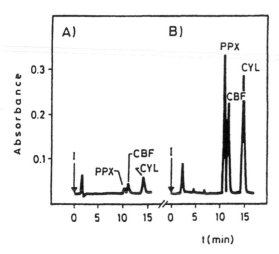

Figure 7 Chromatograms of carbofuran (CBF), propoxur (PPX) and carbaryl (CYL), without (A) and with (B) postcolumn derivatization. Concentration 10 μg/mL.

was subsequently reacted with a mixture of *o*-phthalaldehyde (OPA) and 2-mercaptoethanol (MERC), introduced by a second postcolumn pump, to form a highly fluorescent derivative identified as (1-hydroxyethylthio)-2-methylisoindole (see Figure 8). Krause (1978, 1979) refined the chromatographic and derivatization parameters. This method was rapidly adopted by a large number of researchers for determination of carbamates on a variety of substrates (Blaß, 1991; Lee et al., 1991; Ali et al., 1993; McGarvey, 1994; Yang and Smetena, 1994). Consequently, the U.S. Environmental Protection Agency has established Method 531.1 using this postcolumn derivatization technique for carbamate analysis in food and water. This basic postcolumn derivatization technique was further refined with the introduction of a catalytic solid-phase reactor consisting of a column packed with an anion-exchange resin maintained at 100–120°C for hydrolysis of carbamates (Nondek et al., 1983), and it has been used with various matrices (de Kok et al., 1990; de Kok et al., 1992). The postcolumn reaction technique was simplified by combining the hydrolysis and derivatization steps using a single reagent, OPA-2-mercaptoethanol in 0.01 M KOH, which was delivered by a single postcolumn pump (McGarvey, 1989; Simon et al., 1993). Photolysis of pesticides with subsequent derivatization by OPA-MERC was studied by Patel et al. (1990). Fluorimetric detection has also been used with ion-pair reverse-phase chromatography (Sanchez et al., 1994).

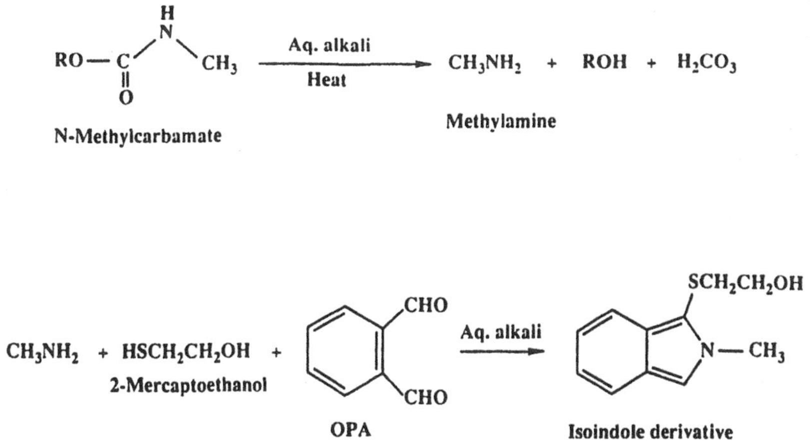

FIGURE 8 Postcolumn derivatization reaction for *N*-methylcarbamate analysis.

Other modes of detection of carbamate insecticide residues include photoconductivity detection, which uses postcolumn photolysis of analytes followed by conductometric detection of the ionic photoproducts (Miles and Zhou, 1990). Miles (1992) also compared 100 analytes in water using postcolumn photolysis followed by fluorescence (PFD), electrochemical (PED) or conductivity (PCD) detection. Carbamates were most readily detected utilizing PFD with OPA-MERC. Howard et al. (1993) developed a thiocarbamate insecticide analysis method by extracting apples with SFE, and detecting the analytes with HPLC/ Sulfur Chemiluminescence Detection (SFD). Carbamate insecticide residues were also detected using electrochemical detection, after hydrolysis to their phenolic derivatives, at a detection limit of 0.7 ppb in water (Díaz et al., 1996).

Several recent investigations deal with optimization of separation in HPLC systems using mathematical modeling. Wang et al. (1993) present the optimization system for preparative liquid chromatography (OS-PLC) and use aldicarb as an example, citing excellent agreement between predicted and experimental results. An optimization procedure by search point (OPSP) based on the Hooke–Jeeves algorithm was developed by Martínez-Vidal et al. (1995), and optimization by orthogonal factorization was used by Sánchez et al. (1995). Rouberty and Fournier (1996) present optimization of HPLC separation of carbamate insecticides (carbofuran, hydroxycarbofuran and aldicarb) by experimental design methodology utilizing isoresponse curves in mathematical modeling.

V. GAS CHROMATOGRAPHY/MASS SPECTROMETRY

Gas chromatography/mass spectrometry has been used as a confirmation procedure for insecticidal carbamates, particularly supporting multiresidue pesticide screening methods (Hong et al., 1993). With GC/MS equipment more readily available in analytical laboratories, quantitation of carbamate pesticides in multiresidue techniques has expanded to all matrices. Direct analysis of insecticidal carbamates in water has been studied with quadrupole MS (Albanis and Hela, 1995), ion trap MS (Benfenati et al., 1990; Muíño et al., 1991; Patsias and Papdopoulou-Mourkidou, 1996), and chemical ionization/ion trap MS (Mattern et al., 1991). Multiresidue screening in fruits and vegetables has been studied by quadrupole MS (Liao et al., 1991; Okihashi et al., 1994) and by ion trap MS (Cairns et al., 1993; Tuinstra et al., 1995). Mattern et al. (1990) have proposed that chemical ionization/ion trap MS could replace the Luke extraction procedure (Luke et al., 1981) for analysis of agricultural products. Lehotay and Eller (1995) successfully used supercritical fluid extraction (SFE) with GC/ion trap MS for analysis of 46 pesticides in fruits and vegetables. Murugaverl et al. (1993a,b) directly connected supercritical fluid chromatography with a benchtop mass spectrometer for carbamate pesticide analysis. Suzuki et al. (1990) used positive and negative ion MS to study carbamate pesticides in human tissues. Determination of carbamates and other pesticides in indoor air and dust was studied by ion trap MS by Roinestad et al. (1993).

For GC/MS of insecticidal carbamates, the problems of thermolability and degradation must still be overcome for consistent quantitative analysis. Climent and Miranda (1996) present a detailed study of photodegradation of carbamate pesticides using GC/MS. Honing et al. (1995a) compared the techniques of desorption chemical ionization (DCI) and flow injection (FIA)-particle beam (PB)-ammonia positive chemical ionization (PCI)-MS on ion formation of carbamates. They found that the ion source pressure and the temperature of analysis can cause irreproducibility of the mass spectra of carbamates and these factors must be carefully monitored for quantitative determination. Other researchers have overcome these problems in carbamate GC/MS analysis by derivatization. Bakowski et al. (1994) studied N-methylcarbamate insecticides in liver using a heptafluorobutyric anhydride derivative to stabilize the ions of interest in a quadrupole MS analysis of liver tissue. Figure 9 is a total ion current (TIC) chromatogram of 10 carbamate-HFB derivatives and depicts the separation achieved using a HP-1 capillary column (Bakowski et al., 1994). Other derivatives used are ammonium with positive ion chemical ionization spectra (Wigfield et al., 1993), flash-heater methylation with trimethylsulfonium hydroxide (Färber and Schöler, 1993), acetic anhydride (Stan and Klaffenbach, 1991) and benzyl derivatization for analysis in

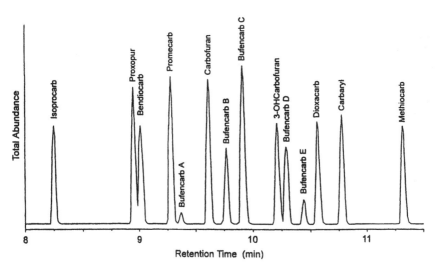

FIGURE 9 Total ion current (TIC) chromatograms of 10 carbamate-HFB derivatives scanned at 50–650 amu.

agricultural products (Akiyama et al., 1995). Temperature degradation has also been overcome by using a programmed temperature evaporation technique where the sample injection was at 39°C and GC/chemical ionization mass spectra analyzed (Vincze and Yinon, 1996).

Many of the problems with carbamate analysis by GC/MS can be overcome with derivatization, attention to thermal programming, and chemical ionization techniques. The coupling of liquid chromatography with mass spectrometry is much more conducive to insecticidal carbamate analysis and is covered below.

VI. LIQUID CHROMATOGRAPHY/MASS SPECTROMETRY

Liquid chromatography with mass spectrometric detection (LC/MS) has been widely accepted as the preferred technique for the identification and quantification of polar and thermally labile compounds such as insecticidal carbamates. The coupling of LC and MS is much less straightforward than that of GC and MS because of the transfer of analytes from the liquid phase into a high-vacuum gas phase. The combination of LC with MS generates more restrictions on the experimental conditions that can be used than with either of them separately. There are two types of commercially available interfaces: nebulization interfaces and analyte-enrichment interfaces. Thermospray (TSP) nebulizes using heat, while electrospray (ES) applies an electric field at the tip of a capillary. Atmospheric pressure chemical ionization (APCI) involves pneumatic nebulization followed by heat-assisted desolvation of the spray droplets, which ionspray (IS) utilizes as pneumatically assisted electrospray nebulization. With the analyte-enrichment type of interface such as particle beam (PB), the analytes are separated from the solvent flow after nebulization. All of these LC/MS interfaces have been used for the determination of carbamate pesticides (Wright, 1982; Honing et al., 1995b).

TSP interface with MS has been used in multiresidue analyses in water, using off-line and on-line enrichment procedures (Bagheri, 1993; Barceló et al., 1993; Chiron et al., 1994; Volmer and Levsen, 1994; Volmer et al., 1994; Sennert et al., 1995) and fruits and vegetables (Lui et al., 1991). An interlaboratory study of TSP for selected N-methylcarbamates showed significant interlaboratory variability, showing that the thermospray tip temperature can play a major role in adduct formation and ion fragmentation in the case of thermally labile carbamate pesticides (Lopez-Avila and Jones, 1993). Durand et al. (1991), Honing et al (1994b) and Vreeken et al. (1994) studied the use of mobile phase additives of ammonium acetate, ammonium formate, triethyl- and triproplyammonium formate, and chloroacetonitrile and how additives affect the adduct and ion formation in carbamate pesticides, thus increasing sensitivity. Volmer et al. (1993, 1995) studied the influence of temperature, salt concentration, and ion source geometry on the mass spectra formation in carbamate pesticides. All of these factors indicate irreproducibility of results from researcher to researcher when TSP is used for confirmation and quantitation. In addition to TSP analysis, DiCorcia et al. (1996) have used electrospray (ES)/MS successfully for analysis of carbamate insecticides in fruits and vegetables using selected ion monitoring (SIM). As with thermospray, ES is a technique able to produce ions at atmospheric pressure but without the need for high temperatures that could decompose labile compounds. The most accepted model of gas-phase ion production involves the formation of charged small droplets by an electrical field, which are more and more shrunken by heat transfer and repeated coulombic explosion until the radius of curvature of the daughter droplets becomes so small that field-assisted ion "evaporation" competes with further droplet disintegration. Nonionic organic species can be electrically charged in the gaseous phase, provided they are able to form stable adducts or complexes with inorganic ions. In addition, in ES, structural information on separated analytes can be easily achieved by collision-induced dissociation with suitable adjustment of the electrical field existing in the desolvation. This can provide structural information very similar to that obtained by MS/MS techniques discussed below. Honing et al. (1996) studied ten carbamate pesticides in water and sediment samples using ionspray (IS)/MS. Because of the "cold" nebulization of the column eluant, 10–50 pg amounts could be quantitated in SIM mode.

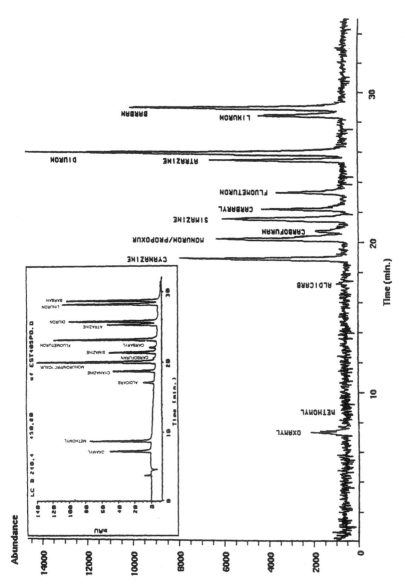

FIGURE 10 RPLC-PB-MS total ion chromatogram of a standard solution of 14 pesticides at 40 µg/mL. The insert shows the RPLC-DAD chromatogram at wavelength 240 nm.

Atmospheric pressure chemical ionization (APCI) techniques using a heated nebulizer interface provided both protonated molecules and abundant, characteristic fragment ions for analysis of *N*-methylcarbamate pesticides (Pleasance et al., 1992). Results indicated the APCI/MS experiments provided the greatest sensitivity, with limits of detection ranging from 0.05 to 0.18 ppm, values that are in most cases ten times lower than those achieved by ISP and TSP (Careri et al., 1996). APCI has been used for carbamate analysis in water (Doerge and Bajic, 1992), serum (Kawasaki et al., 1993), and food (Newsome et al., 1995).

Particle beam (PB)/MS analysis of polar pesticides in water has been studied by Marcé et al. (1995), to which they compared diode array detection (see Figure 10). Detection limits were lower for PB/MS detection; however, they record a distinct matrix effect caused by coeluting compounds acting as carriers. This improves analyte detectability and requires standard addition to be used for quantification. Cappiello et al. (1994) further refined PB/MS analysis using a μL flow rate for column eluant. Miles et al. (1992) used four different nebulizers to study the effect on chromatographic integrity using the PB/MS as a useful qualitative confirmation tool for National Pesticide Survey Analytes.

The inherent power of mass spectrometric analysis is compound specificity, and this is compromised by the "soft" ionization technique typical of thermospray ionization, in which the mass spectra are usually predominated by the relatively intact molecular adduct ions. In tandem spectrometry, LC/MS/MS, carbamate pesticides can be analyzed for structure-specific fragmentations of collisionally activated daughter ions (Chiu et al., 1989). Tandem MS has been used for water analysis with on-line solid-phase extraction and APCI/MS/MS detection (Hogenboom et al., 1996; Slobodník et al., 1996). Volmer et al. (1996) showed that electrospray/MS/MS, with time-scheduled selected ion monitoring, yielded sensitive detection of pesticides at the very low picogram levels in food and water samples. They demonstrated that both in-source collision-induced dissociation and tandem mass spectrometry in the constant neutral-loss and selected retention monitoring modes are excellent means of increasing the sensitivity in analysis of carbamate insecticides and other pesticides.

VII. CONCLUSIONS

Sensitive analytical methods for the detection of insecticidal carbamate residues include the use of thin-layer chromatography (TLC), gas chromatography (GC), high performance liquid chromatography (HPLC), gas chromatography-mass spectrometry (GC/MS), and liquid chromatography-mass spectrometry (LC/MS). Future analytical techniques should include the refinement of LC/MS, with particular attention to the capabilities of LC/MS/MS for quantitative analysis of insecticidal carbamates at trace levels in a variety of matrices.

REFERENCES

Akiyama, Y., Takeda, N., and Adachi, K. (1995), Studies on simple clean-up by the precipitation methods and GC/MS analysis after benzyl derivatization of carbamate pesticides in agricultural products, *Shokuhin Eiseigaku Zasshi 36*(1): 42–49.

Albanis, T. A., and Hela, D. G. (1995), Multi-residue pesticide analysis in environmental water samples using solid-phase extraction discs and gas chromatography with flame thermionic and mass-selective detection, *J. Chromatogr. A. 707*: 283–292.

Albelda, C., Picó, Y., Font, G., and Mañes, J. (1994), Determination of aldicarb, aldicarb sulfoxide, and aldicarb sulfone in oranges by simple gas-liquid chromatography with nitrogen-phosphorus detection, *J. AOAC Intern. 77*(1): 74–78.

Ali, M. S. (1989), Determination of *N*-methylcarbamate pesticides in liver by liquid chromatography, *J. Assoc. Off. Anal. Chem. 72*(4): 586–592.

Ali, M. S., White, J. D., Bakowski, R. S., Stapleton, N. K., Williams, K. A., Johnson, R. C., Phillippo, E. T., Woods, R. W., and Ellis, R. L. (1993), Extension of a liquid chromatographic method for *N*-methylcarbamate pesticides in cattle, swine, and poultry liver, *J. AOAC Intern. 76*(4): 907–910.

Alzaga, R., Durand, G., Barceló, D., and Bayona, J. M. (1994), Comparison of supercritical fluid extraction and liquid-liquid extraction for isolation of selected pesticides stored in freeze-dried water samples, *Chromatographia 38*(7/8): 502–507.

Argauer, R. J. Eller, K. I., Ibrahim, M. A., and Brown, R. T. (1995), Determining propoxur and other carbamates in meat using HPLC fluorescence and gas chromatography/ion trap mass spectrometry after supercritical fluid extraction, *J. Agric. Food Chem. 43*: 2774–2778.

Bagheri, H., and Creaser, C. S. (1991), Determination of carbaryl and 1-naphthol in English apples and strawberries by combined gas chromatography-fluorescence spectrometry, *J. Chromatogr. 547*: 345–353.

Bagheri, H., Brouwer, E. R., Ghijsen, R. T., and Brinkman, U. A. Th. (1993), On-line low-level screening of polar pesticides in drinking and surface waters by liquid chromatography-thermospray mass spectrometry, *J. Chromatogr. 647*: 121–129.

Bakowski, R. S., Sher Ali, M., White, J. D., Phillippo, E. T., and Ellis, R. L. (1994), Gas chromatographic/mass spectrometric confirmation of ten *N*-methylcarbamate insecticides in liver, *J. AOAC Intern. 77*(6): 1568–1574.

Ballesteros, E., Gallego, M., and Valcárcel, M. (1993a), Automatic determination of *N*-methylcarbamate pesticides by using a liquid-liquid extractor derivatization module coupled on-line to a gas chromatograph equipped with a flame ionization detector, *J. Chromatogr. 633*: 169–176.

Ballesteros, E., Gallego, M., and Valcárcel, M. (1993b), Automatic gas chromatographic determination of *N*-methylcarbamates in milk with electron capture detection, *Anal. Chem. 65*: 1773–1778.

Barceló, D., Durand, G., Bouvot, V., and Nielen, M. (1993), Use of extraction disks for trace enrichment of various pesticides from river water and simulated sea water samples followed by liquid chromatography-rapid-scanning UV-visible and thermospray-mass spectrometry detection, *Environ. Sci. Technol. 27*: 271–277.

Barrio, C. S., Asensio, J. S., and Bernal, J. G. (1994), GC-NPD investigation of the recovery of organonitrogen and organophosphorus pesticides from apple samples: the effect of the extraction solvent, *Chromatographia 39*(5/6): 320–324.

Basta, N. T., and Olness, A. (1992), Determination of alachlor, atrazine, and metribuzin in soil by resin extraction, *J. Environ. Qual. 21*: 497–502.

Beauchamp, K. W., Jr., Liu, D. D. W., and Kikta, E. J., Jr. (1989), Determination of carbofuran and its metabolites in rice paddy water by using solid phase extraction and liquid chromatography, *J. Assoc. Off. Anal. Chem. 72*(5): 845–847.

Benfenati, E., Tremolada, P., Chiappetta, L. Frassanito, R., Bassi, G., Di Toro, N., Fanelli, R., and Stella, G. (1990), Simultaneous analysis of 50 pesticides in water samples by solid phase extraction and GC-MS, *Chemosphere 21*(12): 1411–1421.

Berger, T. A., Wilson, W. H., and Deye, J. F. (1994), Analysis of carbamate pesticides by packed column supercritical fluid chromatography, *J. Chromatogr. Sci. 32*: 179–184.

Blaß, W. (1991), Determination of methyl carbamate residues using on-line coupling of HPLC with a post column fluorimetric labeling technique, *Fresenius Z. Anal. Chem. 339*: 340–343.

Bogusz, M., and Erkens, M. (1994), Reversed-phase high-performance liquid chromatographic database of retention indices and UV spectra of toxicologically relevant substances and its interlaboratory use, *J. Chromatogr. A 674*: 97–126.

Brauckhoff, S., and Thier, H. P. (1992), Methyl carbamate insecticides, in: *Manual of Pesticide Residue Analysis, Vol. 2* (H. P. Thier and J. Kirchhoff, eds.), New York: VCH.

Brooks, M. W., Tessier, D., Soderstrom, D., Jenkins, J., and Clark, J. M. (1990), A rapid method for the simultaneous analysis of chlorphyrifos, isofenphos, carbaryl, iprodione, and triadimefon in groundwater by solid-phase extraction, *J. Chromatogr. Science 28*: 487–489.

Burger, K. (1984), DC-PMD, Dünnschicht-chromatographie mit gradienten-elution im vergleich zur saülenflussigkeits-chromatographie, *Frensenius Z. Anal. Chem. 318*: 228–233.

Butz, S., and Stan, H. J. (1995), Screening of 265 pesticides in water by thin-layer chromatography with automated multiple development. *Anal. Chem. 67*: 620–630.

Cabras, P., Spanedda, L., Tuberoso, C., and Gennari, M. (1989), Separation of pirimicarb and its metabolites by high-performance liquid chromatography, *J. Chromatogr. 478*: 250–254.

Cappiello, A., Famiglini, G., and Bruner, F. (1994), Determination of acidic and basic/neutral pesticides in water with a new microliter flow rate LC/MS particle beam interface, *Anal. Chem. 66*: 1416–1423.

Cairns, T., Chiu, K. S., Navarro, D., and Siegmund, E. (1993), Multiresidue pesticide analysis by ion-trap mass spectrometry, *Rapid Commun. in Mass Spectrom. 7*: 971–988.

Careri, M., Mangia, A., and Musci, M. (1996), Applications of liquid chromatography-mass spectrometry interfacing systems in food analysis: pesticide, drug and toxic substance residues, *J. Chromatogr. A 727*: 153–184.

Chaput, D. (1988), Simplified multiresidue method for liquid chromatographic determination of *N*-methyl carbamate insecticides in fruits and vegetables, *J. Assoc. Off. Anal. Chem. 71*(3): 542–546.

Chiron, S., and Barceló, D. (1993), Determination of pesticides in drinking water by on-line solid-phase disk extraction followed by various liquid chromatographic systems, *J. Chromatogr. 645*: 125–134.

Chiron, S., Fernandez Alba, A., and Barceló, D. (1993), Comparison of on-line solid-phase disk extraction to liquid-liquid extraction for monitoring selected pesticides in environmental waters, *Environ. Sci. Technol. 27*(12): 2352–2359.

Chiron, S., Dupas, S., Scribe, P., and Barceló, D. (1994), Application of on-line solid-phase extraction followed by liquid chromatography-thermospray mass spectrometry to the determination of pesticides in environmental waters, *J. Chromatogr. A. 665*: 295–305.

Chiu, K. S., Van Langenhove, A., and Tanaka, C. (1989), High-performance liquid chromatographic/mass spectrometric and high-performance liquid chromatographic/tandem mass spectrometric analysis of carbamate pesticides, *Biomed Environ. Mass Spectrometry 18*: 200–206.

Climent, M. J., and Miranda, M. A. (1996), Gas chromatographic-mass spectrometric study of photodegradation of carbamate pesticides, *J. Chromatogr. A. 738*: 225–231.

Cline, R. E., Todd, G. D., Ashley, D. L., Grainger, J., McCraw, J. M., Alley, C. C., and Hill, R. H., Jr. (1990), Gas chromatographic and spectral properties of pentafluorobenzyl derivatives of 2,4-dichlorophenoxyacetic acid and phenolic pesticides and metabolites, *J. Chromatogr. Sci. 28*: 167–172.

Coats, J. R. (1982), *Insecticide Mode of Action*, Academic Press, New York.

Cohen, I. C., Norcup, J., Ruzicka, J. H. A., and Wheals, B. B. (1970), An electron-capture gas chromatographic method for the determination of some carbamate insecticides as 2,4-dinitrophenyl derivatives of their phenol moieties, *J. Chromatogr. 49*: 215–221.

de Kok, A., and Hiemstra, M. (1992), Optimization, automation and validation of the solid-phase extraction cleanup and on-line liquid chromatographic determination of *N*-methylcarbamate pesticides in fruits and vegetables, *J. AOAC Intern. 75*(6): 1063–1072.

de Kok, A., Hiemstra, M., and Vreeker, C. P. (1990), Optimization of the postcolumn hydrolysis reaction on solid phases for the routine high-performance liquid chromatographic determination of *N*-methylcarbamate pesticides in food products, *J. Chromatogr. 507*: 459–472.

de Kok, A., Hiemstra, M., and Brinkman, U. A. Th. (1992), Low ng/1-level determination of twenty *N*-methylcarbamate pesticides and twelve of their polar metabolites in surface water via off-line solid-phase extraction and high-performance liquid chromatography with post-column reaction and fluorescence detection, *J. Chromatogr. 623*: 265–276.

de la Colina, C., Heras, A. P., Cancela, G. D., and Rasero, F. S. (1993), Determination of organophosphorous and nitrogen-containing pesticides in water samples by solid phase extraction with gas chromatography and nitrogen-phosphorus detection, *J. Chromatogr. A 655*: 127–132.

Díaz, T. G., Guiberteau, A., Salinas, F., and Ortiz, J. M. (1996), Rapid and sensitive determination of carbaryl, carbofuran and fenobucarb by liquid chromatography with electrochemical detection, *J. Liq. Chrom. and Rel. Technol. 19*(16): 2681–2690.

DiCorcia, A., and Marchetti, M. (1991), Multiresidue methods for pesticides in drinking water using a graphitized carbon black cartridge extraction and liquid chromatographic analysis, *Anal. Chem. 63*: 580–585.

DiCorcia, A., and Marchetti, M. (1992), Method development for monitoring pesticides in environmental waters: liquid-solid extraction followed by liquid chromatography, *Environ. Sci. Technol. 26*(1): 66–74.

DiCorcia, A., Crescenzi, C., Laganà, A., and Sebastiani, E. (1996), Evaluation of a method based on liquid chromatography/electrospray/mass spectrometry for analyzing carbamate insecticides in fruits and vegetables, *J. Agric. Food Chem. 44*: 1930–1938.

Doerge, D. R., and Bajic, S. (1992), Analysis of pesticides using liquid chromatography/atmospheric-pressure chemical ionization mass spectrometry, *Rapid Commun. in Mass Spectrom. 6*: 663–666.

Dreyfuss, M. F., Lotfi, H., Marquet, P., Debord, J., Daguet, J. L., and Lachâtre, G. (1994), Analyse de résidus de pesticides dans des miels et des pommes par CLHP et CPG, *Analusis 22*: 273–280.

Driss, M. R., Hennion, M.-C., and Bouguerra, M. L. (1993), Determination of carbaryl and some organophosphorus pesticides in drinking water using on-line liquid chromatographic preconcentration techniques, *J. Chromatogr. 639*: 352–358.

Durand, G., de Bertrand, N., and Barceló, D. (1991), Mobile phase variations in thermospray liquid chromatography-mass spectrometry of pesticides, *J. Chromatography, Biomedical Applic. 562*: 507–523.

Eisert, R., Levsen, K., and Wünsch, G. (1995), Analysis of polar thermally labile pesticides using different solid-phase extraction (SPE) materials with GC and HPLC techniques. *Intern. J. Environ. Anal. Chem. 58*: 103–120.

Färber, H., and Schöler, H. F. (1993), Gas chromatographic determination of carbamate pesticides after flash-heater methylation with trimethylsulfonium hydroxide, *J. Agric. Food Chem. 41*: 217–220.

Frei, R. W., and Lawrence, J. F. (1973), Fluorigenic labelling in high-speed liquid chromatography, *J. Chromatogr. 83*: 321–330.

Frei, R. W., Lawrence, J. F., Hope, J., and Cassidy, R. M. (1974), Analysis of carbamate insecticides by fluorigenic labelling and high-speed liquid chromatography, *Chromatogr. Sci. 12*: 40–44.

Guenu, S., and Hennion, M.-C. (1996), Prediction from liquid chromatographic data of obligatory backflush desorption from solid-phase extraction cartridges packed with porous graphite carbon, *J. Chromatogr. A 725*: 57–66.

Hiemstra, M., and de Kok, A. (1994), Determination of N-methylcarbamate pesticides in environmental water samples using automated on-line trace enrichment with exchangeable cartridges and high-performance liquid chromatography, *J. Chromatogr. A 667*: 155–166.

Hirahara, Y., Narita, M., Okamoto, K., Miyoshi, T., Miyata, M., Koiguchi, S., Hasegawa, M., Kamakura, K., Yamana, T., and Tonogai, Y. (1994), Simple and rapid simultaneous determination of various pesticides of polished rice by gas chromatography, *Shokuhin Eiseigaku Zasshi 35*(5): 517–524.

Hogenboom, A. C., Slobodník, J., Vreuls, J. J., Rontree, J. A., van Baar, B. L. M., Niessen, W. M. A., and Brinkman, U. A. Th. (1996), Single short-column liquid chromatography with atmospheric pressure chemical ionization-(tandem) mass spectrometric detection for trace environmental analysis, *Chromatographia 42*(9/10): 506–514.

Holden, E. R. (1976), Gas chromatographic determination of residues of methylcarbamate insecticides in crops as their 2,4-dinitrophenyl ether derivatives, *J. Assoc. Off. Anal. Chem.* 56: 713–717.

Holden, E. R., Jones, W. M., and Beroza, M. (1969), Determination of residues of methyl- and dimethylcarbamate insecticides by gas chromatography of their 2,4-dinitro-aniline derivatives, *J. Agric. Food Chem.* 18: 56–59.

Holland, P. T., McNaughton, D. E., and Malcolm, C. P. (1994), Multiresidue analysis of pesticides in wines by solid-phase extraction, *J. AOAC Intern.* 77(1): 79–86.

Holstege, D. M., Scharberg, D. L., Tor, E. R., Hart, L. C., and Galey, F. D. (1994), A rapid multiresidue screen for organophosphorus, organochlorine, and *n*-methyl carbamate insecticides in plant and animal tissues, *J. AOAC Intern.* 77(5): 1263–1274.

Hong, J., Eo, Y., Rhee, J., Kim, T., and Kim, K. (1993), Simultaneous analysis of 25 pesticides in crops using gas chromatography and their identification by gas chromatography-mass spectrometry, *J. Chromatogr.* 639: 261–271.

Honing, M., Barceló, D., Ghijsen, R. T., and Brinkman, U. A. Th. (1994a), Optimization of the liquid chromatographic separation of pirimicarb and its metabolites V-VII: application to a soil sample used as a candidate reference material, *Anal. Chim. Acta* 286: 457–468.

Honing, M., Barceló, D., van Baar, B. L. M., Ghijsen, R. T., and Brinkman, U. A. Th. (1994b), Ion formation of *N*-methyl carbamate pesticides in thermospray mass spectrometry: the effects of additives to the liquid chromatographic eluent and of the vaporizer temperature, *J. Am. Soc. Mass Spectrom.* 5: 913–927.

Honing, M., Barceló, D., Jager, M. E., Slobodnik, J., van Baar, B. L. M., and Brinkman, U. A. Th. (1995a), Effect of ion source pressure on ion formation of carbamates in particle-beam chemical-ionisation mass spectrometry, *J. Chromatogr. A.* 712:21–30.

Honing, M., Barceló, D., van Baar, B. L. M., Brinkman, U. A. Th. (1995b), Limitations and perspectives in the determination of carbofuran with various liquid chromatography-mass spectrometry interfacing systems, *Trends in Anal. Chem.* 14(10): 496–504.

Honing, M., Riu, J., Barceló, D., van Baar, B. L. M., Brinkman, U. A. Th. (1996), Determination of ten carbamate pesticides in aquatic and sediment samples by liquid chromatography-ionspray and thermospray mass spectrometry, *J. Chromatogr. A* 733: 283–294.

Howard, A. L., Braue, C., and Taylor, L. T. (1993), Analysis of thiocarbamate pesticides in apples employing SFE and high performance liquid chromatography coupled with sulfur chemiluminescence detection, *Natl. Meet. Am. Chem. Soc., Div. Environ. Chem.* 33(1): 316–319.

Kadenczki, L., Arpad, Z., Gardi, I., Ambrus, A., Gyorfi, L., Reese, G., and Ebing, W. (1992), Column extraction of residues of several pesticides from fruits and vegetables: a simple multiresidue analysis method, *J. AOAC Intern.* 75(1): 53–61.

Kawasaki, S., Nagumo, F., Ueda, H., Tajima, Yu., Sano, M., and Tadano, J. (1993), Simple, rapid and simultaneous measurement of eight different types of carbamate pesticides in serum using liquid chromatography-atmospheric pressure chemical ionization mass spectrometry, *J. Chromatogr. Biomed. Appl.* 620: 61–71.

Kendrick, P. N., Trim, A. J., Atwal, J. K., and Brown, P. M. (1991), Direct gas chroma-tographic determination of carbaryl residues in honeybees (*Apis mellifera* L.) using a nitrogen-phosphorus detector with confirmation by formation of a chemical deriva-tive, *Bull. Environ. Contam. Toxicol.* 46: 654–661.

Koeber, R., and Niessner, R. (1996), Screening of pesticide-contaminated soil by supercritical fluid extraction (SFE) and high-performance thin-layer chromatography with automated multiple development (HPTLC/AMD), *Fresenius' J. Anal. Chem. 354*: 464–469.

Krause, R. T. (1978), Further characterization and refinement of an HPLC post-column fluorometric labeling technique for the determination of carbamate insecticides, *J. Chromatogr. Sci. 16*: 281–288.

Krause, R. T. (1979), Resolution, sensitivity and selectivity of a high-performance liquid chromatographic post-column fluorometric labeling technique for determination of carbamate pesticides, *J. Chromatogr. 185*: 615–624.

Kuhr, R. M., and Dorough, H. W. (1976), *Carbamate Insecticides: Chemistry, Biochemistry and Toxicology*, Cleveland, Ohio: CRC Press.

Lartiges, S. B., and Garrigues, P. (1995), Gas chromatographic analysis of organophosphorus and organonitrogen pesticides with different detectors, *Analusis 23*: 418–421.

Lawrence, J. F. (1979), Practical aspects of chemical derivatization in chromatography, *J. Chromatogr. Sci. 17*: 113–114.

Lawrence, J. F., Lewis, D. A., and McLeod, H. A. (1977), Detection of carbofuran and metabolites directly or as their heptafluorobutyryl derivatives using gas-liquid or high-pressure liquid chromatography with different detectors, *J. Chromatogr. 138*: 143–150.

Lee, S. M., Papthakis, M. L., Feng, H.-M. C., Hunter, G. F., and Carr, J. E. (1991), Multipesticide residue method for fruits and vegetables: California Department of Food and Agriculture, *Fresenius' J. Anal. Chem. 339*: 376–383.

Lehotay, S. J., and Eller, K. I. (1995), Development of a method of analysis for 46 pesticides in fruits and vegetables by supercritical fluid extraction and gas chromatography/ion trap mass spectrometry, *J. AOAC Intern. 78*(3): 821–830.

Liao, W., Joe, T., and Cusick, W. G. (1991), Multiresidue screening method for fresh fruits and vegetables with gas chromatographic/mass spectrometric detection, *J. Assoc. Off. Anal. Chem. 74*(3): 554–565.

Ling, C. F., Melian, G. P., Jimenez-Conde, F., and Revilla, E. (1990), Comparative study of some columns for direct determination of carbofuran by gas-liquid chromatography with nitrogen-specific detection, *J. Chromatogr. 519*: 359–362.

Liska, I., Brouwer, E. R., Ostheimer, A. G. L., Lingeman, H., Brinkman, U. A. Th., Geerdink, R. B., and Mulder, W. H. (1992), Rapid screening of a large group of polar pesticides in river water by on-line trace enrichment and column liquid chromatography, *Intern. J. Environ. Anal. Chem. 47*: 267–291.

Lopez-Avila, V., and Jones, T. (1993), Interlaboratory study of a thermospray-liquid chromatographic/mass spectrometric method for selected *N*-methyl carbamates, *N*-methyl carbamoyloximes, and substituted urea pesticides, *J. AOAC Intern 76*(6): 1329–1343.

Lui, C.-H., Mattern, G. C., Yu, X., Rosen, R., and Rosen, J. (1991), Multiresidue determination of nonvolatile and thermally labile pesticides in fruits and vegetables by thermospray liquid chromatography/mass spectrometry, *J. Agric. Food Chem. 39*: 718–723.

Luke, M. A., Froberg, J. E., Doose, G. M., and Masumoto, H. T. (1981), Improved multiresidue gas chromatographic determination of organophosphorus, organonitrogen, and organohalogen pesticides in produce, using flame photometric and electrolytic conductivity detectors, *J. Assoc. Off. Anal. Chem. 64*(5): 1187–1195.

Machemer, L. H., and Pickel, M. (1994), Chapter 4: Carbamate insecticides, *Toxicology 91*: 29–36.

Maitlen, J. C., and McDonough, L. M. (1980), Derivatization of several carbamate pesticides with methanesulphonyl chloride and detection by gas-liquid chromatography with the flame photometric detector: application to residues of carbaryl on lentil straw, *J. Agric. Food Chem. 28*: 78–82.

Marcé, R. M., Prosen, H., Crespo, C., Calull, M., Borrull, F., and Brinkman, U. A. Th. (1995), On-line trace enrichment of polar pesticides in environmental waters by reversed-phase liquid chromatography-diode array detection-particle beam mass spectrometry, *J. Chromatogr. A 696*: 63–74.

Martínez-Vidal, J. L., Parilla, P., Fernández-Alba, A. R., Carreño, R., and Herrera, F. (1995), A new sequential procedure for the efficient and automated location of optimum conditions in high performance liquid chromatography (HPLC), *J. Liquid Chromatogr. 18*(15): 2969–2989.

Marvin, C. H., Brindle, I. D., Hall, C. D., and Chiba, M. (1990a), Automated high-performance liquid chromatography for the determination of pesticides in water using solid phase extraction, *Anal. Chem. 62*: 1495–1498.

Marvin, C. H., Brindle, I. D., Singh, R. P., Hall, C. D., and Chiba, M. (1990b), Simultaneous determination of trace concentrations of benomyl, carbendazim (MBC) and nine other pesticides in water using an automated on-line pre-concentration high-performance liquid chromatographic method, *J. Chromatogr. 518*: 242–249.

Marvin, C. H., Brindle, I. D., Hall, C. D., and Chiba, M. (1991), Rapid on-line precolumn high-performance liquid chromatographic method for the determination of benomyl, carbendazim and aldicarb species in drinking water, *J. Chromatogr. 555*: 147–154.

Mattern, G. C., Singer, G. M., Louis, J., Robson, M., and Rosen, J. D. (1990), Determination of several pesticides with a chemical ionization ion trap detector, *J. Agric. Food Chem. 38*: 402–407.

Mattern, G. C., Louis, J. B., and Rosen, J. D. (1991), Multipesticide determination in surface water by gas chromatography/chemical ionization/mass spectrometry/ion trap detection, *J. Assoc. Off. Anal. Chem. 74*(6): 982–986.

McGarvey, B. D. (1989), Liquid chromatographic determination of *N*-methylcarbamate pesticides using a single-stage post-column derivatization reaction and fluorescence detection, *J. Chromatogr. 481*: 445–451.

McGarvey, B. D. (1993), High-performance liquid chromatographic methods for the determination of *N*-methylcarbamate pesticides in water, soil, plants and air, *J. Chromatogr. 642*: 89–105.

McGarvey, B. D. (1994), Derivatization reactions applicable to pesticide determination by high-performance liquid chromatography, *J. Chromatogr. B 659*: 243–257.

McGinnis, S. C., and Sherma, J. (1994), Determination of carbamate insecticides in water by C-18 solid phase extraction and quantitative HPTLC, *J. Liquid Chromatogr. 17*(1): 151–156.

Miles, C. J. (1992), Determination of National Survey of Pesticides analytes in groundwater by liquid chromatography with postcolumn reaction detection, *J. Chromatogr. 592*: 283–290

Miles, C. J., and Zhou, M. (1990), Multiresidue pesticide determinations with a simple photoconductivity HPLC detector, *J. Agric. Food Chem. 38*(4): 986–989.

Miles, C. J., Doerge, D. R., and Bajic, S. (1992), Particle beam/liquid chromatography/mass spectrometry of National Pesticide Survey analytes, *Arch. Environ. Contam. Toxicol. 22*: 247–251.

Mora, J. I., Goicolea, M. A., Barrio, R. J., and Gomez de Balugera, Z. (1995), Determination of nematicide aldicarb and its metabolites aldicarb sulfoxide and aldicarb sulfone in soils and potatoes by liquid chromatography with photodiode array detection, *J. Liquid Chromatogr. 18*(16): 3243–3256.

Moye, H. A., Scherer, S. J., and St. John, P. A. (1977), A dynamic fluorogenic labelling of pesticides for high performance liquid chromatography: detection of *N*-methyl carbamates with *o*-phthalaldehyde, *Anal. Lett. 10*: 1049–1073.

Muíño, M. A. F., Gándara, J. S., and Lozano, J. S. (1991), Simultaneous determination of pentachlorophenol and carbaryl in water, *Chromatographia, 32*(5/6): 238–240.

Murugaverl, B., Gharaibeh, A., and Voorhees, K. J. (1993a), Mixed adsorbent for cleanup during supercritical fluid extraction of three carbamate pesticides in tissues, *J. Chromatogr. A 657*: 223–226.

Murugaverl, B., Voorhees, K. J., and DeLuca, S. J. (1993b), Utilization of a benchtop mass spectrometer with capillary supercritical fluid chromatography, *J. Chromatogr. 633*: 195–205.

Nagasawa, K., Uchiyama, H., Ogamo, A., and Shinozuka, T. (1977), Gas chromatographic determination of microamounts of carbaryl and 1-naphthol in natural water as sources of water supplies, *J. Chromatogr. 144*: 77–84.

Narita, M., Miyata, M., Kamakura, K., Hirahaha, Y., Okamoto, K., Hasegawa, M., Koiguchi, S., Miyoshi, T., Yamana, T., Tonogai, Y., and Ito, Y. (1994), Studies on analysis of organophosphorus, carbamate and pyrethroid pesticides as well as determination of residual bromine in raw coffee beans, *Shokuhin Eiseigaku Zasshi 35*(4): 371–379.

Newsome, W. H., Lau, B. P.-Y., Ducharme, D., and Lewis, D. (1995), Comparison of liquid chromatography-atmospheric pressure chemical ionization/mass spectrometry and liquid chromatography-postcolumn fluorometry for determination of carbamates in food, *J. AOAC Intern. 78*(5): 1312–1316.

Nondek, L., Brinkman, U. A. Th., and Frei, R. W. (1983), Band broadening in solid phase reactors packed with catalyst for reactions in continuous-flow systems, *Anal. Chem. 55*: 1466–1470.

Odanaka, Y., Matano, O., and Goto, S. (1991), The use of solid bonded-phase extraction as alternative to liquid-liquid partitioning for pesticide residue analysis of crops, *Fresenius' J. Anal. Chem. 339*: 368–373.

Okihashi, M., Obana, H., Hori, S., and Nishimune, T. (1994), Development of simultaneous analysis for organonitrogen and pyrethroid pesticides with GC/MS, *Journal of the Food Hygienic Society of Japan 35*(3): 258–261.

Page, M. J., and French, M. (1992), Determination of *N*-methylcarbamate insecticides in vegetables, fruits, and feeds using solid-phase extraction cleanup in the normal phase, *J. AOAC Intern. 75*(6): 1073–1083.

Papadopoulou-Mourkidou, E., and Patsias, J. (1996), Development of a semi-automated high-performance liquid chromatographic-diode array detection system for screening pesticides at trace levels in aquatic systems of the Axios River basin, *J. Chromatogr. A 726*: 99–113.

Parrilla, P., Martinez, J. L., Martinez Galera, M., and Frenich, A. G. (1994), Simple and rapid screening procedure for pesticides in water using SPE and HPLC/DAD detection, *Fresenius' J. Anal. Chem. 350*: 633–637.

Patel, B. M., Moye, H. A., and Weinberger, R. (1990), Formation of fluorophores from nitrogenous pesticides by photolysis and reaction with OPA-2-mercaptoethanol for fluorescence detection in liquid chromatography, *J. Agric. Food Chem. 38*: 126–134.

Patil, V. B., and Shingare, M. S. (1993), Thin-layer chromatographic detection of carbaryl using phenylhydrazine hydrochloride, *J. Chromatogr. A 653*: 181–183.

Patsias, J., and Papadopoulou-Mourkidou, E. (1996), Rapid method for the analysis of a variety of chemical classes of pesticides in surface and ground waters by off-line solid-phase extraction and gas chromatography-ion trap mass spectrometry, *J. Chromatogr. A. 740*: 83–98.

Pesticide Analytical Manual. 1979, 1993. Department of Health and Human Services, Food and Drug Administration, Washington, D.C.

Pichon, V., and Hennion, M.-C. (1994), Determination of pesticides in environmental water by automated on-line trace-enrichment and liquid chromatography, *J. Chromatogr. A 665*: 269–281.

Picó, Y., Albelda, C., Moltó, J. C., Font, G., and Mañes, J. (1990), Aldicarb residues in citrus soil, leaves and fruits, *Food Additives and Contaminants 7*(1): S29–S34.

Pleasance, S., Anacleto, J. F., Bailey, M. R., and North, D. H. (1992), An evaluation of atmospheric pressure ionization techniques for the analysis of N-methyl carbamate pesticides by liquid chromatography mass spectrometry, *J. Am. Soc. Mass Spectrom. 3*: 378–397.

Raju, J., and Gupta, V. K. (1991), Thin-layer chromatographic method for the detection of carbaryl using diazotized *p*-nitro-aniline and diazotized *p*-amino-acetophenone, *Fresenius' J. Anal. Chem. 339*: 897.

Rathore, H. S., and Begum, T. (1993), Thin-layer chromatographic behavior of carbamate pesticides and related compounds, *J. Chromatogr. 643*: 321–329.

Rathore, H. S., and Sharma, R. (1992), Sequential thin-layer chromatography of carbaryl and related compounds, *J. Liquid Chromatogr. 15*(10): 1703–1717.

Reuke, S., and Hauck, H. E. (1995), Thin-layer chromatography as a pilot technique for HPLC demonstrated with pesticide samples, *Fresenius' J. Anal. Chem. 351*: 739–744.

Roinestad, K. S., Louis, J. B., and Rosen, J. D. (1993), Determination of pesticides in indoor air and dust, *J. AOAC Intern. 76*(5): 1121–1126.

Rouberty, F., and Fournier, J. (1996), Optimization of HPLC separation of carbamate insecticides (carbofuran, hydroxycarbofuran and aldicarb) by experimental design methodology, *J. Liq. Chrom. and Rel. Technol. 19*(1): 37–55.

Sánchez, F. G., Díaz, A. N., and Pareja, A. G. (1994), Ion-pair reversed-phase liquid chromatography with fluorimetric detection of pesticides, *J. Chromatogr. A 676*: 347–354.

Sánchez, F. G., Díaz, A. N., and Pareja, A. G. (1995), Normal-phase liquid chromatography on amino-bonded-phase column of fluorescence detected pesticides, *J. Liquid Chromatogr. 18*(13): 2543–2558.

Seiber, J. N., Glotfelty, D. E., Lucas, A. D., McChesney, M. M., Sagebiel, J. C., and

Wehner, T. A. (1990), A multiresidue method by high performance liquid chromatography-based fractionation and gas chromatographic determination of trace levels of pesticides in air and water, *Arch. Environ. Contam. Toxicol. 19*: 583–592.

Sennert, S., Volmer, D., Levsen, K., and Wünsch, G. (1995), Multiresidue analysis of polar pesticides in surface and drinking water by on-line enrichment and thermospray LC-MS, *Fresenius' J. Anal. Chem. 351*: 642–649.

Sherma, J. (1986), Thin-layer and paper chromatography, *Anal. Chem. 58*: 69R–81R.

Sherma, J. (1987), Pesticides, *Anal. Chem. 59*: 18R–31R.

Siebers, J., Koehle, H., and Nolting, H. G. (1992), Carbosulfan, carbofuran, in: *Manual of Pesticide Residue Analysis, Vol. 2* (H. P. Thier and J. Kirchhoff, eds.), New York: VCH.

Simon, V. A., Pearson, K. S., and Taylor, A. (1993), Determination of *N*-methylcarbamates and *N*-methylcarbamoyloximes in water by high-performance liquid chromatography with the use of fluorescence detection and a single *o*-phthalaldehyde post-column reaction, *J. Chromatogr. 643*: 317–320.

Slobodnik, J., Groenewegen, M. G. M., Brouwer, E. R., Lingeman, H., and Brinkman, U. A. Th. (1993), Fully automated multi-residue method for trace level monitoring of polar pesticides by liquid chromatography, *J. Chromatogr. 642*: 359–370.

Slobodník, J., Hogenboom, A. C., Vreuls, J. J., Rontree, J. A., van Baar, B. L. M., Niessen, W. M. A., and Brinkman, U. A. Th. (1996), Trace-level determination of pesticide residues using on-line solid-phase extraction-column liquid chromatography with atmospheric pressure ionization mass spectrometric and tandem mass spectrometric detection, *J. Chromatogr. A 741*: 59–74.

Sparacino, C. M., and Hines, J. W. (1976), High-performance liquid chromatography of carbamate pesticides, *J. Chromatogr. Sci. 14*: 549–556.

Specht, W., Pelz, S., and Gilsbach, W. (1995), Gas-chromatographic determination of pesticide residues after clean-up by gel-permeation chromatography and mini-silica gel-column chromatography, *Fresenius' J. Anal. Chem. 353*: 183–190.

Stafford, S. C., and Lin, W. (1992), Determination of oxamyl and methomyl by high-performance liquid chromatography using a single-stage postcolumn derivatization reaction and fluorescence detection, *J. Agric. Food Chem. 40*: 1026–1029.

Stan, H.-J., and Klaffenbach, P. (1991), Determination of carbamate pesticides and some urea pesticides after derivatization with acetic anhydride by means of GC-MSD, *Fresenius' J. Anal. Chem. 339*: 151–157.

Strait, J. R., Thornwal, G. C., and Ehrich, M. (1991), Sensitive high-performance liquid chromatographic analysis for toxicological studies with carbaryl, *J. Agric. Food Chem. 39*: 710–713.

Suzuki, O., Hattori, H., Liu, J., Seno, H., and Kumazawa, T. (1990), Positive- and negative-ion mass spectrometry and rapid clean-up of some carbamate pesticides, *Forensic Sci. Intern. 46*: 169–180.

Tena, M. T., Luque de Castro, M. D., and Valcárcel, M. (1992a), Sensitivity enhancement by using an HPLC flow-through sensor for determination of pesticide mixtures, *J. Liquid Chromatogr. 15*(13): 2373–2383.

Tena, M. T., Linares, P., Luque de Castro, M. D., and Valcárcel, M. (1992b), Total and individual determination of carbamate pesticides by the use of an integrated flow-injection/HPLC system, *Chromatographia 33*(9/10): 449–453.

Tjan, G. H., and Jansen, J. T. A. (1979), Gas-liquid chromatographic determination of thiabendazole and methyl 2-benzimidazole carbamate in fruits and crops, *J. Assoc. Off. Anal. Chem. 62*(4): 769–773.

Tuinstra, L. G. M. Th., Van de Spreng, P., and Gaikhorst, P. (1995), Ion trap detection for development of a multiresidue/multimatrix method for pesticide residues in agricultural products, *Intern. J. Environ. Anal. Chem. 58*: 81–91.

Vincze, A., and Yinon, J. (1996), Analysis of thermally labile pesticides by gas chromatography/mass spectrometry and gas chromatography/tandem mass spectrometry with a temperature-programmed injector, *Rapid Commun. in Mass Spectrom. 10*: 1638–1644.

Volmer, D., and Levsen, K. (1994), Mass spectrometric analysis of nitrogen- and phosphorus-containing pesticides by liquid chromatography-mass spectrometry, *J. Am. Soc. Mass Spectrom. 5*: 655–675.

Volmer, D., Preiss, A., Levsen, K., and Wünsch, G. (1993), Temperature and salt concentration effects on the ion abundances in thermospray mass spectra, *J. Chromatogr. 647*: 235–259.

Volmer, D., Levsen, K., and Wünsch, G. (1994), Thermospray liquid chromatographic-mass spectrometric multi-residue determination of 128 polar pesticides in aqueous environmental samples, *J. Chromatogr. A. 660*: 231–248.

Volmer, D., Levsen, K., Honing, M., Barceló, D., Abian, J., Gelpí, E., van Baar, B. L. M., and Brinkman, U. A. Th. (1995), Comparative study of different thermospray interfaces with carbamate pesticides: influence of the ion source geometry, *J. Am. Soc. Mass Spectrom. 6*: 656–667.

Volmer, D. A., Vollmer, D. L., and Wilkes, J. G. (1996), Multiresidue analysis of pesticides by electrospray LC-MS and LC-MS-MS, *LC-GC 14*(3): 216–224.

Vreeken, R. J., Van Dongen, W. D., Ghijsen, R. T., and Brinkman, U. A. Th. (1994), The use of mobile phase additives in the determination of 55 (polar) pesticides by column liquid chromatography-thermospray-mass spectrometry, *Intern. J. Environ. Anal. Chem. 54*: 119–145.

Wang, Q.-S., Gao, R.-Y., and Yan, B.-W. (1993), Optimization of the separation of technical aldicarb by semi-preparative reversed-phase high-performance liquid chromatography, *J. Chromatogr. 628*: 127–132.

Wigfield, Y. Y., Grant, R., and Snider, N. (1993), Gas chromatographic and mass spectrometric investigation of seven carbamate insecticides and one metabolite, *J. Chromatogr. A. 657*: 219–222.

World Health Organization (1986), *Carbamate Pesticides: A General Introduction*, Environmental Health Criteria, No. 64, Geneva: World Health Organization, International Programme on Chemical Safety.

Wright, L. H. (1982), Combined liquid chromatographic/mass spectrometric analysis of carbamate pesticides, *J. Chromatogr. Science 20*: 1–6.

Yang, S. S., and Smetena, I. (1994), Determination of aldicarb, aldicarb sulfoxide and aldicarb sulfone in tobacco using high-performance liquid chromatography with dual post-column reaction and fluorescence detection, *J. Chromatogr. A 664*: 289–294.

Zoun, P. E. F., and Spierenburg, Th. J. (1989), Determination of cholinesterase-inhibiting pesticides and some of their metabolites in cases of animal poisoning using thin-layer chromatography, *J. Chromatogr. 462*: 448–453.

10

Toxic Carbonyl Compounds

HELEN C. YEO, HAROLD J. HELBOCK,
AND BRUCE N. AMES
University of California, Berkeley, California

I. INTRODUCTION

Carbonyl compounds such as aldehydes and ketones are reactive species in which the chemical and biological properties are governed by the aldehydic moiety. They occur in food systems, naturally as flavorants or synthetically as food additives or contaminants. Most of the low molecular weight carbonyls are volatile, and some are emitted into the environment as pollutants by industrial activities. Other sources of carbonyls in the environment include cigarette smoke [1] and automobile exhaust [2].

Toxic aldehydes are generally categorized into three classes: saturated (simple) aldehydes, α,β-unsaturated aldehydes, and substituted aldehydes (see structures in Fig. 1). The vapors from simple aldehydes such as formaldehyde and acetaldehyde have been shown to cause carcinomas in rats [3,4]. α,β-Unsaturated aldehydes such as acrolein and crotonaldehyde are known to cross-link with DNA [5]. 4-Hydroxynonenal is one of the modified aldehydes that is formed during lipid peroxidation. *In vitro* studies show that it binds irreversibly to protein and has

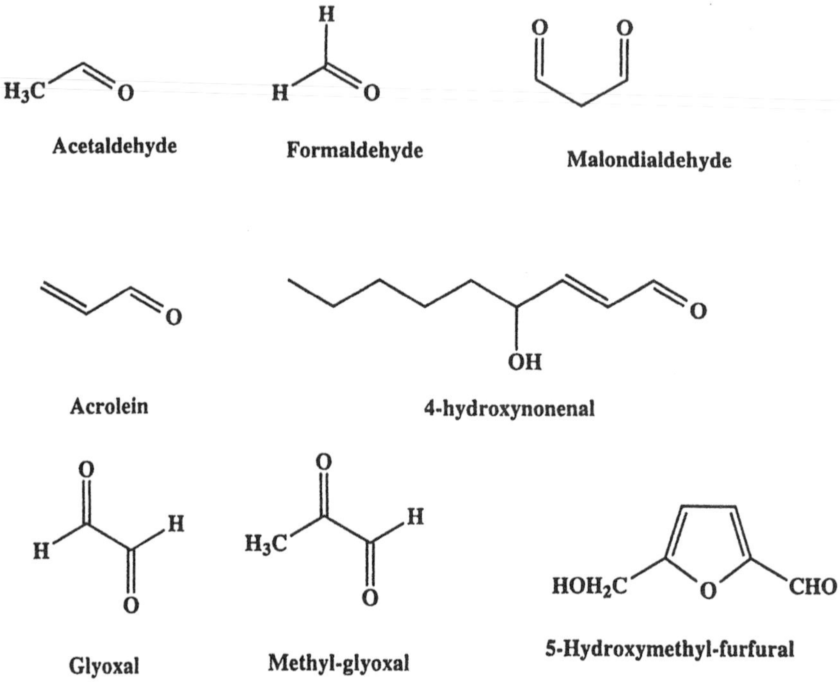

FIGURE 1   Chemical structures of some toxic carbonyls.

cytotoxic and genotoxic properties. Ketones that are known to be toxic are usually diketones such as methylglyoxal, a known mutagen found in coffee [6,7].

Malondialdehyde (MDA) is a well-known compound in rancid food and is a product of lipid oxidation and prostaglandin biosynthesis. A commonly used method employs thiobarbituric acid (TBA), which has been broadly criticized for the ambiguity in the results obtained. In-depth reviews of the TBA assay have been published elsewhere [8,9,10] and therefore will not be covered here. In this paper, recent work on the MDA assay using PFPH will be discussed in the GC section.

Quantitative analysis of carbonyls, both in the environment and in foods, is a challenge because these compounds are water soluble and form polymers in aqueous solution. These properties make extraction into the organic solvents difficult. Due to the high volatility of these species, direct analyses have sometimes led to irreproducible and unreliable results. Various types of derivatizing agents have been employed in an attempt to form stable adducts with carbonyls for chromatographic analyses. This chapter will discuss derivatization, chromatographic separation, and measurement of carbonyl compounds in food and environmental systems.

II. GAS CHROMATOGRAPHIC TECHNIQUES

A. Hydrazine

1. Pentafluorophenyl Hydrazine (PFPH)

A gas chromatographic method for MDA was developed by Tomita and coworkers by converting the highly volatile and aqueous soluble MDA using pentafluorophenyl hydrazine (PFPH) into its corresponding pyrazole [81]. Hydrazines are one of the most popular reagents used for the quantitation of carbonyls due to their high reactivity toward carbonyls. Halogen containing hydrazines, such as PFPH, allow for enhanced sensitivity on detectors such as the electron capture detector (ECD) and gas chromatography–mass spectrometry with negative chemical ionization (GC-MS-NCI) capabilities. Tomita and colleagues optimized the reaction conditions for MDA and found that the derivatization is complete in 30 min at room temperature using pH conditions between pH 4 and 5. Figure 2 shows the reaction between MDA and PFPH to form the highly stable *N*-pentafluorophenyl pyrazole. This method was applied to urine samples resulting in recoveries of 58% and a detection limit of 36 nM. The calibration curve was linear up to 2 μM. Low recoveries suggest that the MDA may be conjugated to proteins and other reactive biomolecules with resulting potential toxicity in biological systems.

The GC-ECD technique using PFPH was later tested on a GC-MS using the NCI mode [12]. The latter technique allows for increased sensitivity and selectivity when measuring MDA in complex systems. This is because the background noise on the MS-NCI is lower than that on the ECD. Further, the MS detector provides mass spectral information specific to the compound of interest. This technique was applied to biological systems including plasma, tissue homogenate, and brain cells [12,13]. Briefly, sulfuric acid and sodium tungstate were added to the sample (blood plasma or tissue homogenate—10%) to hydrolyze the MDA and precipitate the protein. The PFPH reagent was then added and allowed to react for 30 min at room temperature, after which the MDA-PFPH derivative is extracted into the hexane or isooctane phase and analyzed by GC-MS. The sample preparation is simple and straightforward and thus lends itself to studies of environmental and clinical samples where large sample loads are typical. The linear range was between 5 pmol/mL and 10 nmol/mL with the limit of detection in biological samples of 5 nM on-column. This method was developed on a DBWAX capillary column, but other commonly used columns such as the DB-1 and DB-5 work just as well.

In addition to the enhanced sensitivity and selectivity of this technique, the other advantage is the use of a stable isotope internal standard for accurate

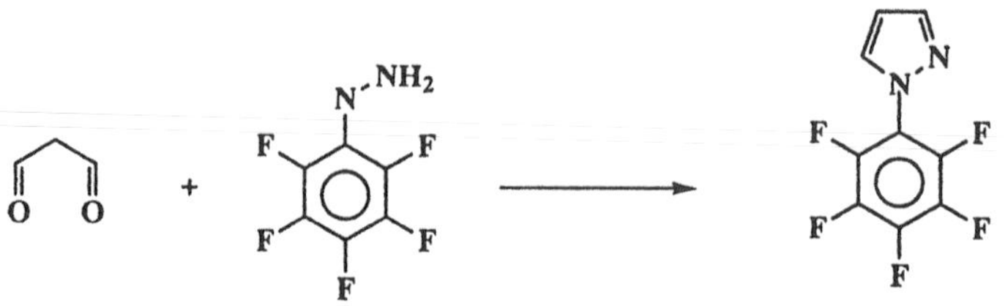

FIGURE 2 Derivatization of malondialdehyde by PFPH.

quantitation and to correct for derivatization and extraction efficiencies. The disadvantage of this method is the high cost of instrumentation and maintenance of the MS. However, bench top models of the GC-MS with NCI capabilities are now commercially available at considerably lower cost.

The PFPH assay has also been expanded to measure aliphatic aldehydes. Unlike MDA, which forms a stable pyrazole adduct with PFPH, the aliphatic aldehydes form the unstable syn and anti isomers of the corresponding hydrazones. These isomers split the signal of the derivative into two peaks on the gas chromatogram, causing a decrease in sensitivity. Efforts were made to stabilize the hydrazone and increase the signal-to-noise ratio by reducing the Schiff base of the hydrazone with sodium cyanoborohydride (Yeo, unpublished observation). By using deuterated acetaldehyde as an internal standard, it was found that the aldehyde derivatives were efficiently reduced to the more stable form. Figure 3 shows the mass spectrum of acetaldehyde-PFPH before and after reduction. The molecular ion at $m/z = 224$ in the reduced form is the base peak, while the molecular ion in the unreduced form is only 20% of the base peak. Cyanoborohydride treatment resulted in a fivefold increase in the signal when monitoring at $m/z = 224$. Interferences from contaminating formaldehyde and acetaldehyde present in the PFPH solution also reduces the sensitivity of the analysis. Efforts were made to remove the hydrazones in the PFPH solutions by extracting it with carbon tetrachloride [14,15]. Significant reduction in the background signal was observed with this procedure.

2. *N*-Methyl hydrazine (NMH)

In 1988, Umano and coworkers [16] developed an assay using NMH to detect α,β-unsaturated aldehyde (such as acrolein) and β-dicarbonyl (such as malondialdehyde) using a gas chromatograph equipped with a nitrogen-phosphorus detector (GC-NPD). These carbonyls react with the NMH to form *N*-methyl pyrazole, which is highly stable yet volatile [see Fig. 4]. The derivatization is complete at room temperature in 30 min with a yield of 86% for MDA. Chromatographic separation was achieved using a 30 meter polar capillary column such as

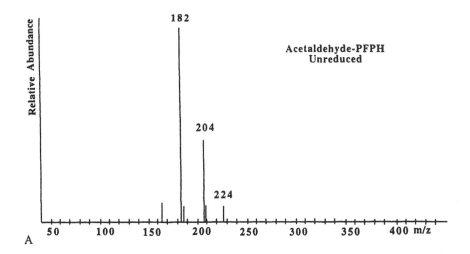

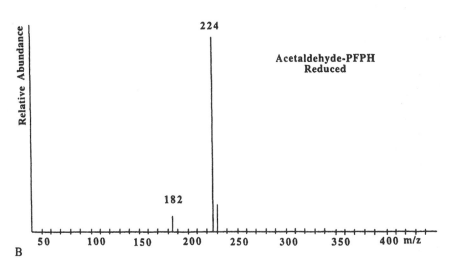

FIGURE 3 Mass spectrum of acetaldehydes-PFPH adduct in the (A) before and (B) after reduction by sodium cyanoborohydride.

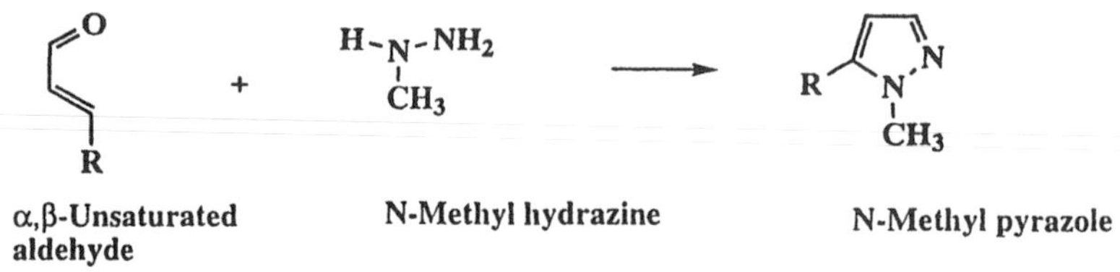

FIGURE 4 Derivatization of α,β-unsaturated aldehydes by NMH.

the DBWAX. In this study, various fats (beef fats and corn oil) were photoirradi-ated, and aldehyde levels were determined by extraction of the corresponding pyrazole into the hexane layer.

Similar studies were conducted by Yasuhara et al. [17,18] to measure acrolein in various oils and lard. In these studies, aldehydes in the headspace of the heated fats were swept with air into two impingers containing the derivatizing agent, NMH. The derivatives were further extracted into the dichloromethane layer for GC analysis using a nitrogen phosphorus detector and a bonded phase DBWAX column. The detection limit using this method for acrolein was 5.9 pg, and recoveries ranged from 98 to 100%. The levels of acrolein in 26 commercial brands of cigarettes were analyzed using the same derivatizing agent [19]. Ciga-rette smoke was first introduced into an evacuated separating funnel containing NMH. Acrolein was then extracted into the dichloromethane phase using a continuous liquid-liquid extractor and analyzed as described above. Recoveries of acrolein by this method were above 99%. The levels of acrolein found in the various cigarettes ranged between 124 and 337 μg/cigarette. The high recoveries are credited to the use of a continuous liquid-liquid extractor. However, to accom-modate a large sample load in routine analysis, a simple liquid-liquid extractor procedure may be adequate due to the high sensitivity of this method.

3. 2,4-Dinitrophenyl Hydrazine (DNPH)

One of the most widely used derivatizing agent for aldehydes and ketones is 2,4-dinitrophenyl hydrazine. Conversion to the derivative is quantitative, and analysis of the derivative may be performed on the HPLC or GC. However, due to the relatively low volatility of the DNPH derivative compared to hydroxylamine derivatives, HPLC is the preferred technique. The GC technique is sometimes employed to enhance the sensitivity of the analysis by taking advantage of the nitro group for an electron capture detection.

Buckley et al. [20,21] developed a method to detect trace levels of formalde-hyde in milk from animals consuming formaldehyde-treated feedstuff, using a GC equipped with an electron capture detector. Aliquots of milk were homogenized with 2,4-DNPH and allowed to react for 1 hour at room temperature (Fig. 5).

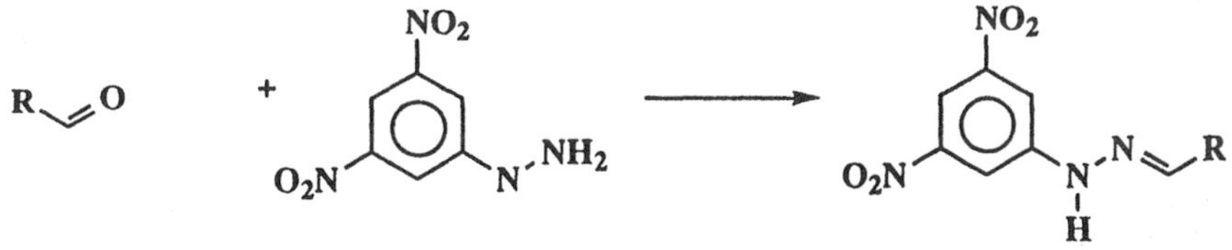

Aldehyde | 2,4-Dinitrophenyl Hydrazine | 2,4-Dinitrophenyl Hydrazone

FIGURE 5 Derivatization of aldehydes by 2,4-DNPH.

The resulting derivative was then extracted with "formaldehyde-free" toluene. Formaldehyde-free solvents are important as exemplified by dichloromethane, which has been shown to contain between 2.3–14 ppm of formaldehyde [22]. The recoveries of formaldehyde from 12 milk samples averaged 96%, and the detection limit was 26 μg/kg.

The 2,4-DNPH derivative has also been applied to measure formaldehyde in waste water [23]. Gas chromatographic methods (using ECD and FID) were compared to spectrophotometric methods (using chromotropic acid (CTA) and 3-methyl-2-benzothiazolone hydrazone (MBTH)). The detection limits were 0.41, 0.06, 0.72, and 0.06 mg/L using the CTA, MBTH, GC-ECD, and GC-FID methods, respectively. The recoveries were found to range between 87 and 101%.

Use of 2,4-DNPH to determine formaldehyde concentrations in air was recently reported by Dalene et al. [24]. Air was drawn into 2,4-DNPH impregnated glass at a flow rate of 1 lit/min. The derivative was then extracted with acetonitrile and passed through a cation-exchange column to remove the excess reagent. The eluate is further evaporated to dryness and resuspended in toluene for GC-TSD analysis. The limit of detection was estimated to be 10 $\mu g/m^3$. The overall recovery for formaldehyde (92%) was good considering the number of steps involved. The linear range was between 1 and 6 μg.

B. Hydroxylamine

1. *O*-(2,3,4,5,6-Pentafluorophenyl)methylhydroxylamine Hydrochloride (PFBHA)

Since its discovery in the 1970s [25], PFBHA has become one of the most versatile derivatizing agents for aldehyde analyses. This reagent has been reported under different names and abbreviations: *O*-(2,3,4,5,6-pentafluorophenyl)methylhydroxylamine hydrochloride, pentafluorobenzyloxylamine, PFBHA, PFBOA, PFBHOX [26]. In this paper, it is referred to as PFBHA. The reactions between

PFBHA and carbonyls occur over a wide pH range and are complete within an hour at room temperature. The derivatives are more volatile than those formed with DNPH. The formation of syn and anti isomers is less prominent, thereby allowing more precise quantitation at low aldehyde levels. The presence of fluoro groups gives rise to greater sensitivity on the ECD and GC-MS in the NCI mode, compared to the nonfluorinated reagent. It was found that low pg quantities of aldehydes were easily detected in drinking water, using the GC-MS methods [27].

Vidal et al. [28] employed PFBHA to determine carbonyl levels in cognac at ppb levels. The carbonyls were extracted into the pentane phase and allowed to react with PFBHA for 30 min at 65°C (Fig. 6). The resulting Schiff base on the oxime was selectively reduced with pyridine:borane reagent to convert the anti and syn isomers into one major compound and one peak on the GC. The organic layer was concentrated and an aliquot was analyzed by GC-MS, using acetophenone as an internal standard. Aliphatic aldehydes, methylketones, and cyclic aldehydes were detected in this study with emphasis on the quantitation of methylketones. The average recoveries of methylketones from cognac was 113% with detection at the ppb levels and a linear range between 10 and 400 µg/L.

The advantage of the above technique is the selective reduction of the oxime to form a more stable compound, thereby allowing for better reproducibility. This method could potentially be used to detect trace levels of carbonyls in food and environmental systems by using a mass detector operating in the NCI mode. The disadvantages are the high levels of formaldehyde in the blanks and the lack of a commercially available oxime.

PFBHA has been applied to various environmental systems including form-

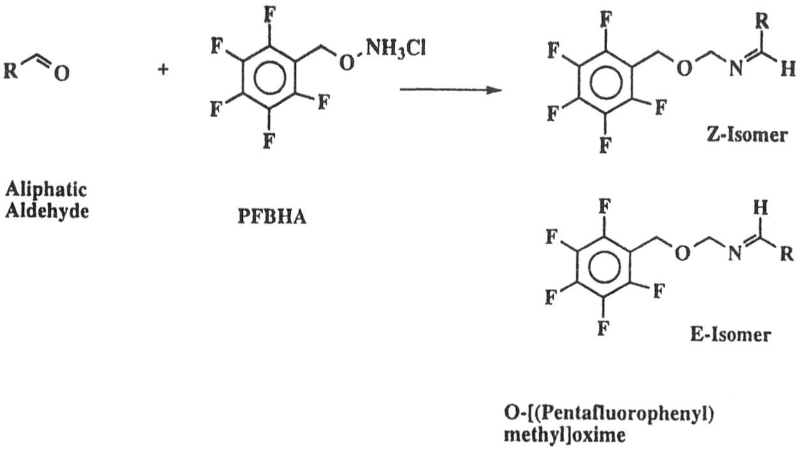

O-[(Pentafluorophenyl)
methyl]oxime

FIGURE 6   Derivatization of aldehydes by PFBHA.

aldehyde in air [29]. The air sample of interest is adsorbed in distilled water and reacted with PFBHA. The reaction mixture was allowed to react for 40 min and then extracted into hexane for analysis on the GC-ECD. Recovery of formaldehyde was 94% and the linear range was between 10 and 80 ng. The levels of formaldehyde detected in 5 liters of air was in the low ppb range.

PFBHA has also been successfully used to measure aldehydes (including MDA and 4HNE) in urine, plasma, and tissue homogenate. GC-MS has proven to be a powerful tool with detection limits of 50 to 100 fmol on column [30]. With the exception of 4-HNE, recoveries of aldehydes from these biological samples were over 85%. The recoveries of 4HNE were below 80% due in part to binding. 4HNE is tightly bound to proteins and other molecules with nucleophilic groups. As a result, recoveries tend to be lower than those obtained with simple carbonyls.

2. *O*-Benzylhydroxylamine Hydrochloride (BOA)

BOA has also been extensively used to measure aldehydes in environmental samples such as cigarette smoke [31,1]. Cigarette smoke is drawn into a column and trapped on silica gel, after which it is eluted with water. The aldehyde is converted to the oxime by the addition of BOA to the aqueous sample, and the oximes are extracted with diethyl ether for GC-NPD analysis (Fig. 7). The reaction is complete after 1 h at 65–70°C. Recoveries and detection limits by this method were not reported. Some of the drawbacks include the high background interference from acetaldehyde contamination, poor reproducibility between runs, formation of syn and anti isomers, and susceptibility to decomposition of the derivative in the GC column. The latter problem was solved by using a short (12 meter) column to suppress the decomposition. Similar studies measuring volatile aldehydes in air and unleaded gasoline were reported by Levine et al. [32]. Using a 2.4 m packed glass column, decomposition of the derivative was not observed.

Magin [31] performed gas chromatographic characterization of benzyloximes and *p*-nitrobenzyloximes of low molecular weight carbonyls and found some interesting trends using the 12 m FFAP capillary column:

> Iso-branching of the alkyl chain of aldehydes and ketones decreases the retention time.

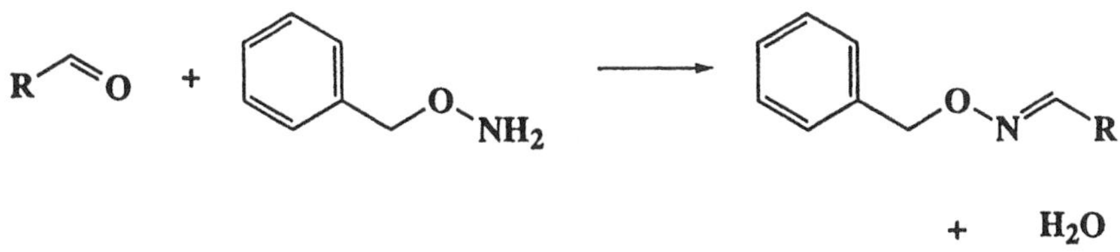

FIGURE 7 Derivatization of aldehydes by BOA.

Unsaturation along the alkyl chain of aldehydes and methyl ketones prolongs retention time.

Shifting the keto group inward toward the center of the alkyl chain also
prolongs retention time.

Aromaticity in the molecule increases retention time.

Ring closure of the alkyl side chain of the ketones to form cycloketones
prolongs retention time.

The above information is valuable when evaluating unknowns in complex mixtures.

3. Bromination of Oxime

Thus far we have only discussed the utility of oximes with saturated aldehydes.
Several studies have improved upon the oxime technique for the determination of
unsaturated aldehydes such as acrolein by brominating the carbon-carbon double
bond on the aldehyde [33,34]. The reaction mechanism is shown in Figure 8.

Nishikawa et al. [33] applied this technique to measure acrolein and crotonaldehyde in the exhaust gas of automobiles. The gaseous sample was bubbled
through impingers containing ethanol. BOA or *o*-methoxylamine hydrochloride
(MOA) reagent was added to the sample and reacted for 20 min at room temperature. After the formation of the oxime, bromination was initiated by adding
potassium bromate and potassium bromide, which were allowed to react for 15
min at room temperature. Excess bromine was quenched with thiosulfate solution,
and the final derivative was extracted with diethyl ether for GC-ECD analysis.

The bromination step is quantitative, with greater than 96% overall recovery
for acrolein and crotonaldehyde using BOA. With MOA, the overall recoveries
were lower—92% and 79% for acrolein and crotonaldehyde, respectively. The
linear range was between 0 and 6 μg for acrolein and between 0 and 5 μg for
crotonaldehyde. The limits of detection were 0.2 μg and 0.15 μg for acrolein and
crotonaldehyde, respectively. Bromination of MOA derivatives was used to detect

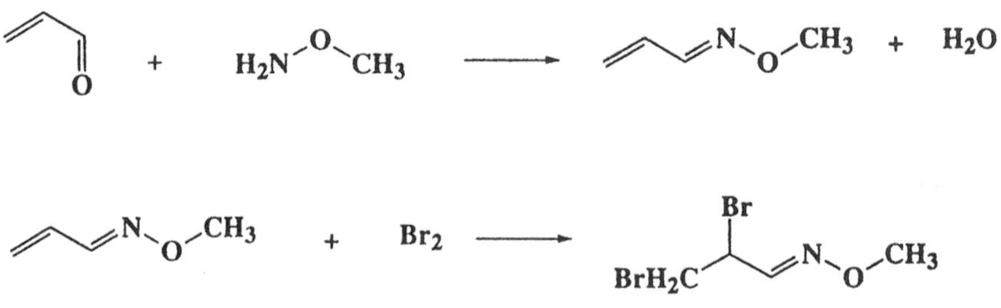

FIGURE 8 Derivatization of α,β-unsaturated aldehydes by MOA, followed by
bromination.

acrolein in rain water with recoveries above 90% and a detection limit of 0.4 ng/ mL sample [35].

4. Hydroxylamine/*N*-Methyl-*N*-*tert*-butyldimethylsilyl trifluoroacetamide (MTBSTFA)

Hydroxylamine has been used extensively for aldehyde analysis, especially for bound aldehydes, because it competes with the nucleophilic sites on biomolecules. One such bound aldehyde is 4HNE. There is growing evidence that 4HNE is implicated in biological damage associated with oxidative stress. It is therefore not surprising that considerable effort has been expended to develop methods to measure 4HNE in biological samples [36].

Kinter in 1996 [37] reported a GC-MS technique to detect 4HNE in blood and tissues using hydroxylamine, with sensitivity in the low pmol range. Oximated reagent of hydroxylamine hydrochloride in acetate buffer was added to the sample of interest, containing hydroxynonanal as the internal standard. The reaction mixture was allowed to react for 1 hour at 70°C. The 4HNE-oxime was recovered in methanol and the solvent evaporated to dryness for further derivatization with DMF and MTBSTFA. The method involves tedious sample preparation and does not address the complications created by the propensity of 4HNE to form Michael addition products with cysteine, histidine, and lysine residues [38]. Similarly, most chromatographic methods for the analysis of 4HNE measure only free and Schiff base forms of the aldehyde [37]. Given this limitation, these methods are likely to be of little value in assessing total, and therefore meaningful, 4HNE activity. Bruenner et al. [38] have attempted to address the adduct issue by using electrospray mass spectrometry to study protein aldehyde adducts of 4HNE. These authors used this technique successfully in an elegant *in vitro* study of hemoglobin and lactoglobulin, confirming the importance of 4HNE adducts. Alternatively, Toyokuni et al. [39] approached the problem by applying polyclonal antibody that recognizes 4HNE, including the adduct forms. With the exception of these two reports, plausibly valid methods for measurement of 4HNE in biological systems are not available.

C. Cysteamine

The use of cysteamine as a derivatization reagent for carbonyls was first reported by Hayashi et al. in 1986 [22]. Cysteamine reacts with monocarbonyls to form thiazolidines, stable 5-membered ring compounds (Fig. 9). The reaction is complete in 1 h at room temperature and pH 8. Analyses are usually performed on a GC equipped with a nitrogen-phosphorus detector or a flame photometric detector to take advantage of the presence of the nitrogen and sulfur groups, respectively. Chromatographic separations are usually done on a DBWAX column [18,40] or DB-17 [41] and its equivalent.

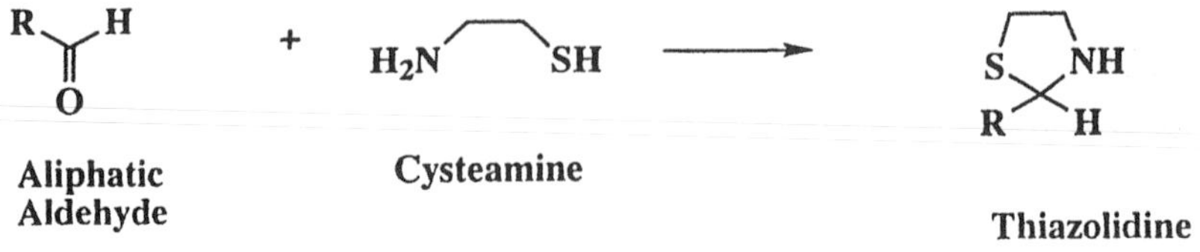

FIGURE 9 Derivatization of aldehydes by cysteamine.

The cysteamine method has been used to determine levels of formaldehyde in coffee [22], short chain monocarbonyls in fish flesh [42], low molecular weight aldehydes in automobile exhaust [2], methyl gyloxal in beverages [43], formaldehyde in air [44], and headspace carbonyls of heated pork fat [45]. To quantitate formaldehyde levels in air, samples were bubbled into impingers containing aqueous cysteamine solutions [44]. The thiazolidine formed was then extracted into the chloroform layer for GC analysis.

Chloroform was used as an extraction solvent because methylene chloride contains detectable levels of formaldehyde. Recovery using paraformaldehyde as a standard was greater than 90%. The lowest detection limit for thiazolidine was 17.2 pg, which is equivalent to 5.8 pg of formaldehyde. The advantage of this method is that it does not form syn and anti isomers. Instead, it forms only one peak for each aldehyde, thereby increasing the sensitivity of the method. In addition, the derivatization step is carried out under mild conditions, thus suppressing artifactual breakdown that can occur at more extreme pH and temperature settings. A minor disadvantage is the time-consuming continuous liquid-liquid extraction procedure.

The cysteamine technique was improved by Kataoka et al. in 1995 [41] to include detection of both saturated and unsaturated aldehydes. This modified technique was applied to various cooking oils, cheese, eggs, milk, chocolate, and potato chips with overall recoveries 82–111%. The limit of detection at a signal-to-noise ratio of 3 was from 4 to 100 pg injected with a linear range of 20 to 2500 ng. The primary improvements of this method are the replacement of the tedious continuous liquid-liquid extraction, thus allowing for larger sample load for routine analysis, and the reduced derivatization time. A drawback of this method, and probably of most aldehyde techniques described here, is the increased variability due to interference from background acetaldehyde that is present in the reagents and surrounding air.

D. Other Techniques

1. o-Phenylenediamine (PDA)

α-Dicarbonyls such as diacetyl are present in cigarette smoke and play a significant role in the aroma of smoke. These aldehydes are also found in milk [46] and coffee [47]. The measurement of diacetyl and acetoin are important in the area of fermentation technology, food chemistry, and toxicology. Moree-Testa and Saint-Jalm [48] measured dicarbonyls in cigarette smoke by trapping the smoke in an aqueous solution of PDA (Fig. 10). The derivatives were extracted with chloroform for GC analysis, and portions of the aqueous samples were used for HPLC analysis, which will be discussed later. In addition to diacetyl, this method was able to detect 2,3-pentanedione, glyoxal, pyruvic aldehyde, and 2-oxobutanal, with sensitivities in the nanogram range. This derivative has also been used successfully to measure glyoxal in fish liver oil using a GC-NPD [49].

2. 4,5-Dichloro-1,2-diaminobenzene (DCDB)

Otsuka and Ohmoro [50] developed a GC-ECD method by converting diacetyl into 6,7-dichloro-2,3-dimethyl quinoxaline using DCDB. The reaction mechanism is shown in Figure 11. Acetoin may also be measured simultaneously using this method by first oxidizing it into diacetyl with Fe^{3+} in 1 M perchloric acid. The derivatization step is carried out at 40°C for 90 min, after which the derivative is extracted with benzene for analysis. The detection limit by this method, for both diacetyl and acetoin, was 10 fmol/μg extract. This sensitive method was applied to normal human urine or blood (using only 100 μL samples), alcoholic beverages, rat liver, kidney, and brain. Recoveries of these aldehydes were greater than 90%. Previous methods for the detection of α-dicarbonyls include gravimetric and colorimetric techniques that are tedious and insensitive. The method of Otsuka

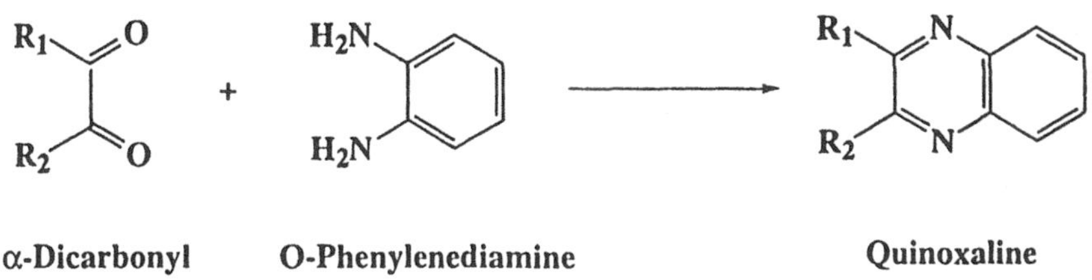

α-Dicarbonyl   O-Phenylenediamine   Quinoxaline

Carbonyl	R_1	R_2
Glyoxal	H	H
Methylglyoxal	CH_3	H
Diacetyl	CH_3	CH_3

FIGURE 10 Derivatization of α-dicarbonyls by the *o*-PDA.

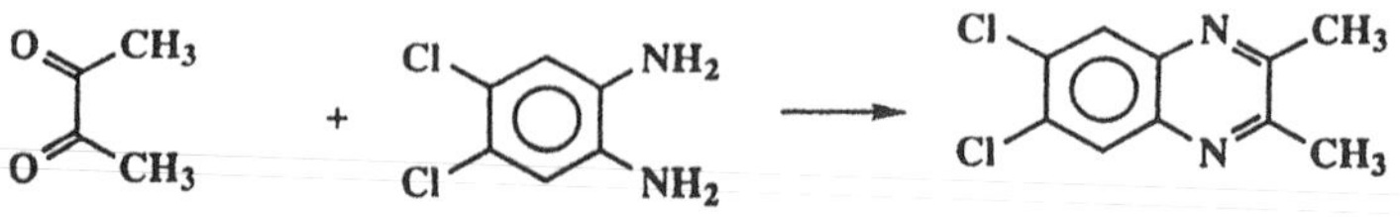

FIGURE 11   Derivatization of diacetyl by DCDB.

and Ohmori is a versatile technique that allows for simultaneous analysis of various dicarbonyls, is sensitive, and is simple to use. The only drawback of this method is the use of the strong carcinogenic benzene as an extraction solvent.

3. Direct Analysis

Volatility, high background signal, and poor reproducibility have made the direct measurement of low molecular weight carbonyl compounds a challenge. Direct analysis may be necessary, however, to understand the chemical profile of the sample, as in the case of flavor studies of complex mixtures. Careri et al. [51] and Barbieri et al. [52] studied the flavor characteristics of Parmesan cheese using two sampling methods: simultaneous distillation-extraction (SDE) and dynamic headspace techniques (DHT). In SDE, a modified Likens–Nickerson extractor was used to isolate the volatiles using liquid-liquid continuous extraction with dichloromethane. The extract was then analyzed by GC-MS. Using DHT, volatiles from the cheese samples were swept through a Tenax GC trap where they were adsorbed. The volatiles were then thermally desorbed and eluted onto a GC column equipped with a mass spectrometer. Hawthorne and Miller [53] also used the DHT to detect aldehydes in whisky. These authors also reported an on-column technique where samples, such as lemon oil, were injected into the GC column directly.

III. LIQUID CHROMATOGRAPHIC TECHNIQUES

A. 2,4-Dinitrophenyl Hydrazine

Methods for aldehyde analyses utilizing 2,4-DNPH are commonly performed on HPLC due to the relatively low volatility of the DNP hydrazone derivative. Typically, the derivatizing step is performed at acidic pH, and the derivative is extracted into an organic solvent such as methanol or actonitrile. This technique has been applied to numerous food and environmental systems including automotive exhaust [54,55], waste and drinking water [56,57], air [58,59], rainwater [60], and fried foods [61].

Acrolein has been demonstrated in the vapor phase of heated fats using 2,4-

DNPH. Lane and Smathers [61] pursued measurements of acrolein and other low molecular weight aldehydes in deep-fried foods such as codfish and doughnuts. A homogenate of the sample was codistilled with steam, and the vapors were condensed into a solution of 2,4-DNPH and isooctane. The reaction mixture was stirred for 30 min and allowed to separate, after which the derivative in the isooctane layer was extracted with acetonitrile for HPLC analysis. Recovery of acrolein from doughnut homogenates was 90%. Concentration of acrolein, formaldehyde, and acetaldehyde in the fish coatings and fried doughnuts ranged from 0.1 to 1.9 ppm; the detection limit was not reported.

Even though this technique is simple, the UV detector does not provide the specificity and sensitivity normally found on an ECD. Lack of specificity was clearly demonstrated in a study by Puputti and Lehtonen [62], where aldehydes and ketones are measured in whiskies. It was found that the acetonitrile/water solvent system was more efficient in separating the carbonyls than a methanol/water system. Using a C18 column, iso and *n*-butyraldehyde were not resolved under any of the conditions used. While increasing the column temperature to 60°C decreased the analysis time and improved the separation of acrolein, furfural, and propionaldehyde, other sources of error were noted. UV spectra obtained from authentic diacetyl were vastly different from that of the standard even though their retention times coincided. Table 1 shows the λ max of the various

Table 1 UV Maximum of Dinitrophenyl Hydrazones

Carbonyl	λ max	Reference
Formaldehyde	351	Coutrim et al., 1993 [55]
Acetaldehyde	361	Coutrim et al., 1993
Acrolein	371	Coutrim et al., 1993
Acetone	365	Coutrim et al., 1993
Propionaldehyde	362	Coutrim et al., 1993
Methyl ethyl ketone	366	Coutrim et al., 1993
Butyraldehyde	362	Coutrim et al., 1993
Benzaldehyde	381	Coutrim et al., 1993
5-Hydroxymethylfurfural	396	Puputti and Lehtonen, 1986 [62]
Diacetyl	360	Puputti and Lehtonen, 1986
Furfural	390	Puputti and Lehtonen, 1986
Isobutaraldehyde	364	Puputti and Lehtonen, 1986
Isovaleraldehyde	364	Puputti and Lehtonen, 1986
2-Methylbutyraldehyde	364	Puputti and Lehtonen, 1986
Caproaldehyde	362	Puputti and Lehtonen, 1986
2,4-Pentanedione	400	Puputti and Lehtonen, 1986
Caprylaldehyde	362	Puputti and Lehtonen, 1986
Glyoxal	436	Puputti and Lehtonen, 1986

dinitrophenyl hydrazones of carbonyl compounds obtained by several researchers. Clearly, identification on retention time alone is unreliable, and this method needs further validation.

To detect aldehydes in the air, several cartridges and impingers have been tested to determine their collection efficiencies. Grosjean and Fung [58] undertook a study comparing the trapping efficiencies of microimpingers and cartridges consisting of glass beads impregnated with DNPH reagent. They found that collection efficiency was influenced by two factors: humidity and the introduction of organic solvent into the aqueous DNPH reagent. For formaldehyde trapping, humid air was more efficient than dry air in the cartridges. Humidity, however, did not affect the trapping efficiencies in the impingers. Organic solvent is added to the DNPH solution for three reasons: to decrease the liquid surface tension and thereby improve the mixing of the air stream and the reagent solution, to improve solubility of the carbonyl, and to displace the reaction step in favor of hydrazone formation by extracting the hydrazone. Cyclohexane and isooctane were found to be suitable solvents.

A year later, Kuwata et al. [63,64] developed a more sensitive and less tedious method to measure low molecular weight volatiles in air. Using solid phase cartridges impregnated with DNPH, they were able to trap and detect sub-ppb levels of carbonyls in the atmosphere, and in industrial and incinerator emissions. The technique was further simplified by eluting the derivative with acetonitrile, thereby allowing the sample to be directly injected onto the HPLC. Good separation was achieved using a Develosil ODS-3 analytical column with an acetonitrile/water solvent system. The detection limits were 0.1 ng and 0.2 ng for C1–C3 aldehydes and C4 aldehydes, respectively. The linear range was between 0.2 and 10 ng for C1–C3 and 0.5 and 20 ng for C4. Recoveries of these aldehydes exceeded 94%. The efficiency of the solid phase cartridges in trapping aldehydes was not affected by humidity and was stable for over 6 weeks. This simpler and more sensitive method is suitable for routine analysis, especially field studies where controlling humidity may not be practical.

Aldehydes are one of the major output components of automobile exhaust and wood-fired furnaces. Lehotay and Halmo [65] and Lehotay and Hromulakova [57] further simplified the Kuwata technique by compressing the sample preparation steps. The sample collection and derivatization were performed directly on a column containing Chromosorb P impregnated with DNPH. The derivative was then eluted with methanol and the extract injected directly onto the HPLC system. Separation was accomplished on a C18 column using a methanol/water solvent system. Formaldehyde, acetaldehyde, propanal, and butanal were detected in the waste gas of a furnace with a detection limit below 1 mg/m^3 for a 10 liter sample. Geng and coworkers [54] also developed a simplified method of aldehyde analysis by bubbling automobile exhaust gas into a methanol or acetonitrile solution of 2,4-DNPH. An aliquot of the resulting mixture was then injected into the LC system. Recoveries of the same four most volatile aldehydes were above 85%. The detection limits range was from 12 to 57 ppb. Other studies monitoring aldehydes generated by automobile exhaust include those by Lipari and Swarin [66,67] and Olson and Swarin [68]. Tobacco smoke was studied by Manning and coworkers [69].

B. *O*-Phenylenediamine (PDA)

O-Phenylenediamine is another reagent that forms derivatives of α-carbonyls that are both GC and HPLC compatible. However, as Bernaski and coworkers [70] pointed out, the PDA derivative is more amenable to HPLC than to GC analysis due to the polar nature of the derivative. Moree-Testa and Saint Jalm [48] studied the α-dicarbonyl profile in cigarette smoke using both GC and HPLC techniques. They concluded that the HPLC method is more suitable than the GC method because of its simplicity, specificity, and short analysis time. Interferences from nicotine and the need for preconcentration makes the GC-NPD method unattractive.

Typically, for the analysis of α-dicarbonyl in food samples, an aqueous sample is reacted with PDA at pH 8 and 25°C for 4 h [70]. The chemical reaction is shown in Fig. 9. The derivative is then extracted with chloroform after acidifying the aqueous solution to pH 3 so that the excess reagent remains in the aqueous phase. The chloroform layer is evaporated to dryness and the residue is resuspended in methanol for HPLC analysis. Even though the analysis time is only 13 min, the sample preparation step is rather tedious for routine analysis.

A variation of the PDA reagent is 1,2-diamino-4,5-methylenedioxybenzene (DMB). Figure 12 shows the reaction of α-carbonyls with DMB. Yamaguchi and coworkers [71] developed a method to measure α-carbonyls (such as glyoxal, methylglyoxal, diacetyl, and 2,3 pentanedione) in fermented foods where they are contributors to food deterioration. A methanol solution of the food (yogurt or wine) was reacted with DMB in seal tubes at 60°C for 40 min in the dark. Upon cooling, an aliquot of the reaction mixture was injected directly onto the HPLC. Detection is achieved with a fluorescence detector operating at excitation and emission wavelengths of 350 nm and 390 nm, respectively. Recoveries of the α-carbonyls from food samples were in excess of 94%. The limits of detection for

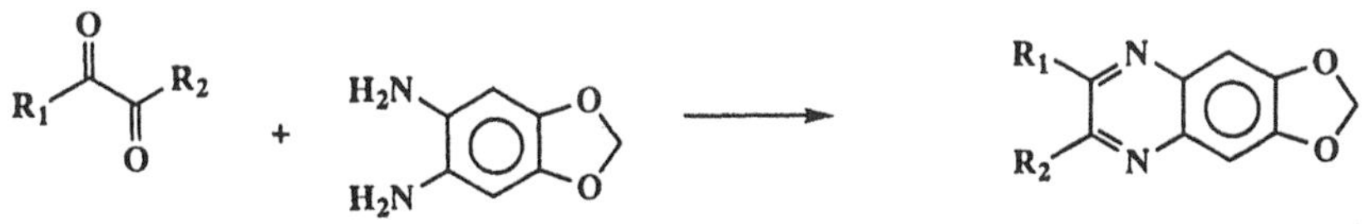

FIGURE 12 Derivatization of α-dicarbonyl by DMB.

glyoxal, methyl glyoxal, diacetyl, and 2,3-pentanedione were 260, 270, 400, and 330 pmol/g, respectively, which are adequately sensitive for most food systems. In this study, the concentrations of α-carbonyls in yogurt, beer, and wine were in the low nmol/g levels. The advantage of this method is the minimum interference from other common food constituents such as sugars, carboxylic acids, alcohols, amino acids, phenols, and even simple aldehydes and ketones because they do not produce fluorescent derivatives. α-Keto acids, on the other hand, generate fluorescent derivatives but with excitation and emission wavelengths of 367 and 446 nm, respectively.

C. 1,3-Cyclohexanedione (CHD)

A derivatizing agent such as 1,3-cyclohexanedione (also known as dihydroresorcinol) is employed for aldehyde analysis because of the fluorescent nature of the derivative, acridine. Figure 13 shows the conversion of a carbonyl into a fluorescent derivative. Holley and coworkers [72] developed an HPLC assay to detect *n*-alkanals and 4-hydroxynonenal in biological samples based on this property. Samples containing the aldehydes were incubated in the CHD reagent at 60°C for 1 h in sealed tubes. After the reaction is complete, methanol was added to precipitate the protein. The resulting supernatant is passed through a Sep Pak C18 cartridge and the aldehydes were then eluted with methanol for HPLC analysis.

Separation of the decahydroacridine derivative is achieved by using a 5 μM LiChrosphere RP18 column with an elution gradient of aqueous THF from 5% to 50% over 40 min. It was found that THF is crucial for baseline resolution of the aldehydes. Fluorescent detection is accomplished at excitation and emission wavelengths of 380 nm and 445 nm, respectively. The detection limit by this method was 50 fmol per injected aldehyde, which corresponded to 1 nM concentration in plasma. This is 20-fold more sensitive than methods using DNPH. The

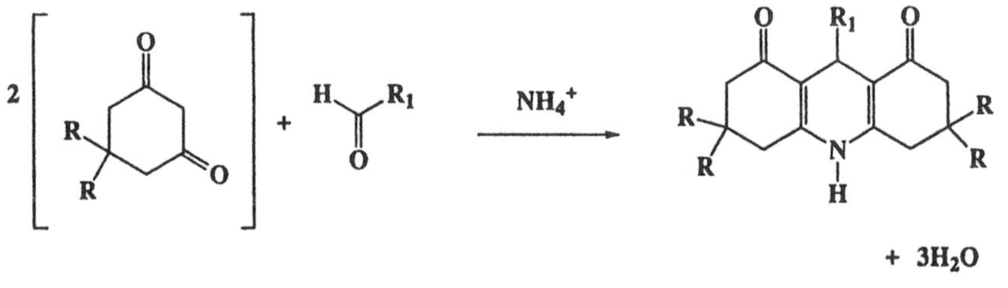

R = H   CHD
R = CH₃   Dimedone

FIGURE 13 Derivatization of aldehydes by CHD.

recoveries of the individual aldehydes are dependent on chain length. The recoveries of the less hydrophobic aldehydes such as acetaldehyde were found to be 95%, and that for the more hydrophobic aldehydes, such as octanal and 4-HNE, were only 8%. One of the major drawbacks of this sensitive assay is the background contamination from the CHD reagents and solvents. The CHD reagent contains high levels of aldehydes, especially acetaldehyde, propanal, and butanal. Recrystalization of the reagent from ethyl acetate was shown to remove only 65% of the aldehydes. Another technique used to remove these aldehydes involves heating the reagent solution at 60°C for 1 hour and then passing it through C18 cartridges [73]. High levels of aldehydes were also detected in commercial HPLC-grade methanol, water, and chloroform [72]. One way to remove most of the aldehydes in solvents involves pretreatment with DNPH.

The CHD reagent has several superior qualities to dimedone, an analogous reagent. CHD is more soluble in water, allowing the preparation of a more concentrated stock solution. Derivatization with CHD is accomplished at 60°C for 1 h, as opposed to 100°C with dimedone. The lower temperature would potentially reduce artifactual formation of aldehydes during sample workup. The CHD derivatives are better resolved on HPLC, with methanol as the eluting solvent, than dimedone derivatives. One drawback of the CHD reagent is the elaborate cleanup step, as described above, to remove background contamination, whereas in dimedone, an organic solvent extraction will suffice.

D. Anthrone

In complex mixtures such as foods and biological systems a selective derivatizing agent would greatly improve analytical capabilities. Nonspecific agents often lead to overlapping or poorly resolved peaks on the HPLC, resulting in ambiguity in data. Anthrone is a reagent that selectively derivatizes α,β-unsaturated aldehydes such as acrolein, crotonaldehye, and methacrolein to form fluorescent benzathrone derivatives. Figure 14 shows the condensation of unsaturated aldehydes with anthrone. Miller and Danielson [74] developed a method to measure unsaturated aldehydes in alcoholic beverages based on this property.

To the aqueous samples of interest, anthrone in concentrated sulfuric acid is added and the mixture is allowed to react for 10 min at room temperature. Acetonitrile is added to reduce the acidity and viscosity of the final reaction mixture, which is then injected directly onto the HPLC system. Care should be taken with the amount of acid present in the sample, as it can alter the wavelength maxima on the fluorimeter. Separation of the benzanthrones is accomplished with a Nova-PAK C18 cartridge and detection is at wavelengths 405 and 480 nm for excitation and emission, respectively. Linearity range by this method was reported to be between 0.02 and 14 ppm, and the detection limit was 5 ppb.

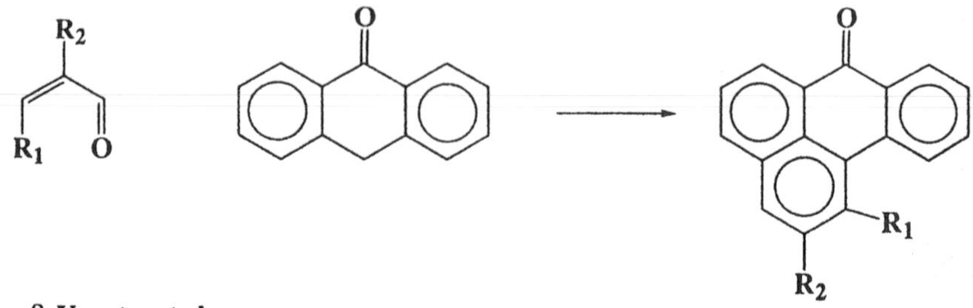

α,β-Unsaturated Aldehyde | Anthrone | Benzanthrone

Aldehyde	R_1	R_2
Acrolein	H	H
Crotonaldehyde	CH_3	H
Methacrolein	H	CH_3

Figure 14 Derivatization of α,β-unsaturated aldehyde by anthrone.

E. Fluoral-P

As discussed above, the use of reagents that form fluorogenic derivatives with aldehydes has become popular for the analysis of aldehydes on the HPLC. Another such reagent is 4-amino-3-penten-2-one, also known as Fluoral-P (see structure in Fig. 15). Formaldehyde, for example, condenses with Fluoral-P to produce 3,5-diacetyl-2,6-dihydrolutidine, which fluoresces [75]. In 1994, Tsuchiya et al. [76] reported a method with flow injection capabilities for the determination of formaldehyde using this reagent in beverages. Aldehydes are separated on a column prior to reacting with Fluoral-P, on-line, at pH 2. The resulting derivatives are detected at the excitation and emission wavelengths of 410 nm and 510 nm, respectively. Acidic pH condition is needed for the condensation of

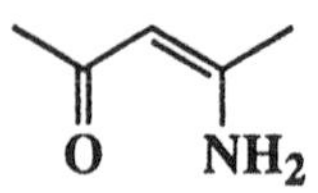

4-Amino-3-penten-2-one (Fluoral-P)

Figure 15 Chemical structure of Fluoral-P.

Fluoral-P and formaldehyde. However, to achieve enhanced sensitivity on the fluorescence detector, a neutral or weakly alkaline condition is essential. This is because Fluoral-P hydrolyzes to 2-4-pentanedione in an acidic environment. In order to prevent decomposition, the reagent was prepared in acetonitrile, and the acidic carrier solution was introduced by flow injection using a separate pump. pH 2 gave the most sensitive detection [76].

This method was successfully applied to various alcoholic beverages such as beer, wine, and sake by directly injecting the sample onto the column without pretreatment. The linear calibration range was between 0.5 and 100 nmol/mL. Recoveries of formaldehyde were above 97%, and the quantitative limit of this method was 25 pmol per injection.

F. Luminarin

For the detection of 5-hydroxymethyl-furfural (HMF) and β-dicarbonyls such as MDA and acetylacetone, luminarin has been employed in food systems [77,78]. Luminarins have quinolizinocoumarin moieties as fluorophores and carboxylic acid hydrazide as the reacting group (Fig. 16). An acidified sample is allowed to react with luminarin hydrazide at room temperature for 30 min. The pH of the solution is adjusted to 7.0 with sodium hydrogencarbonate. Dichloromethane is added to extract the derivative, and the pooled extract is evaporated to dryness. The residue is resuspended in acetonitrile for reverse phase HPLC analysis, or in dichloromethane for normal phase chromatography. It was found that less polar solvents resulted in more sensitive detection, which suggested that normal phase chromatography may be superior. However, reverse phase chromatography was used in this study because of its ease of use and popularity. The detection limit for the carbonyls was in the sub-pmol range. The linear range was between 0.1 and 10 nmol with recoveries of 98% or more.

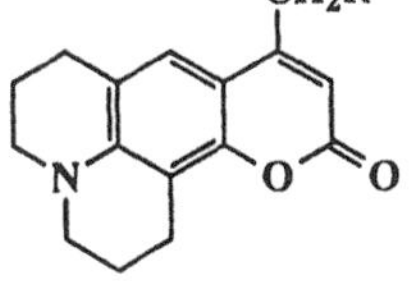

Luminarin	R
3	$CONHNH_2$
11	$(CH_2)_3CONHNH_2$
12	$O(CH_2CH_2O)_2CH_2CONHH_2$

FIGURE 16 Chemical structure of luminarin.

IV. CONCLUDING REMARKS

Over the last 15 years, improvements in detector sensitivity and enhanced resolving power of columns have resulted in less tedious techniques. One such improvement is the elimination of the preconcentration steps. With increased sensitivity, however, background contamination is often detectable and may interfere with sample analysis. Common sources of contamination in aldehyde analyses are solvents, the surrounding air, and derivatizing reagents. The latter trap carbonyls from the atmosphere to form their corresponding derivatives. Therefore proper storage of the reagents is required, and it may prove necessary to recrystalize or extract the reagents prior to use. Several extraction methods have been attempted and found suitable for specific solvents, reagents, and detectors. These methods include extraction with carbon tetrachloride [14], use of 5-methoxytryptamine [79], and passing the reagent through a C18 Sep Pak cartridge [80].

A proper internal standard is essential if carbonyls are to be measured quantitatively. The standard should have similar chemical (reaction efficiency) and physical (solubility, stability) properties as the analyte of interest. It must be stable over the course of the reaction and not interfere with the sample workup. Ideally, the internal standard should elute close to the analyte of interest. For MS analyses, a stable-isotope labeled (SIL) internal standard is crucial, as variation of the fragmentation pattern and therefore ion intensity of the analyte can make quantitation almost impossible. By using a SIL internal standard, any change in the intensity of the analyte is reflected in the intensity of the standard, resulting in accurate quantitation. An external standard technique is sometimes used if a proper internal standard is not available. This method is simple and valid provided that the injection volume is constant and the sample preparation is reproducible. This means that changes in extraction efficiencies due to variation in sample matrix, spills, and evaporation of volatile compounds (most carbonyls) can invalidate the data.

The most appropriate method for a given application depends on several factors. Large numbers of samples require a rugged and simple workup. This would generally eliminate methods where preconcentration with cartridges or continuous extractions are employed. Techniques requiring short analysis time would also improve sample turnover. For applications where the researcher is interested in recovering all the volatile aldehydes in a complex mixture, a procedure using continuous liquid-liquid extraction is beneficial. Choosing a derivatizing agent that selectively reacts with carbonyls is advantageous in complex mixtures (see Tables 2 and 3). Selective reagents coupled with a specific detector (such as NPD, ECD, and MS) can improve analyses in complex food, environmental, and biological samples. In air, water, and other environmental applications where only trace levels are present, preconcentration is usually necessary. For food and biological systems, where care is needed to prevent further oxidation of

Table 2 Chromatographic Conditions for Selected Carbonyl Analysis Using the Gas Chromatograph

Reagent	Sensitivity	Chromatographic conditions	Reference
Saturated and unsaturated aldehydes; Ketones			
PFBHA	50–100 fmol/μL	DB-5 column; MS operating in the negative chemical ionization mode. Measures saturated and unsaturated aldehydes and 4-hydroxyonenal and MDA. Internal standard benzaldehyde. Application to biological systems.	Luo et al., 1995 [30]
PFBOA	10 ppb	CPSIL 5 CB column; MS. Oximes are reduced by pyridine/borans reagent to form stable derivatives. Reaction at 65°C for 30 min. Application to cognac.	Vidal et al., 1993 [28]
BOA and MOA	0.2 μg	OV-17 column; ECD. Reaction at room temperature for 20 min. After oxime formation, the unsaturated bond of the aldehyde is brominated before ECD analysis. Applications for unsaturated aldehyde in automobile exhaust. Detection limits are 0.2 and 0.15 μg of acrolein and crotonaldehyde, respectively, in 2 mL of absorption solution.	Nishikawa and Hayakawa, 1987 [33]
	ppb	Application for acrolein in air.	Nishikawa and Hayakawa, 1986 [34]
	0.4 ppb	Application for acrolein in rain water.	Nishikawa et al., 1987 [33]
	μg/cigarette	FFAP column. NPD. Measures saturated and unsaturated aldehydes.	Magin, 1980 [1]
Cysteamine	Low pg	DBWAX column; NPD. Internal standard *N*-methyl acetamide. Reaction at room temperature for 1 h at pH 8. Application in headspace of heated food oils.	Yasuhara and Shibamoto, 1989, 1991 [40,44]

Table 2 Continued

Reagent	Sensitivity	Chromatographic conditions	Reference
Saturated and unsaturated aldehydes; Ketones (continued)			
Cysteamine (continued)	Low ppb	Application to automobile exhaust.	Yasuhara and Shibamoto, 1994 [2]
	5.8 pg	Application to headspace of fish flesh.	Yasuhara and Shibamoto, 1995 [42]
	ppb	Application to air samples.	Yasuhara and Shibamoto, 1989 [44]
Cysteamine	Low ppm	DB-17 or DB-210 or OV-1 columns; FPD Internal standard phenyl sulfide. Reaction at room temperature for 10 min. Application for aliphatic and unsaturated aldehyde in salad, sesame, and corn oil.	Kataoka et al., 1995 [41]
Cysteamine	Low pg	DBWAX column; NPD or FPD Internal standard N-methyl acetamide Application for formaldehyde in coffee.	Hayashi et al., 1985 [43]
DNPH	Low ppm	DB-5 column, ECD For formaldehyde analysis in water.	Velikonja et al., 1995 [23]
DNPH	0.1 ppb	CP-Sil 8 CB column, TSD Formaldehyde in air is collected in miniature glass fibers impregnated with DNPH. Internal standard is isobutyl chloroformate derivative of di-n-butylamine.	Dalene et al., 1992 [24]
DNPH	20 µg/kg	SP-2250 packed column. ECD For formaldehyde analysis in milk.	Buckley and Fisher, 1986 [20]
Direct	n/a	OV-351 column; MS. Carbonyls were extracted using dynamic headspace technique. Also include analysis of diones in whisky.	Hawthorne and Miller, 1986 [53]
Direct	n/a	DBWAX column; MS. Two types of extraction procedure were used: simultaneous distillation extraction and dynamic headspace extraction. Also include analysis of aromatic ketones in cheese.	Careri et al., 1994 [51]

Compound / Derivative	Detection limit	Method / Notes	Reference
α-carbonyl DCDMQ	10 fmol/μL	OV-17 column; ECD. Reaction at 40°C for 90 min. Application for diacetyl and acetoin in alcoholic beverages.	Otsuka and Ohmori, 1992 [50]
PDA	n/a	Carbowax 20M; MS; For the analysis of diacetyl, 2,3-pentanedione, glyoxal, pyrivic acid, and 2-oxo-butanal in cigarette smoke.	Moree-Testa and Saint-Jalm, 1981 [48]
PDA	nmol	DBWAX column, NPD. Reaction at room temperature for 1 h at pH 7.5. Inernal standard indole. Application for glyoxal in fish liver oil.	Nishiyama et al., 1994 [49]
Malondialdehyde			
PFPH	n/a	2% OV-17 packed column; ECD: Reaction at room temperature for 30 min	Tomita et al., 1990 [11]
PFPH	5 nM	HP-5 or DBWAX columns; MS operating in negative ionization mode. Applications to blood plasma, tissues homogenate, sperm cells. Reaction at room temperature for 1 h. Internal standard ^{2}H2MDA. Amenable to large sample analyses.	Yeo et al., 1994 [12]
NMH	Low pg	DBWAX; NPD, Reaction at room temperature for 30 min. Internal standard N-methylacetamide. Application to corn oil and beef fat.	Umano et al., 1988 [16]
PFBHA	50–100 fmol/μL	See Luo et al. above.	Luo et al., 1995 [30]
Acrolein			
NMH	Low ppb	DBWAX; NPD, Reaction at room temperature for 30 min. Internal standard N-methylacetamide. Application to heated vegetable oil.	Yasuhara et al., 1989 [18]
			Yasuhara and Shibamoto, 1991 [17]
	μg/cigarette	Application to cigarette smoke.	Miyake et al., 1995 [19]

Table 3 Chromatographic Conditions for Selected Carbonyl Analysis Using the High-Performance Liquid Chromatograph

Reagent	Sensitivity	Chromatographic conditions and applications	Reference
Aliphatic aldehydes, vinyl aldehydes, and ketones			
CHD	nM	5 µM LiChrosphere RP18; Mobile phase: THF/water; Fluorescence 380 nm (ex), 445 nm (em); For analysis of alkanals and hydroxyalkenals in biological systems.	Holley et al., 1993 [72]
DNPH	1 ppb	5 µMTessek SGX C18 column; Mobile phase: MeOH/water, UV at 355 nm; Application to emissions from wood fired furnace, aldehyde is collected in tube containing Chromosorb P impregnated with DNPH.	Lehotay and Halmo, 1994 [65]
DNPH	Low ppb	3 µM Develosile ODS-3 column; Mobile phase: ACN/water; UV at 365 nm. Application to air, aldehyde is collected with SP-cartridge impregnated with DNPH.	Kuwata et al., 1983 [64]
DNPH	30 ppb	Zorbax ODS or Supelcosil LC-PAH; Mobile phase: MeOH/water, UV; 365 nm; Conditions may also be applied to ketone analysis in automotive exhaust.	Coutrim et al., 1993 [55]
DNPH	51 ppb	5 µM ODS C18 column; Mobile phase: MeOH/water; UV 360·nm; Detection limit of 51, 57, 12, 14 ppb for formaldehyde, acetaldehyde, propionaldehyde, and butyraldehyde, respectively. Applications to highly volatile carbonyls in automotive exhaust.	Geng and Chen, 1992 [54]
DNPH	Low ppm	5 µm C18 column; Mobile phase: ACN/water; UV at 254 nm; Applications to fried foods for analysis of formaldehyde, acetaldehyde, and acrolein.	Lane and Smathers, 1991 [61]

Reagent	Detection limit	Method	Reference
DNPH	n/a	Nucleosil 5C18 and Spherisorb S5 ODS-2. Mobile phase: ACN/water, UV; 375 nm	Puputti and Lehtonen, 1986 [62]
DNPH	10 µg/cigarette	Zorbax ODS column. Mobile phase: MeOH/water, UV at 365 nm. Other aldehydes detected by this method include acetaldehyde, acroelein, propioaldehyde, and acetone. Sensitivity reported is acrolein.	Manning et al., 1983 [69]
DNPH	Low ppb	5 µM Zorbax-ODS column. Mobile phase: ACN/water; UV at 365 nm. Detection limit of 20, 10, 5, and 4 for formaldehyde, acetaldehyde, acrolein, and benzaldehyde, respectively.	Lipari and Swarin, 1982 [66]
Anthrone	5 ppb	Nova-PAK C18 Radial-PAK cartridge. Mobile phase: ACN/water. Fluorescence 405 nm (ex) and 480 nm (em). For analysis of vinyl aldehydes such as acrolein, crotonaldehyde, and methacrolein.	Miller and Danielson, 1988 [74]
Fluoral-P	Low µM	5 µM NS-Gel C18 column. Mobile phase: ACN/PO4. Fluorescence 410 nm (ex), 510 nm (em). Formaldehyde analysis in beverages.	Tsuchiya et al., 1994 [76]
Luminarin	68 fmol	*Normal phase:* 5 µM Nucleosil silica. Mobile phase: Diisopropylether/ethyl acetate; Fluorescence 368 nm (ex), 416 nm (em). *Reverse phase:* 5 µM Nucleosil ODS. Mobile phase: ACN/Imidazole nitrate buffer, pH 7.5. Fluorescence 387 (ex), 450 (em).[a]	Traore et al., 1993 [77]
α-dicarbonyls			
DAMDB	12–14 fmol	L-column, 5 µM ODS. Mobile phase: ACN/0.5 M ammonium acetate; Fluorescence 350 nm (ex), 390 nm (em). For analysis of glyoxal, methylglyoxal, diacetyl in fermented food.	Yamaguchi et al., 1994 [71]

Table 3 Continued

Reagent	Sensitivity	Chromatographic conditions and applications	Reference
α-dicarbonyls			
o-PDA	ppm	Supercosil LC-18; UV at 254 nm Mobile phase: MeOH/water For analysis of glyoxal, methylglyoxal, diacetyl	Bednarski et al., 1989 [70]
o-PDA	μg/cigarette	Partisil 10 ODS 3 column Mobile phase: ACN/water, UV; 312 nm For analysis of diacetyl, 2,3-pentanedione, glyoxal, pyruvic aldehyde, and 2-oxo-butanal.	Moree-Testa and Saint-Jalm, 1981 [48]
Malondialdehyde			
Luminarin	159 fmol	5 μM Nucleosil silica Mobile phase: Diisopropylether/ethyl acetate; Fluorescence 387 nm (ex), 444 nm (em) Reaction at room temperature for 30 min.	Traore et al., 1993 [77]
TBA	15 pg	10 μM C18 uBondapak column Mobile phase: ACN/THF in 5 mM phosphate buffer pH 7. Fluorescence 515 nm (ex) and 550 nm (em). Reaction at 100°C for 30 min.	Draper et al., 1993 [10]
Hydroxyalkenals			
CHD	nM	5 μM LiChrosphere RP18; Mobile phase: THF/water Fluorescence 380 nm (ex), 445 nm (em)	Holley et al., 1993 [72]
Hydroxymethylfufural			
Luminarin	0.1–10 nmol	5 μM Nucleosil silica Mobile phase: Diisopropylether/ethyl acetate; Fluorescence 387 nm (ex), 444 nm (em) Reaction at room temperature for 30 min.	Traore et al., 1993 [77]

[a]The wavelength selected is dependent on the aldehyde of interest and the luminarin used as the derivatizing reagent. The excitation and emissions wavelengths reported above are for the analysis and separation of malondialdehyde and acetylacetone using Luminarin 11.

the sample, a technique with mild reaction conditions is essential to avoid artifactual formation of carbonyls.

REFERENCES

1. Magin, D. F., Gas chromatography of simple monocarbonyls in cigaret whole smoke as the benzyloxime derivatives, *J. Chromatogr. 202*: 255–261 (1980).
2. Yasuhara, A., and Shibamoto, T., Gas chromatographic determination of trace amounts of aldehydes in automobile exhaust by a cysteamine derivatization method, *J. Chromatogr., A 672*: 261–266 (1994).
3. Kerns, W. D., Pavkov, K. L., Donofrio, D. J., Gralla, E. J., and Swenberg, J. A., Carcinogenicity of formaldehyde in rats and mice after long-term inhalation exposure, *Cancer Res. 43*: 4382–4392 (1983).
4. Woutersen, R. A., Appelman, L. M., Van Garderen-Hoetmer, A., and Feron, V. J., Inhalation toxicity of acetaldehyde in rats. III. Carcinogenicity study, *Toxicology 41*: 213–231 (1986).
5. Marnett, L. J., DNA adducts of α,β-unsaturated aldehydes and dicarbonyl compounds, *IARC Sci. Publ. 125*: 151–163 (1994).
6. Nagao, M., Fujita, Y., Sugimura, T., and Kosuge, T., Methylglyoxal in beverages and foods: its mutagenicity and carcinogenicity, *IARC Sci. Publ. 70*: 283–291 (1986).
7. Nagao, M., Fujita, Y., Wakabayashi, K., Nukaya, H., Kosuge, T., and Sugimura, T., Mutagens in coffee and other beverages, *EHP, Environ. Health Perspect. 67*: 89–91 (1986).
8. Janero, D. R., Malondialdehyde and thiobarbituric acid-reactivity as diagnostic indexes of lipid peroxidation and peroxidative tissue injury, *Free Radical Biol. Med. 9*: 515–540 (1990).
9. Hoyland, D. V., Chemical methods for assessing lipid oxidation in food, 5662 (1990).
10. Draper, H. H., Squires, E. J., Mahmoodi, H., Wu, J., Agarwal, S., and Hadley, M., A comparative evaluation of thiobarbituric acid methods for the determination of malondialdehyde in biological materials, *Free Radical Biol. Med. 15*: 353–363 (1993).
11. Tomita, M., Okuyama, T., Watanabe, S., and Kawai, S., Free malondialdehyde levels in the urine of rats intoxicated with paraquat, *Arch. Toxicol. 64*: 590–593 (1990).
12. Yeo, H. C., Helbock, H. J., Chyu, D. W., and Ames, B. N., Assay of malondialdehyde in biological fluids by gas chromatography–mass spectrometry, *Anal. Biochem. 220*: 391–396 (1994).
13. Liu, J., Wang, X., Yeo, H. C., Shigenaga, M. K., Mori, A., and Ames, B. N., Immobilization stress causes oxidative damage to lipid, protein, and DNA in the brain of rats, *FASEB J. 10*: 1532–1538 (1996).
14. Kieber, R. J., and Mopper, K. Determination of picomolar concentrations of carbonyl compounds in natural waters, including seawater, by liquid, *Environ. Sci. Technol. 24*: 1477–1481 (1990).
15. Zhou, X., and Mopper, K., Measurement of sub-parts-per-billion levels of carbonyl compounds in marine air by a simple cartridge trapping procedure followed by liquid chromatography, *Environ. Sci. Technol. 24*: 1482–1485 (1990).

16. Umano, K., Dennis, K. J., and Shibamoto, T., Analysis of free malondialdehyde in photoirradiated corn oil and beef fat via a pyrazole derivative, *Lipids 23*: 811–814 (1988).

17. Yasuhara, A., and Shibamoto, T., Determination of acrolein evolved from heated vegetable oil by *N*-methylhydrazine conversion, *Agric. Biol. Chem. 55*: 2639–2640 (1991).

18. Yasuhara, A., Dennis, K. J., and Shibamoto, T., Development and validation of new analytical method for acrolein in air, *J. Assoc. Off. Anal. Chem. 72*: 749–751 (1989).

19. Miyake, T., Yasuhara, A., and Shibamoto, T., Gas chromatographic analysis of acrolein as 1-methyl-2-pyrazoline in cigarette smoke, *J. Environ. Chem. 5*: 569–573 (1995).

20. Buckley, K. E., Fisher, L. J., and MacKay, V., Electron capture gas chromatographic determination of traces of formaldehyde in milk as the 2,4-dinitrophenylhydrazone, *J. Assoc. Off. Anal. Chem. 69*: 655–657 (1986).

21. Buckley, K. E., Fisher, L. J., and MacKay, V. G., Levels of formaldehyde in milk, blood, and tissues of dairy cows and calves consuming formalin-treated whey, *J. Agric. Food Chem. 36*: 1146–1150 (1988).

22. Hayashi, T., Reece, C. A., and Shibamoto, Gas chromatographic determination of formaldehyde in coffee via thiazolidine derivative, *J. Assoc. Off. Anal. Chem. 69*: 101–105 (1986).

23. Velikonja, S., Jarc, I., Zupancic-Kralj, L., and Marsel, J., Comparison of gas chromatographic and spectrophotometric techniques for the determination of formaldehyde in water, *J. Chromatogr. A, 704*: 449–454 (1995).

24. Dalene, M., Persson, P., and Skarping, G., Determination of formaldehyde in air by chemisorption on glass filters impregnated with 2,4-dinitrophenylhydrazine using gas chromatography with thermionic specific detection, *J. Chromatogr. 626*: 284–288 (1992).

25. Koshy, K. T., Kaiser, D. G., and VanDerSlik, A. L., *O*-(2,3,4,5,6-Pentafluorobenzyl)-hydroxylamine hydrochloride as a sensitive derivatizing agent for the electron capture gas liquid chromatographic analysis of keto steroids, *J. Chromatogr. Sci. 13*: 97–104 (1975).

26. Cancilla, D. A., and Que Hee, S. S., *O*-(2,3,4,5,6-Pentafluorophenyl)methylhydroxylamine hydrochloride: a versatile reagent for the determination of carbonyl-containing compounds, *J. Chromatogr. 627*: 1–16 (1992).

27. Cancilla, D. A., Chou, C. C., Barthel, R., and Hee, S. S. Q., Characterization of the *O*-(2,3,4,5,6-pentafluorobenzyl)-hydroxylamine hydrochloride (PFBOA) derivatives of some aliphatic mono- and dialdehydes and quantitative water analysis of these aldehydes, *J. AOAC Int. 75*: 842–854 (1992).

28. Vidal, J. P., Estreguil, S., and Cantagrel, R., Quantitative analysis of cognac carbonyl compounds at the ppb level by GC-MS of their *O*-(pentafluorobenzyl amine) derivatives, *Chromatographia 36*: 183–186 (1993).

29. Nishikawa, H., and Sakai, T., Derivatization and chromatographic determination of aldehydes in gaseous and air samples, *J. Chromatogr., A 710*: 159–165 (1995).

30. Luo, X. P., Yazdanpanah, M., Bhooi, N., and Lehotay, D. C., Determination of aldehydes and other lipid peroxidation products in biological samples by gas chromatography-mass spectrometry, *Anal. Biochem. 228*: 294–298 (1995).

31. Magin, D. F., Preparation and gas chromatographic characterization of benzyloximes and *p*-nitrobenzyloximes of short-chain (C1–C7) carbonyls, *J. Chromatogr. 178*: 219–227 (1979).

32. Levine, S. P., Harvey, T. M., Waeghe, T. J., and Shapiro, R. H., *O*-alkyloxime derivatives for gas chromatographic and gas chromatographic-mass spectrometric determination of aldehydes, *Anal. Chem. 53*: 805–809 (1981).

33. Nishikawa, H., Hayakawa, T., and Sakai, T., Determination of acrolein and crotonaldehyde in automobile exhaust gas chromatography with electron-capture detection, *Analyst 112*: 859–861 (1987).

34. Nishikawa, H., Hayakawa, T., and Sakai, T., Determination of micro amounts of acrolein in air by gas chromatography, *J. Chromatogr. 370*: 327–332 (1986).

35. Nishikawa, H., Hayakawa, T., and Sakai, T., Gas chromatographic determination of acrolein in rain water using bromination of *O*-methyloxime, *Analyst 112*: 45–48 (1987).

36. Esterbauer, H., Schaur, R., and Zollner, H., Chemistry and biochemistry of 4-hydroxynonenal, malonaldehyde and related aldehydes, *Free Radical Biology Medicine 11*: 81–128 (1991).

37. Kinter, M., Analytical technologies for lipid oxidation products analysis, *J. Chromatogr. B 671*: 223–236 (1995).

38. Bruenner, B., Jones, A. D., and German, J. B., Direct characterization of protein adducts of the lipid peroxidation product 4-hydroxy-2-nonenal using electrospray mass spectrometry, *Chem. Res. Toxicol. 8*: 552–559 (1995).

39. Toyokuni, S., Uchida, K., Okamoto, K., Hattori-Nakakuki, Y., Hiai, H., and Stadman, E., Formation of 4-Hydroxy-2-nonenal-modified proteins in the renal proximal tubules of rats with renal carcinogen, ferric nitrilotriacetate, *Proc. Natl. Acad. Sci. 94*: 2616–2620 (1994).

40. Yasuhara, A., and Shibamoto, T., Determination of volatile aliphatic aldehydes in the headspace of heated food oils by derivatization with 2-aminoethanethiol, *J. Chromatogr. 547*: 291–298 (1991).

41. Kataoka, H., Sumida, A., Nishihata, N., and Makita, M., Determination of aliphatic aldehydes as their thiazolidine derivatives in foods by gas chromatography with flame photometric detection, *J. Chromatogr. A 709*: 303–311 (1995).

42. Yasuhara, A., and Shibamoto, T., Quantitative analysis of volatile aldehydes formed from various kinds of fish flesh during heat treatment, *J. Agric. Food Chem. 43*: 94–97 (1995).

43. Hayashi, T., and Shibamoto, T., Analysis of methyl glyoxal in foods and beverages, *J. Agric. Food Chem. 33*: 1090–1093 (1985).

44. Yasuhara, A., and Shibamoto, T., Formaldehyde quantitation in air samples by thiazolidine derivatization: factors affecting analysis, *J. Assoc. Off. Anal. Chem. 72*: 899–902 (1989).

45. Yasuhara, A., and Shibamoto, T., Analysis of aldehydes and ketones in the headspace of heated pork fat, *J. Food Sci. 54*: 1471–1472, 1484 (1989).

46. Urbach, G., Dynamic headspace gas chromatography of volatile compounds in milk, *J. Chromatogr. 404*: 163–174 (1987).

47. Harada, K., Nishimura, O., and Mihara, S., Rapid analysis of coffee flavor by gas chromatography using a pyrolyzer, *J. Chromatogr. 391*: 457–460 (1987).

48. Moree-Testa, P., and Saint-Jalm, Y., Determination of α-dicarbonyl compounds in cigaret smoke, *J. Chromatogr. 217*: 197–208 (1981).

49. Nishiyama, T., Hagiwara, Y., Hagiwara, H., and Shibamoto, T., Formation and inhibition of genotoxic glyoxal and malonaldehyde from phospholipids and fish liver oil upon lipid peroxidation, *J. Agric. Food Chem. 42*: 1728–1731 (1994).

50. Otsuka, M., and Ohmori, S., Simple and sensitive determination of diacetyl and acetoin in biological samples and alcoholic drinks by gas chromatography with electron-capture detection, *J. Chromatogr. 577*: 215–220 (1992).

51. Careri, M., Manini, P., Spagnoli, S., Barbieri, G., and Bolzoni, L., Simultaneous distillation-extraction and dynamic headspace methods in the gas chromatographic analysis of Parmesan cheese volatiles, *Chromatographia 38*: 386–394 (1994).

52. Barbieri, G., Bolzoni, L., Careri, M., Mangia, A., Parolari, G., Spagnoli, S., and Virgili, R., Study of the volatile fraction of Parmesan cheese, *J. Agric. Food Chem. 42*: 1170–1176 (1994).

53. Hawthorne, S. B., and Miller, D. J., Water chemical ionization mass spectrometry of aldehydes, ketones, esters, and carboxylic acids, *Appl. Spectrosc. 40*: 1200–1211 (1986).

54. Geng, A. C., Chen, Z. L., and Siu, G. G., Determination of low-molecular-weight aldehydes in stack gas and automobile exhaust gas by liquid chromatography, *Anal. Chim. Acta 257*: 99–104 (1992).

55. Coutrim, M. X., Nakamura, L. A., and Collins, C. H., Quantification of 2,4-dinitrophenylhydrazones of low molecular mass aldehydes and ketones using HPLC, *Chromatographia 37*: 185–190 (1993).

56. Takeda, S., Wakida, S.-I., Yamane, M., and Higashi, K., Analysis of lower aliphatic aldehydes in water by micellar electrokinetic chromatography with derivatization to 2,4-dinitrophenylhydrazones, *Electrophoresis 15*: 1332–1334 (1994).

57. Lehotay, J., and Hromulakova, K., HPLC determination of trace levels of aliphatic aldehydes C1–C4 in river and tap water using online preconcentration, *J. Liq. Chromatogr. 17*: 579–588 (1994).

58. Grosjean, D., and Fung, K., Collection efficiencies of cartridges and microimpingers for sampling of aldehydes in air as 2,4-dinitrophenylhydrazones, *Anal. Chem. 54*: 1221–1224 (1982).

59. Grosjean, D., Formaldehyde and other carbonyls in Los Angeles ambient air, *Environ. Sci. Technol. 16*: 254–262 (1982).

60. Matsumoto, M., Nishikawa, Y., Murano, K., and Fukuyama, T., Determination of aldehydes in rainwater by HPLC, *Bunseki Kagaku 36*: 179–183 (1987).

61. Lane, R. H., and Smathers, J. L., Monitoring aldehyde production during frying by reversed-phase liquid chromatography, *J. Assoc. Off. Anal. Chem. 74*: 957–960 (1991).

62. Puputti, E., and Lehtonen, P., High-performance liquid chromatographic separation and diode-array spectroscopic identification of dinitrophenylhydrazone derivatives of carbonyl compounds from whiskies, *J. Chromatogr. 353*: 163–168 (1986).

63. Kuwata, K., Uebori, M., and Yamasaki, Y., Determination of aliphatic and aromatic aldehydes in polluted airs as their 2,4-dinitrophenylhydrazones by high performance liquid chromatography, *J. Chromatogr. Sci. 17*: 264–268 (1979).

64. Kuwata, K., Uebori, M., Yamasaki, H., Kuge, Y., and Kiso, Y., Determination of aliphatic aldehydes in air by liquid chromatography, *Anal. Chem.* *55*: 2013–2016 (1983).

65. Lehotay, J., and Halmo, F., Determination of aliphatic aldehydes C1–C4 in waste gas by HPLC, *J. Liq. Chromatogr.* *17*: 847–854 (1994).

66. Lipari, F., and Swarin, S. J., Determination of formaldehyde and other aldehydes in automobile exhaust with an improved 2,4-dinitrophenylhydrazine method, *J. Chromatogr.* *247*: 297–306 (1982).

67. Lipari, F., and Swarin, S. J., 2,4-Dinitrophenylhydrazine-coated Florisil sampling cartridges for the determination of formaldehyde in air, *Environ. Sci. Technol.* *19*: 44–48 (1985).

68. Olson, K. L., and Swarin, S. J., Determination of aldehydes and ketones by derivatization and liquid chromatography-mass spectrometry, *J. Chromatogr.* *333*: 337–347 (1985).

69. Manning, D. L., Maskarinec, M. P., Jenkins, R. A., and Marshall, A. H., High-performance liquid chromatographic determination of selected gas phase carbonyls in tobacco smoke, *J. Assoc. Off. Anal. Chem.* *66*: 8–12 (1983).

70. Bednarski, W., Jedrychowski, L., Hammond, E. G., and Nikolov, Z. L., A method for the determination of α-dicarbonyl compounds, *J. Dairy Sci.* *72*: 2474–2477 (1989).

71. Yamaguchi, M., Ishida, J., Zhu, X. X., Nakamura, M., and Yoshitake, T., Determination of glyoxal, methylglyoxal, diacetyl, and 2,3-pentanedione in fermented foods by high-performance liquid chromatography with fluorescence detection, *J. Liq. Chromatogr.* *17*: 203–211 (1994).

72. Holley, A. E., Walker, M. K., Cheeseman, K. H., and Slater, T. F., Measurement of *n*-alkanals and hydroxyalkenals in biological samples, *Free Radical Biol. Med.* *15*: 281–289 (1993).

73. Stahovec, W. L., and Mopper, K., Trace analysis of aldehydes by precolumn fluorigenic labeling with 1,3-cyclohexanedione and reversed-phase high-performance liquid chromatography, *J. Chromatogr.* *298*: 399–406 (1984).

74. Miller, B. E., and Danielson, N. D., Derivatization of vinyl aldehydes with anthrone prior to high-performance liquid chromatography with fluorometric detection, *Anal. Chem.* *60*: 622–626 (1988).

75. Compton, B. J., and Purdy, W. C., Fluoral-P, a member of a selective family of reagents for aldehydes, *Anal. Chim. Acta* *119*: 349–357 (1980).

76. Tsuchiya, H., Ohtani, S., Yamada, K., Akagiri, M., Takagi, N., and Sato, M., Determination of formaldehyde in reagents and beverages using flow injection, *Analyst* *119*(6): 1413–1416 (1994).

77. Traore, F., Farinotti, R., and Mahuzier, G., Determination of malonaldehyde by coupled high-performance liquid chromatography-spectrofluorometry after derivatization with luminarin 3, *J. Chromatogr.* *648*: 111–118 (1993).

78. Traore, F., Pianetti, G. A., Dallery, L., Tod, M., Chalom, J., Farinotti, R., and Mahuzier, G., Determination of picomole amounts of carbonyls as luminarin hydrazones by high-performance liquid chromatography with fluorescence detection, *Chromatographia* *36*: 96–104 (1993).

79. Bosin, T. R., Holmstedt, B., Lundman, A., and Beck, O., Presence of formaldehyde in

biological media and organic solvents: artifactual formation of tetrahydro-β-carbo-lines, *Anal. Biochem. 128*: 287–293 (1983).

80. Kieber, D. J., and Mopper, K., Reversed-phase high-performance liquid chromato-graphic analysis of α-keto acid quinoxalinol derivatives. Optimization of technique and application to natural samples, *J. Chromatogr. 281*: 135–149 (1983).

81. Tomita, M., Okuyama T., Kawai, S., Determination of malonaldehyde in oxidized biological materials by high-performance liquid chromatography, *J. Chromatogr. 515*: 391–397 (1990).

Index

9 780367 400576